Hydrogeochemical Evaluation and Groundwater Quality

Md. Shajedul Islam

Hydrogeochemical Evaluation and Groundwater Quality

Springer

Md. Shajedul Islam
Department of Chemistry
Govt. Hosain Shahid Shohrawardi College
Ministry of Education
Magura, Bangladesh

ISBN 978-3-031-44303-9 ISBN 978-3-031-44304-6 (eBook)
https://doi.org/10.1007/978-3-031-44304-6

This Springer imprint is published by the registered company Springer Nature Switzerland AG
The registered company address is: Gewerbestrasse 11, 6330 Cham, Switzerland

Paper in this product is recyclable.

To
Parents

Preface

Groundwater is the largest supply of fresh water and a vital resource for the living and food security of billions of people, largely in a prosperous global agricultural economy. Approximately 97% of all unfrozen freshwater is found underneath the Earth's surface, and globally, among the total uses, it offers separately about 50% of potable water supplies, 40% of the industrial demand, and 35% of the water used for agricultural purposes. Though, for potable purposes, some areas were completely dependent on groundwater. For instance, in Pacific regions and Southeast Asia, an average of 60% of households in village areas and 66% of families in municipal areas depend on groundwater for drinking purposes. Surface water is becoming insufficient and contaminated due to unplanned industrialization and urbanization, and groundwater is considered safer to use for every purpose. The maximum population in Southeast Asia relies on groundwater for domestic, industrial, and irrigational uses. Water quality issues in Bangladesh have become serious concerns due to excess water mining, industrial development, agricultural diversity, urbanization, and weak water management. Most of the rural people of this country and some African and Asian countries drink untreated raw groundwater, and more than 85% of agricultural water is supplied from this resource. Recently, groundwater quality and quantity have drastically deteriorated worldwide due to overexploitation of water, excessive use of agrochemicals, urbanization, and industrialization. Massive agrochemical leaching, salinization, topsoil contamination, landfill, and vast flooding have worsened the groundwater quality. Moreover, geogenic sources have also enormously deteriorated the groundwater quality in some parts of the world during the last 30 years. Due to sea level rises, another global calamity can unfit groundwater for every use through seawater intrusion in coastal areas. Thus, the proper evaluation of groundwater water quality is very important to ensure safe water for all-purpose uses. This book aims to evaluate the hydrogeochemical process controlling the water quality variables and assess the water type of groundwater to find its suitability for drinking, irrigation, and industrial purposes.

This book guided the improvement of the management of groundwater resources on a zonal scale and may be put on another part with analogous topography. Accordingly, it is expected that the findings of this book will deliver a useful understanding for future groundwater monitoring and management. Also, the contents of this book deliver a guide for agricultural activists, policymakers, public health departments, and water managers.

The contents of this book are composed through the output of Ph.D. works and fundamental concepts of water chemistry and geochemical processes. This book contains five main parts: (1) natural and anthropogenic geochemical processes at the aquifer level, (2) water–rock interaction courses, (3) sources and mobilization of trace metals in groundwater, (4) assessment techniques of groundwater quality, and (5) groundwater management. The water data were collected from groundwater samples of hand and engine tube wells in the pre-monsoon, monsoon, and post-monsoon seasons, 2019–2020 and 2020–2021, in Kushtia District (upper Bengal basin), Bangladesh. I think not only graduate students but also academics, researchers, and professional persons will equally desire this book.

Chapters 1 and 2 of this book present introductory materials, including the nature of the aquifer, geochemistry, geochemical processes, groundwater chemistry and quality, contamination, groundwater availability, etc. The main contents of this book are the evaluation of hydrogeochemical characteristics and the assessment of water quality, which essentially depend on precise water data. To produce exhaustive water data, the analytical techniques of water analysis are described in Chap. 3. Chapter 4 is a very important section of the book. For the identification of hydrogeochemical processes, this chapter describes the various geological formations and hydrogeological settings, water chemistry, geostatistical modelling, spatiotemporal variations in water characteristics, pollution sources, water–rock interactions, water facies, source rock identification, thermodynamic equilibrium, CO_2 chemistry, etc. Later, the occurrence, mobilization, and sources of trace metals, especially Fe, Mn, and As, in groundwater are described methodically in Chap. 5. In addition, the assessment methods of drinking, irrigation, and industrial water quality are presented in Chaps. 6–8. Finally, Chap. 9 describes climate change and groundwater management.

Magura, Bangladesh

Md. Shajedul Islam

Acknowledgments

This book arose from my Ph.D. works at the Water Lab, Institute of Environmental Science, University of Rajshahi, Bangladesh. I thank the research supervisor (Prof. M. G. Mostafa, D.Eng.) and professors of this Institute. I gratefully acknowledge this Institute for granting me a fellowship for this research work.

I am thankful to the Ministry of Education, Bangladesh, for sanctioning the deputation to complete this study. I convey my special thanks to all officers and staff of Central Science Laboratory, University of Rajshahi, and DPHE Laboratory, Dhaka, for their cooperation in sample analysis. I wish to express thanks to all research fellows and colleagues for their friendly cooperation.

Also, I would like to thank the staff at Springer, in particular Mr. Aaron Schiller and Mr. Herbert Moses, for their help and support.

Md. Shajedul Islam

Contents

1 Introduction to Hydrogeochemical Processes 1

2 Groundwater: Sources, Functions, and Quality 17

3 Water Analysis 37

4 Evaluation of Hydrogeochemical Processes 65

5 Trace Metals in Groundwater: *Sources and Mobilization* 135

6 Drinking Water Quality 187

7 Irrigation Water Quality 223

8 Industrial Water Quality 281

9 Climate Change and Groundwater Management 301

Appendices 309

Index 325

List of Figures

Fig. 1.1 Subfield of geochemistry 3
Fig. 1.2 (**a**) Recharge and discharge area in aquifers and (**b**) confined and unconfined aquifer 6
Fig. 1.3 Confined and unconfined beds and various types of wells 10

Fig. 2.1 Groundwater profiles beneath the Earth's surface 18
Fig. 2.2 Groundwater beneath the surface 19
Fig. 2.3 Groundwater in the water cycle 20
Fig. 2.4 The average depth pattern of the global groundwater table. The white areas indicate the regions where a stable shallow groundwater table may not exist 24
Fig. 2.5 Factors involved in groundwater chemistry 25
Fig. 2.6 Pathway of groundwater contamination 29
Fig. 2.7 Sources of groundwater contamination 30
Fig. 2.8 Worldwide pesticide consumption in 2019 33

Fig. 3.1 Flowchart of methods to evaluate the geochemical processes and water quality 38
Fig. 3.2 Map of the study area and sampling points 39
Fig. 3.3 Sampling location and piezometric lines (RS2) in the study area 40
Fig. 3.4 Hydrogeologic formation of the study area 41
Fig. 3.5 Pump sampling instruments 44
Fig. 3.6 Depth groundwater sampling 45
Fig. 3.7 Sampling equipment 47
Fig. 3.8 Various types of water quality analyser (**a**) potable digital multimeter (**b**) UV–visible spectrophotometer (**c**) AAS 49

Fig. 4.1 Average values of (**a**) temp. (°C) and pH. (**b**) EC (μS/cm), TDS (mg/L), and TH (mg/L) in the different sampling seasons 69

Fig. 4.2 Average levels of (**a**) major cations (mg/L) and (**b**) anions (mg/L) in the different sampling seasons 73

Fig. 4.3 Presence of major (**a**) cations and (**b**) anions of total concentration (mg/L) in groundwater of the study area 74

Fig. 4.4 Comparable loads of Na^+ and Ca^{2+} in groundwater of different sampling regions 76

Fig. 4.5 Biplots of groundwater depth vs. (**a**) EC (μS/cm), (**b**) TDS (mg/L), (**c**) Ca (mg/L), and (**d**) HCO_3^- (mg/L) 77

Fig. 4.6 Boxplot diagrams of the upper and lower quartiles (box); median (black line inside the box); 1.5× interquartile range, IQR (cross mark within the box); and outliers (circles) for nominated variables in the three sampling periods PRM (green), MON (sky blue), and POM (red). As stated by the Mann–Whitney U test, the letters 'a', 'b', and 'c' in each panel indicate provocatively different data distributions at $p < 0.05$ (two-tailed) 83

Fig. 4.7 Biplot (axes PC1 and PC2) of robust principal component analysis for (**a**) pre-monsoon (PRM), (**b**) monsoon (MON), and post-monsoon (POM) sampling seasons 92

Fig. 4.8 Hierarchical cluster analysis of 40 sampling sites for the PRM, MON, and POM sampling rounds. The dendrogram of Q-mode hierarchical cluster analysis (HCA) displays connotations between samples from different regions of the hydrologic system. Figures A1, B1, and C1 denote the sample ID-based clusters, and Figures A2, B2, and C2 represent the water parameter-based clusters 95

Fig. 4.9 Morphological changes in the Padma River (Bangladesh) in (**a**) 1988 and (**b**) 2018: decrease the river size and increase the density of green plants. (Due to the Green Revolution in the country from 1985) 97

Fig. 4.10 Spatial variation in clusters and recharge/discharge zones in the study area for different sampling periods 98

Fig. 4.11 The concentration/value variation of water variables in a separate cluster of the PRM and POM sampling seasons. The maximum values of the variable are higher in cluster I than in cluster II in both seasons 99

Fig. 4.12 Sample-by-sample values of BCI in the three different sampling seasons 100

Fig. 4.13 Bivariate plots of most important ions against TDS values (square, triangle, and round shapes denote the pre-monsoon. monsoon and post-monsoon sampling seasons, respectively) 103

Fig. 4.14 Piper diagram for groundwater classification 106

Fig. 4.15 Chadha's plot for groundwater classification 108

Fig. 4.16 Gibbs diagrams for groundwater samples of the study area 110

Fig. 4.17 Bivariate plot of (**a**) Ca/Na vs. HCO_3/Na and (**b**) Ca/Na vs. Mg/Na to categorize the mineral weathering of groundwater in the study zone. Previous observations have shown that the major cations and anions result from rock weathering rather than other processes ... 112
Fig. 4.18 Bivariate plots of (**a**) Ca^{2+} + Mg^{2+} vs. HCO_3^-, (**b**) Na^+ + K^+ vs. HCO_3^-, (**c**) Ca^{2+} + Mg^{2+} vs. total cations, (**d**) Ca^{2+} + Mg^{2+} vs. total anions, (**e**) Na^+ + K^+ vs. total cations, and (**f**) Na^+ vs. Cl^- ... 113
Fig. 4.19 (**a**) Bivariate plot of Cl^- correlated (Na^+ + K^+) and (Ca^{2+} + Mg^{2+}) correlated (HCO_3^- + SO_4^{2-}) to assess the cation exchange of water in the study area; (**b**) concentration (mEq/L) of earth metals over basic metals: bivariate plot of (Ca + Mg) vs. (Na + K) to determine the inter-dominance of major cations ... 117
Fig. 4.20 Bivariate plot of Cl^- vs. Cl^-/HCO_3^- to determine saltwater intrusion ... 118
Fig. 4.21 Test of cation exchange: CAI1 vs. CAI2 in the three sampling rounds ... 120
Fig. 4.22 Bivariate correlation between Ca^{2+} and Mg^{2+} concentrations in the samples ... 121
Fig. 4.23 Geochemical relationship of (**a**) pH vs. $\log_{10}(pCO_2)$ and (**b**) $[HCO_3^-]$ vs. $\log_{10}(pCO_2)$ in groundwater ... 125
Fig. 5.1 Average metal concentrations (mg/L) of three sampling seasons. The concentrations of Fe, Mn, Cu, Zn, B, and Ni are denoted by 100 × mg/L ... 139
Fig. 5.2 Robust PC of metals and other selected water parameters in the water samples ... 152
Fig. 5.3 Hierarchical cluster analysis of 40 sampling sites for the average value (mg/L) of trace metals and other selected parameters in three sampling seasons (PRM, MON, and POM). Dendrogram of Q-mode hierarchical cluster analysis (HCA) presenting connotations between samples from different regions of the hydrologic system ... 153
Fig. 5.4 Bivariate diagram of TDS vs. trace metals ... 154
Fig. 5.5 Projection of the factor loading for the different components (PC1 and PC2) for the robust PCA ... 159
Fig. 5.6 Main lithotypes and soil formations present in the study area and locations of the monitoring points ... 160
Fig. 5.7 Fe and Mn concentrations in different sampling seasons of the Gangetic alluvial (*Ga*) and Deltaic alluvial (*Da*) platforms ... 161
Fig. 5.8 Bar diagram of the clay thickness of sample well and average Fe and Mn concentrations in the PRM and POM sampling periods ... 163

Fig. 5.9 Fe vs. Mn concentration in the pre-monsoon and post-monsoon seasons 165
Fig. 5.10 Bivariate diagrams of Fe vs. (**a**) pH, (**b**) alkalinity (HCO_3^-), (**c**) DO, and (**d**) DOC 168
Fig. 5.11 Log-normal probability plots of Fe (**a**, **b**) and Mn (**c**, **d**). Proof of individuality of the slope changes consistently to the natural background level, NBL (red line) 172
Fig. 5.12 Rust-coloured leaves due to high iron in the soil 173
Fig. 5.13 Distribution of arsenic in groundwater of Bangladesh 175
Fig. 5.14 Symptoms of arsenic-affected human skin in Bangladesh 179
Fig. 5.15 Global picture of As contamination in groundwater 179

Fig. 6.1 Theoretical model of the Canadian Water Quality Index 192
Fig. 6.2 Index values of CWQI and WWQI of three sampling seasons with the mean values 204
Fig. 6.3 The pair difference statistics between CWQI (drinking) and WWQI values for all water samples 209
Fig. 6.4 Total hazard index (HI_{total}) value (via oral and dermal absorption) for different trace metals of the (**a**) POM, (**b**) MON, and (**c**) POM sampling periods 212
Fig. 6.5 $ILCR_{total}$ values for Cr, Ni, Cd, and Pb metals in the three sampling periods 216

Fig. 7.1 Box plots of irrigation water suitability parameters as %Na, SSP, RSBC, PI, and MAR (*Y*-axis in the above figure denotes the value of the corresponding parameter) 240
Fig. 7.2 Box plots of irrigation water quality parameters as %Na, SSP, RSBC, PI, and MAR (*Y*-axis in the above figure denotes the value of the corresponding parameter) 241
Fig. 7.3 Box plots of irrigation water suitability parameters such as SAR, KR, Mg:Ca, and Na:Ca (the *Y*-axis in the above figure denotes the value of the corresponding parameter) 241
Fig. 7.4 The pair difference statistics between CWQI (irrigation) and MWQI values for all water samples 247
Fig. 7.5 US salinity hazard diagram 248
Fig. 7.6 Wilcox diagram for the assessment of irrigation water quality 249
Fig. 7.7 Permeability index (PI) diagram and Wilcox diagram for the assessment of irrigation water quality 250
Fig. 7.8 Sampling locations of sodic-type groundwater 270
Fig. 7.9 (**a**) EC vs. $IIWQ_{Index}$ and (**b**) SAR vs. $IIWQ_{Index}$ of groundwater for different regions 275
Fig. 8.1 Study area and sampling sites 282
Fig. 8.2 Location of major industries in Kushtia (denoted by numerical numbers) 283

Fig. 8.3 Percentage of samples exceeding various water quality parameters: (**a**) wood, (**b**) food, (**c**) paper, (**d**) polymer, (**e**) sugar, and (**f**) canned fruit industries 291
Fig. 8.4 Percentage of the samples crossing the (**a**) scale formation and (**b**) corrosive tendency in the pre-monsoon (PRM) and post-monsoon (POM) sampling seasons 294

Fig. 9.1 Rate of (1960–2010) worldwide groundwater demand, abstraction, and water table depletion 303

List of Tables

Table 1.1 Differences between confined and unconfined aquifers 11

Table 2.1 Total amount consumed of chemical fertilizer and pesticides in several countries 32
Table 2.2 Common sources of groundwater pollution from industry 35

Table 3.1 Sampling descriptions and possible pollution sources of the sampling sites 43
Table 3.2 Groundwater sampling log 46
Table 3.3 Sample container types and preservation 49
Table 3.4 List of studies and parameters used in each investigation 63

Table 4.1 Summary of the analysed data of the groundwater in the pre-monsoon (PRM), monsoon (MON), and post-monsoon (POM) of 2019–2020 and 2020–2021 sampling campaign, with basic statistics 67
Table 4.2 Drinking and irrigation water quality standards 70
Table 4.3 Pearson's correlation matrix (PCM) of water parameters in the (a) PRM, (b) MON, and (c) POM sampling seasons in the study zone 86
Table 4.4 Principal component (five components extracted) loadings of the analysed parameters in the groundwater samples (sorted by size) 90
Table 4.5 Application of PCA for groundwater in different geographical areas 93
Table 4.6 The sample ID included water parameters in two different clusters of three sampling periods 96
Table 4.7 Average water variable values of two clusters according to Q-mode hierarchical analysis for three sampling periods 101
Table 4.8 Chloro-alkaline index (CAI 1 and CAI 2) value of groundwater for the three sampling seasons 119

Table 4.9 CAI−1 and CAI−2 values for groundwater from different places 121
Table 4.10 Statistical data of saturation indices (*SI*s) of minerals using the PHREEQC−3v programme in groundwater samples 123

Table 5.1 Statistical analysis of trace metals in the groundwater samples of the pre-monsoon (PRM), monsoon (MON), and post-monsoon (POM) sampling seasons in 2019–2020 and 2020–2021 138
Table 5.2 Drinking and irrigation water quality standards 140
Table 5.3 Statistics of trace metal concentrations in groundwater during the PRM and POM sampling rounds 141
Table 5.4 Trace metal concentration in groundwater of different regions 142
Table 5.5 pH, EC (μS/cm), TDS (mg/L), and TH (mg/L) values in three different sample seasons 147
Table 5.6 The average concentration of analysed trace metals and some selected parameters in the groundwater samples of the three sampling seasons in 2019–2020 and 2020–2021, with basic statistics 148
Table 5.7 Pearson's correlation matrix for trace metals and physicochemical parameters of groundwater samples 150
Table 5.8 Principal component analysis of detected water parameters (sorted by size) 151
Table 5.9 Statistical data of the Fe and Mn with some other parameters in groundwater of the pre-monsoon (PRM) and post-monsoon (POM) sampling periods 156
Table 5.10 Pearson's correlation matrix of selected water variables in both the pre-monsoon (PRM) and post-monsoon (POM) periods 158
Table 5.11 Principal component loadings of the analysed parameters in the study areas (sorted by size) 159
Table 5.12 Statistics of arsenic concentration in Bangladesh 176
Table 5.13 A complete scenario of As concentration in the groundwater of Bangladesh 177

Table 6.1 Standards (S_i), weights (w_i), and relative weights (W_i) used for WQI calculation 195
Table 6.2 Drinking water suitability classification based on the classical WQI method 195
Table 6.3 Standard values (S_i), ideal values (I_i), and maximum admissible concentrations (all are in mg/L unit) for the analysed metal elements 198
Table 6.4 Crucial factors for calculating the health risks for drinking water through oral and dermal pathways 199

Table 6.5 Values of RfD_{oral}, CSF_{oral}, and GI_{ABS} for analysed metals 200
Table 6.6 Statistical summary of WQI values of collected water samples 203
Table 6.7 WQI designations and summarized results of samples 205
Table 6.8 Paired sample *t*-test between CWQI–WWQI methods for comparative study 208
Table 6.9 Summarized results of heavy metal contamination indices 210
Table 6.10a Noncarcinogenic health risks of trace elements in samples by oral and dermal pathways in the PRM period 213
Table 6.10b Noncarcinogenic health risks of trace elements in samples by oral and dermal pathways in the MON period 213
Table 6.10c Noncarcinogenic health risks of trace elements in samples by oral and dermal pathways in the POM period 214
Table 6.11 Summary results of noncarcinogenic cataloguing based on HI_{total} values 214
Table 6.12 Carcinogenic health risks ($ILCR_{total}$) of heavy metals by oral and dermal pathways in the different sampling seasons 215
Table 6.13 Judgement of carcinogenic and noncarcinogenic risk values of the present study with earlier studies in some different countries 217

Table 7.1 Crop variety and rate of production in the study area 226
Table 7.2 Categorization for IWQ index parameters 230
Table 7.3 The limiting values of water parameters for quality measurement (q_i) calculation 232
Table 7.4 Limits of important parameter indices for rating groundwater quality and its suitability in irrigation use 234
Table 7.5 Destructive statistics of irrigation water suitability parameters in groundwater throughout PRM, MON, and POM and per cent of suitability for irrigation use 238
Table 7.6 Statistical results of the Simsek irrigation water quality index ($SIWQ_{index}$) 243
Table 7.7 Statistical summary of WQI (CWQI and MWQI) values for irrigation of collected water samples 244
Table 7.8 WQI descriptions and summarized results of water samples 244
Table 7.9 Paired sample *t*-test (CWQI vs. MWQI) for comparative study 246
Table 7.10 Indexing model of irrigation water quality evaluation 253
Table 7.11 The scoring and weight factor of hazard classes and involved parameters in each class 254
Table 7.12 Cataloguing of irrigation water quality parameters with the rating value (r) 255

Table 7.13 Irrigation water classification according to the overall IWQ_{Index} 258
Table 7.14 Scoring of hazard class and rating of parameters in the proposed new indexing model for the assessment of irrigation water suitability 261
Table 7.15 Proposed irrigation water category according to $IIWQ_{Index}$ 266
Table 7.16 List of parameters used in each hazard class and calculation results of $IIWQ_{Index}$ for calcite water 268
Table 7.17 List of parameters used in each hazard class and calculation results of $IIWQ_{Index}$ for sodic water 271
Table 7.18 Calculated values of $IIWQ_{Index}$ of groundwater samples in different geographical places 273

Table 8.1 Guideline values for industrial water quality indices 285
Table 8.2 Descriptive statistics of physicochemical parameters in groundwater samples during the PRM and POM sampling rounds 288
Table 8.3 Industrial water quality requirements 290
Table 8.4 Cooling water standard for industrial purposes 292
Table 8.5 Statistical illustration of water quality indices for industrial purposes 295

Table 9.1 Groundwater stock, rate of abstraction, depletion, and water level decline of most vulnerable countries and regions 304

About the Author

Md. Shajedul Islam is specialized in Geochemistry, Hydrology, Water Chemistry and Quality, Water Quality Indexing, Climate Change, Risk Assessment, and Geostatistics.

He has a strong academic and research background in environmental science. Dr. Islam graduated with first class honours and first class master's degree from the Department of Chemistry, University of Chittagong, and was awarded a doctoral degree in Environmental Science from IES, University of Rajshahi, Bangladesh in the year of 2022. He has a rich publication record that includes books, book chapters, and numerous famous journal papers. From 1999, he engaged in teaching, administration, research, etc. as a member of Bangladesh Civil Servies (BCS). Dr. Islam attends several national and international conferences and delivered his research works.

Chapter 1
Introduction to Hydrogeochemical Processes

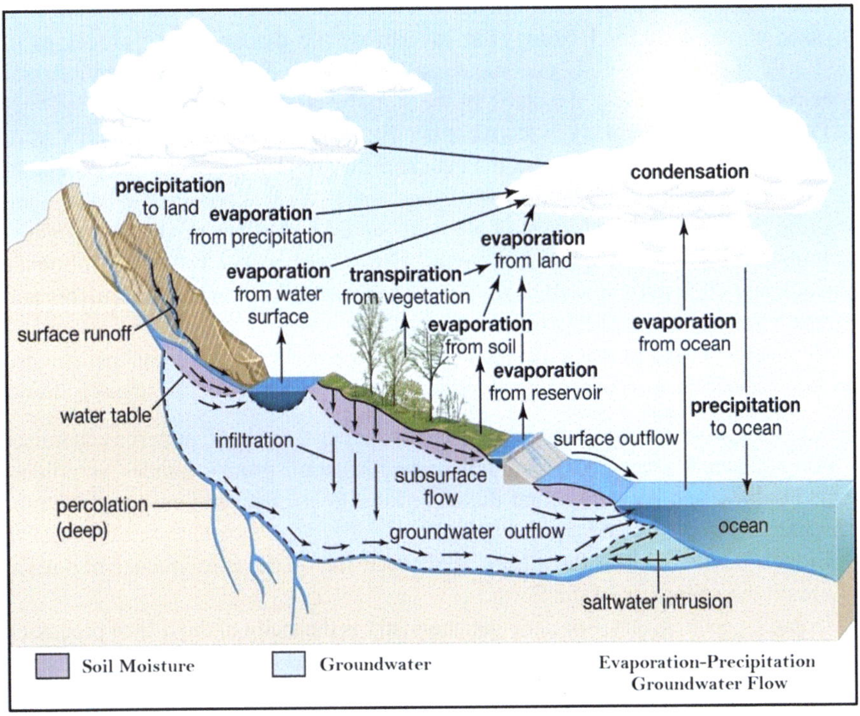

Abbreviations

BGS	British Geological Survey
BWS	Basement Water Sand
KSA	Kingdom of Saudi Arabia

M. S. Islam, *Hydrogeochemical Evaluation and Groundwater Quality*,
https://doi.org/10.1007/978-3-031-44304-6_1

PBM Padma–Brahmaputra–Meghna
SAGD Steam-assisted gravity drainage
USGS United States Geological Survey

1.1 Geochemistry

The word 'geochemistry' is obtained from the etymological terms 'geo' and 'chemistry', which mean Earth and chemical principles, respectively. So it is that subject of science that put on chemical ideologies to understand the Earth system. This ideology is used to clarify the mechanisms behind the main geological systems, such as Earth's crust, seas, rocks, fluids, biosphere and climate, water–rock interactions, gases, and biotic reactions, that exchange substances and energy over a range of time eras. It integrates the principles of several subjects, such as chemistry, physics, climatology, and biology, to investigate the dynamics and processes of the Earth. Besides, the study encompasses the Earth on all sides of the entire solar system (i.e. other planets), the study of the sun, the origin of rocks, etc.

Practically, geochemistry is conventionally alienated into the geochemistry of elements (mainly metal elements), the geochemistry of processes, and the geochemistry of systems. The geochemistry of elements is very important to study and describe the behaviour and loads of elements under conditions of several geogenic or human-caused processes and systems. How geochemical processes influence groundwater chemistry as well as the level of the ionic form of metal elements and groups is the key topic of this book.

Geochemistry is the interconnected study of chemistry, mineralogy, and in vitro geology. The numerous common subfields of geochemistry (Fig. 1.1) are as follows:

- Aqueous geochemistry – this geochemistry studies the role of numerous chemical components in watersheds, including metal ions, ionic groups, halides, sulphides, and oxides, and how elemental fluxes or ions are exchanged between the atmosphere, lithosphere, and aquatic systems.
- Biogeochemistry – it is the area of study focusing on the consequence of natural life on the chemistry of the Earth.
- Organic geochemistry – organic geochemistry is the study of the role of processes and components that result from living organisms.
- Geochemistry of trace elements – in the subject, a trace element (e.g. Fe, Mn, Cr, Si, Pb, Hg, Cu, Zn, etc.) is one whose level is less than 1000 mg/L or 0.1% (W/W) of the rock composition. This term is utilized largely in igneous mineralogy.
- Isotope geochemistry – it comprises the determination of the comparative and complete levels of the elements and their isotopes in the inner Earth, Earth surface, and aquifer sediments. In this case, generally, stable isotope analysis with δ^2H, $\delta^{18}O$, and $^{87}Sr/^{86}Sr$ is measured, and these values indicate the main source of groundwater and rock sources.

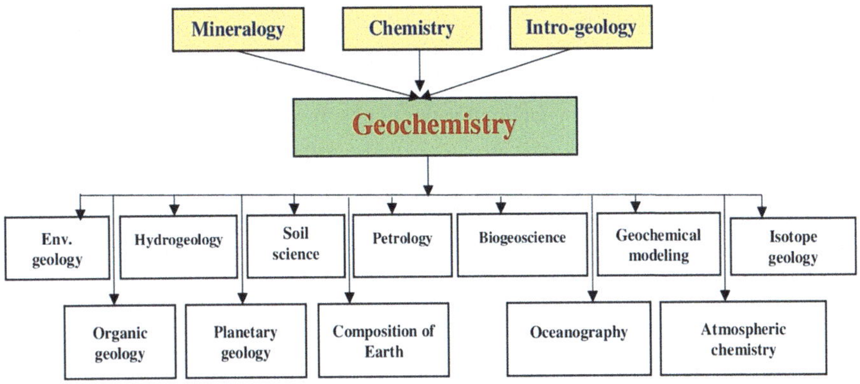

Fig. 1.1 Subfield of geochemistry

- Rock geochemistry – it is an integral section of research into geologic processes, where precise major, minor, and trace element chemistry is essential.
- Regional geochemistry – it comprises application to environmental, hydrological, and rock investigation studies.
- Photogeochemistry – it is the study of solar-induced photochemical reactions that happen or may occur between natural components of the surface of Earth.

History of Geochemistry

C. F. Schönbein, the Swiss-German chemist, first used the term *geochemistry* in 1838, though, for the rest of the century, the more general term was *chemical geology*, and there was little interaction among chemists and geologists.

Then geochemistry appeared as a distinct subject after some sophisticated laboratories were established, initiated with the United States and British Geological Survey (US-GS and BGS) in 1884, which started systematic studies of the geochemistry of minerals and rocks.

In the early 1900s, O. C. Farrington hypothesized that, while there were differences, the comparative composition of meteorites should still be the same. This was the initial stage of the field of cosmochemistry and has donated much of what we know about the formation of the solar system and Earth.

At the beginning of the twentieth century, M. von Laue and W. L. Bragg showed that X-ray scattering could be utilized to regulate the structures of crystals. Between 1920 and 1930, V. Goldschmidt and associates at Oslo University used these approaches for many common rocks and formulated a set of guidelines for how elements are grouped.

From 1960 to 2002, the investigation of M. Schidlowski was concerned with the biochemistry of the early Earth with a focus on isotopic biogeochemistry and the indication of the earliest life processes in the Precambrian age. Recently, several researchers engaged in the field of geochemistry related to groundwater quality following several geostatistical modellings, modern computer programmes, and isotope dilution methods.

1.2 Hydrogeochemistry

Natural water is a very dilute solution of naturally added chemical elements or ions. The groundwater chemistry, specifically the connection between the chemical features and suitability of waters and the local and regional geological formations, is a complicated and complex subject. Instead, the term hydrochemistry, i.e., water chemistry, mentions to the concentrations of the numerous chemical components present in a specific water source. Chemically, water is a simple chemical compound like H_2O, but aquifer water contains an extensive varies of dissolved chemical solutes and ions from the neighbouring solid environment like rocks, minerals, sediments, sands, clay, stone, etc. Maximum of these above mentioned dissolved solutes happen naturally, and many of these remain only in small concentrations. Characteristically, only six cations and anions make up approximately 95% of the total ions present in most groundwater [1, 2]. These are sodium (Na^+), calcium (Ca^{2+}), magnesium (Mg^{2+}), bicarbonate (HCO_3^-), sulphate (SO_4^{2-}), and chloride (Cl^-). Additionally, other chemical ions known as minor and trace quantities typically exist in very little concentrations, as do specific isotopes. In coastal groundwater, Na^+ and Cl^- are abundant components (over 500 mg/L), but in upland groundwater, Ca^{2+}, Mg^{2+}, HCO_3^-, and SO_4^{2-} are very common and present in large amounts. Other ions, such as NH_4^+, NO_3^-, NO_2^-, PO_4^{3-}, and F^-, are present in water at minor levels. On the other hand, groundwater contains very low amounts of heavy metals, which are very toxic to the environment and human health at specific concentrations. Some physical features of water, such as acidic or basic nature (pH), total dissolved solids (TDS), electrical conductivity (EC), turbidity, etc.; Earth's crust tracers, such as deuterium ($_1H^3$), carbon-13 ($_6C^{13}$), and oxygen-18 ($_8O^{18}$); and radioactive isotopes, for example, carbon-14 ($_6C^{14}$) and chloride-36 ($_{17}Cl^{36}$), are also giving important information about groundwater geochemistry. These chemical components and physical parameters can give us a lot of information about the origin of the water, its suitability, and its probable utilizations.

The chemical features of water of any groundwater sample have a different chemical sign. This sign replicates the aggregation of all natural and anthropogenic procedures that influenced the water from the time it started as precipitation, penetrated the topsoil overhead the water table, passed into the aquifers (discussed in the next section), and moved over long detachments and depths to the sampling station or release from the aquifers. These processes that impact hydrochemistry can deliver a significant tool for tracking groundwater movement paths, identifying solute sources, and sustainably managing groundwater.

The hydrochemistry of a particular groundwater can help us to answer many queries, such as:

- What are the actual sources of the dissolved chemical component in water?
- What is the mechanism of adding solutes from rock and minerals?
- How and how much do pollutants mix with groundwater?
- Can this water be used for domestic, industrial, agricultural, and drinking needs?

- Where does the aquifer water come from, i.e. the origin of groundwater?
- How long previously was the groundwater refilled?

Several researches have been conducted in diverse demographic areas of the world. The hydrochemical data varied with a variety of different types of geological settings, local climates, seasonal and spatial changes, etc. Different geochemical classes are primarily the chemical characteristics of groundwater and are well documented by numerous investigators. The weathering of minerals from the neighbouring rocks in the water is the leading influencing factor controlling the chemical features of these waters. According to Garrels and MacKenzie [3], some minerals, such as metal carbonates, are quickly dissolved and suggestively altered the chemical arrangement of the water, whereas other minerals or rocks, such as silicate rocks or mixed rocks, are dissolved more gradually and thus have fewer consequence on the chemical arrangement of groundwater. In this case, temperature also has a vigorous role in the chemical constituents of groundwater. Maximum studies, such as Adams et al. [4], Alberto et al. [5], and Hartman et al. [6], have shown that groundwater chemistry is mostly a function of the interaction between water and the mineralogical arrangement of the aquifer. The hydrochemical processes happening within this system are weathering, dissolution, precipitation, percolation, and cation ion exchange. These phenomena arise along with the groundwater movement route and are contingent on the residence time, which regulate the progression of the chemical composition of groundwater.

In Bangladesh, very high sodic groundwater with elevated levels of Na^+ and Cl^- (water facies is Na–Cl class) and silicate weathering is the main geochemical course in the coastal zone. But groundwater in the upper part of this country is very hard, with a high concentration of Earth metal and bicarbonate ions (water type is Ca–HCO_3), and dissolution of carbonate minerals is the key geochemical factor. The same findings were obtained from the Indian semiarid area, China river basin area, and Mexico City area, but in the Talensi District of Ghana, both carbonate and silicate mineral and rock dissolution were recognized as the major controls on groundwater geochemistry. Also, reverse ion exchange played an important role. On the other hand, it was found that the groundwater of the Asir region of KSA was oversaturated by the calcite and dolomite minerals but unsaturated with high halite and gypsum, which suggested that the weathering of gypsum and halite as a foremost process resulted in high chloride ion (Cl^-) content in aquifer water (see Chap. 4). This study accomplishes that the collective technique of a multivariate statistical approach, multivariate diagrams, plots, computer programmes, and geological modelling is operative techniques in defining the factors controlling aquifer water chemistry.

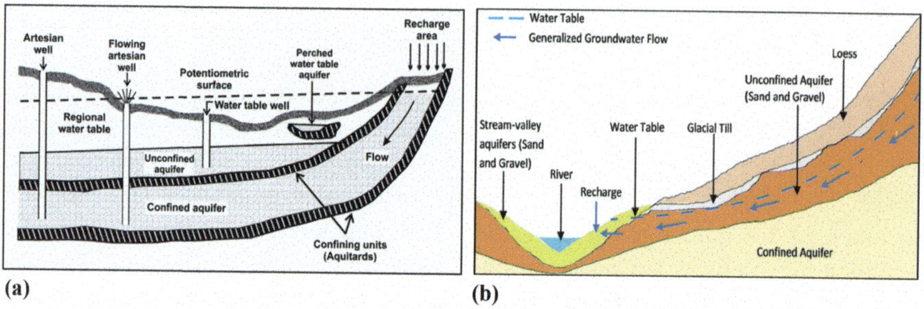

Fig. 1.2 (**a**) Recharge and discharge area in aquifers and (**b**) confined and unconfined aquifer

1.3 Aquifers and Hydrogeology

Geochemistry and hydrogeochemistry are both influenced by the local aquifer conditions. The aquifer is water-bearing underground layers, which consist of porous rock, rock breakages, or unconsolidated constituents such as gravel, silt, or sand. Groundwater can be taken out from an aquifer using a hand tube well. Aquifers differ significantly in their appearances and nature. Hydrogeology is defined as the study of water movement, the direction of water movement, the degree of recharge/discharge in aquifer, and the classification of aquifers. In regard to aquifer, two geologic terms comprise *aquitard*, which is a divan of low penetrability along an aquifer, and *aquiclude*, which is a hard, solid, waterproof zone overlying or underlying an aquifer, the pressure of which could generate a confined aquifer. Figure 1.2 shows the general sketch of aquifer conditions that included the confined and unconfined zones, confined bedrock, water table, and the age of aquifers.

1.3.1 Properties of Aquifers

To appropriately manage any aquifer, its characteristics must be understood. Several characteristics of aquifers should be recognized to identify:

- How does an aquifer respond to precipitation, contamination, scarcity, and propelling?
- Wherever and how much water arrives in the groundwater from precipitation?
- What direction and how fast does the groundwater move?
- How much water volume leaves the aquifer as springs?
- How much volume of water can be sustainably propelled out?
- How rapidly will a pollution event spread through a spring or well?

Computer and statistical programmes can be utilized to assess how exactly the understanding of the aquifer possessions matches the actual aquifer presentation.

Environmental guidelines require sites with possible sources of pollution to determine that the hydrology has been categorized.

In Fig. 1.2a, two aquifers with one *aquitard*, a confining bed and water-resistant layer, among them, enclosed by the bedrock *aquiclude*, which is in interaction with a stream, are displayed. An *aquitard* is a place inside the Earth that confines the movement of groundwater from one aquifer to another aquifer. It can occasionally, if impervious, be known as an *aquiclude*. Aquitards are made of deposits of whichever fine silty clay or nonporous rock with very low hydraulic conductivity. On the other hand, the aquiclude is made of solid hard rock with almost zero hydraulic conductivity (Fig. 1.2b).

The zone nearly below the top surface, which comprises water and air in the open places or pores, is called an unsaturated zone (Fig. 1.2b). On the other hand, a region in which all the openings and rock ruptures are occupied with water lies underneath the unsaturated neighbourhood and is called the saturated zone (Fig. 1.2b). The top of the saturated region is called the water table. Aquifers are situated from near the top surface of the Earth to deeper than 9 km. Those nearer to the surface are not only more likely to be used for water supply for drinking and irrigation but are also more possible to be refilled by local precipitation. Nevertheless, aquifers are infrequently considered underground river or lake, and they are permeable rocks saturated with water.

The aquifer condition varies with distinct geographic forms, and every formation has separate characteristics of the aquifer. A very rich aquifer exists in Bangladesh, which has two discrete shallow and deep aquifers. Geologically, this country falls under the foredeep portion of the Ganges delta. It is positioned in a floodplain of the Padma–Brahmaputra–Meghna (PBM) River and characterized by dynamic Ganga floodplain deposits with low liberation, crisscrossed by rivers and canals, and bounded by alluvial very fertile land. The lithological surface of the country consists of sedimentary sand in the south and deltaic silt and sand in the northern parts of the country. The sedimentary sands are a portion of the active floodplain of the Padma (Ganga) River and overstep the deltaic sand and slit deposits, which crop out further south (see Chap. 3). The lithology includes gravel and coarse-grained sand, which occurs at the base, and fine-grained sand and silt in the uppermost region of the study zone. The movement track of groundwater in the country is typically from northern point to southern point. Groundwater recharge happens from precipitation and floodwaters all through the rainy monsoon period, resulting in groundwater level increase. After the rainy period, share of the water is recharged back from the river, pond, stream, and wetland areas. The groundwater table drops throughout the summer season owing to overwater withdrawal for irrigation in farmland with a low specific yield.

Many desert regions of the world have limestone hills inside them or close to them that can be exploited as groundwater resources. Some part of the North Africa's Atlas Mountains, Jebel Akhdar in Oman, Lebanon and Anti-Lebanon ranges among Lebanon and Syria, and the Sierra Nevada and adjacent areas in the Southwest USA have shallow aquifers that are mining for their water needed. Excess groundwater mining can lead to the surpassing of the practical sustained produce, i.e. additional water is taken out than can be recharged. Along the coastal part of some countries, like Libya, Bangladesh, India, and Israel, augmented water demand accompanied by population overgrowth has caused a dropping of the water table and the following pollution of the groundwater with more saline water from the ocean.

Some big freshwater aquifers were identified in 2013 under the continental shelves of both North and South Africa, Australia, and China. These aquifers hold an assessed 5×10^6 km^3 of very low saline water that could be commercially treated into drinking water. These water reserves formed when sea levels were lower and fresh rainwater made its way into the ground in land regions that were not water-logged up until the Ice Age ended 20,000 years ago. The volume of aquifer water is estimated to be 100 times the amount of water mined from other aquifers since 1900.

Groundwater recharge or deep filtration is a hydrogenic procedure, where water flows down from the surface to the underground. Recharge is the principal technique through which water goes into an aquifer. This process typically happens in the vadose zones or unsaturated areas (Fig. 1.2b) below the plant roots and is often stated as a flux to the water table surface. Water recharge also includes water flowing away from the water table beyond the saturation zone. Water recharge happens together naturally (by water cycle) and through man-made courses, i.e. nonnatural groundwater recharge, where precipitation and/or domestic water is directed to the subsurface and aquifer.

1.3.2 Classification of Aquifers

The categorization of aquifers is stated as follows:

- Saturated and unsaturated aquifers
- Aquitard versus aquifer
- Confined and unconfined aquifer
- Isotropic and anisotropic aquifers
- Porous, fractured, or karst aquifer
- Transboundary aquifer

1.3.2.1 Saturated and Unsaturated Aquifer

Groundwater can be found close to every small portion in the shallow subsurface of Earth to some degree, while aquifers do not essentially hold freshwater. The Earth crust can be separated into two areas: the *saturation zone* (e.g. aquifers and aquitards), wherever all accessible places are occupied with water, and the *unsaturated zone* or the vadose zone, where there are still pouches of air that cover little amount of water but can contain more water.

A saturated state means the vapour pressure of the water is larger than normal air pressure (gauge pressure > 0), and at the groundwater table, this pressure head is equal to air pressure (gauge pressure = 0). *Unsaturated* conditions happen overhead the water table wherever this gauge pressure is negative, and the water that partially blocks up the pores of the aquifer solids is under mechanical suction process. The water contented in the unsaturated section is detained in place by surface tension, and it increases above the water table by the action of capillary to saturate a small zone above the 'phreatic surface', the location where the pore water pressure is less than air pressure. The water contented in a capillary periphery declines with cumulative distance from the phreatic surface and the capillary head depending on the size of the soil pore. In the sandy or gravel soils with greater pores, the head will be fewer than in clay-type soils with very minor pores.

1.3.2.2 Aquifer Versus Aquitard

Aquifers are characteristically saturated zones of the subsurface that yield an economically viable volume of water to any type of well or natural spring and are made of gravel, sand, silt, and/or cracked bedrock. An aquitard is a region inside the Earth that confines the movement of groundwater from one aquifer to another aquifer. A wholly water-resistant aquitard is defined as an aquiclude, which was discussed earlier. Aquitards contain layers of clay and nonporous rocks with small hydraulic conductivity.

In hilly areas, the primary aquifer is usually unconsolidated alluvial sediment, collected of typically horizontal layers of ingredients deposited by water courses like rivers and channels, which in cross sections seem to be layers of irregular rough and fine substances. Coarse or rough substances, because of the elevated energy required to interchange them, tend to originate closer to the source (mountain fronts or streams), though the fine sand will make it beyond the source (called the pressure area). Meanwhile, there are fewer fine sand deposits adjacent to the source, which is a place wherever the aquifer is often unconfined (i.e. forebay zones) or in hydraulic connectivity with the terrestrial soil layer.

1.3.2.3 Confined and Unconfined Aquifer

Generally, there are three end fragments in a variety of kinds of aquifers: confined, semiconfined, and unconfined. An unconfined aquifer is where the rock is just open at the permeable surface of the ground and groundwater is recharged directly by rainfall, streams, or snowmelt. Sometimes this is also called the water table or phreatic or shallowest aquifer. A confined aquifer is extended beyond by a confining layer, often made up of a clay impermeable layer. The constraining layer might offer some defence from surface pollution. The semiconfined aquifer is positioned between the confined and unconfined layers. Figure 1.3 displays both confined and unconfined aquifers.

At times, geologically, the variance between confined and unconfined aquifers is not fully understood. In this condition, the value of storability gives back from an

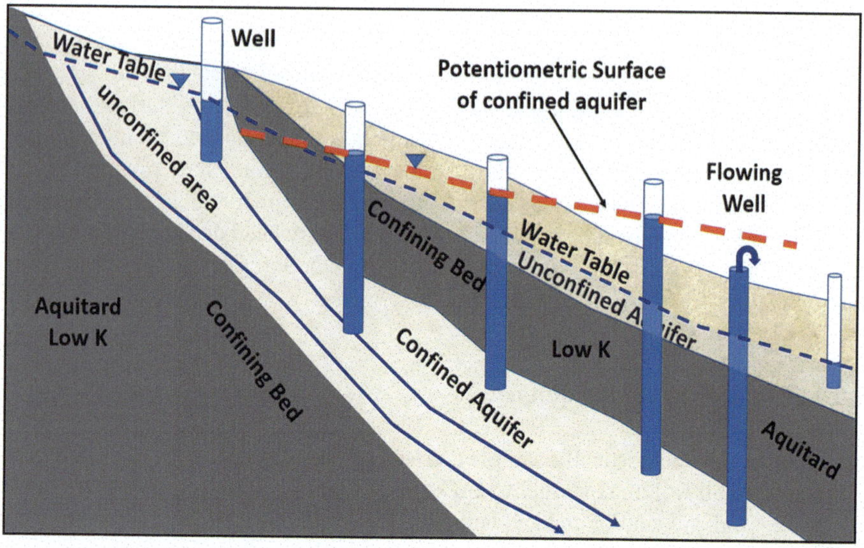

Fig. 1.3 Confined and unconfined beds and various types of wells

aquifer test can be used to fix it. The confined aquifer has very low storability values, which means that the aquifer is storage water utilizing the technique of aquifer matrix extension and the squeezability of water, which characteristically are both relatively fewer amounts. The unconfined aquifer has a storability greater than 0.01, i.e. 1% of the total volume. It releases water from storage by basically demanding the pores of the aquifer, discharging fairly large volumes of water. The dissimilarities between confined and unconfined aquifers are explained in Table 1.1.

Not only dissimilarities but also some similarities between confined and unconfined aquifers are as follows:

- They are found underground where they serve as a storage house for numerous uses.
- Both have an underlying impervious inside layer.
- The filtration process is the main operator of water flows through them.
- Both aquifers can artificially recharge to refill both when their water content is very low.
- Hydrogeology is defined as the study of groundwater flow in both aquifers.

1.3.2.4 Isotropic and Anisotropic Aquifers

The hydraulic conductivity is identical for water movement in all directions in the isotopic aquifers, while in the anisotropic aquifers, it fluctuates, specifically in vertical and horizontal lines. A semiconfined aquifer with one or more aquitards

Table 1.1 Differences between confined and unconfined aquifers

Category	Confined aquifers	Unconfined aquifers
Definition	It is the aquifer with a deposit of a porous substance above it	This aquifer is overlaid by a layer of penetrated rock or fine sandy clay
Closeness of surface	It is situated close to the Earth's surface	It typically occurs at a considerable depth below the Earth's surface
Formation of aquifer	These aquifers form when water infiltrates the soil and permeates down to the water table over the layer of porous and penetrable materials	It forms when water seepages between layers of permeated rocks under the slow effect of gravity and pressure
Pollution	The restricted layer in this aquifer offers some protection from surface pollution or contamination	At a more risk of pollution from outside sources like rivers or drains and from weakly managed human events like septic leaks and unnecessary uses of pesticides and herbicides in agriculture
Rate of recharge	Take a long time to become recharged through the aquifer	The rate of recharge is faster than in confined aquifers
Driving rate	The flow rate is very high even when it is retarded by an aquitard	The flow rate is very low due to the presence of an impervious inside layer
Neighbouring layers	Characteristically protected by an aquitard and an aquiclude placed above and below it	Constrained by an aquiclude above and below it
Degree of storability	It has storability values greater than 0.01	It has storability values much less than 0.01
Hydraulic conductivity	It has higher hydraulic conductivity	It has very low hydraulic conductivity
Alternative names	Another name is water table aquifer	Common name is artesian aquifer

functions as an anisotropic aquifer, even when the distinct layers are isotropic since the values of the vertical and horizontal hydraulic conductivity are dissimilar. When scheming movement to drains or well in an aquifer, the anisotropy must be taken into account in case the resultant structure of the drainage systems may be broken down.

1.3.2.5 Porous, Karst, or Fractured Aquifer

Porous Aquifers

Generally, this kind of aquifer occurs in sand and sandstone basements. The properties of porous aquifers depend on the deposited alluvial basements and lithologic cementation of the sand particles. The horizontal and vertical orientation of the sand particles, as well as the spread of shale layers, is controlled by the aquifer basements where a sand body was deposited. Groundwater movement is significantly impeded by these thin shale layers. These elements affected the penetrability and porosity of sandy groundwater. In areas where the aquifer is close to the surface, precipitation enters the groundwater levels. Potentiometric surface maps of the water levels in springs and wells can be used to identify the directions of groundwater movement. The porous aquifers have been found to have a groundwater discharge rate of 0.3 m/d. Although porosity is significant, it does not control the rock's capacity to serve as a reservoir on its own. The Deccan Traps in West Central India is an excellent example of a basaltic lava formation with high porosity but low penetrability, making it a poor aquifer. Like the Upper Cretaceous Chalk Group in southeast England, which is microporous and has a low grain permeability despite having a logically high porosity, the excellent water-yielding characteristics are primarily the result of micro-fracturing and micro-fissuring.

Karst Aquifers
Underlying water reserves in karstified rocks are known as aquifers. They are fuelled by water from waterways and precipitation, just like all aquifers. This water seeps through the Earth and can be accessed by wells for human use. The water eventually hits a location at a lower elevation where springs cause it to resurface. The complexity of this aquifer type is unmatched anywhere else in the globe. Most water is produced by karst aquifers, but they are also the most prone to pollution. In the USA, karst aquifers are a major source of groundwater. In this nation, karst aquifers provide about 50% of the groundwater used for household purposes. Australia, China, and other nations are other areas of the globe with significant karst terrain. Australia, China, Europe, and the Caribbean are some other parts of the globe that have significant karst terrain.

Fractured Aquifers
If a low-porosity rock unit is heavily fractured, it can also function well as an aquifer (via fracture flow), providing the rock has a sufficiently high hydraulic conductivity to allow water to flow through it easily.

1.3.2.6 Transboundary Aquifer

Transboundary aquifers are those that support multiple nations or independent states. So, the management of these resources depends on international collaboration, and they must be thoroughly understood to ensure that they are used sustainably. The aquifer's physical characteristics are successfully overridden by the socioeconomic and political contexts, which also increase its geostrategic value. The criteria outlined in this approach make an effort to enumerate and quantify all possible factors that could be involved in determining an aquifer's transboundary nature and its multidimensional boundaries.

1.4 Human Use of Aquifer Water

Using groundwater presents difficulties such as drafting (pulling out the groundwater outside the equilibrium production of the aquifer), land subsidence brought on by groundwater, groundwater turning salty, and groundwater pollution. Groundwater dependence will only expand as a result of rising water demand across all sectors and changing rainfall patterns brought on by climate change. Globally, from 1960 to 2000, the rate of global aquifer water depletion rose from 126 (±32 km^3/y) to 283 (±40 km^3/y) km^3 [7]. According to the nation or continent, the overall groundwater depletion and water table decline in Libya, Australia, India, and its neighbouring countries, Bangladesh, Mexico, and the North China Plain, are extremely risky situations. However, modern methods of managing groundwater, such as artificial recharge and surface water injection throughout periodic wet and dry times, have increased the lifespan of numerous freshwater aquifers, particularly in the USA. The Great Artesian Basin in Australia may be the world's biggest groundwater aquifer (over 20,000 km^3). It contributes significantly to the water sources for Queensland and some isolated regions of South Australia. But from 2000 to 2008, 6 km^3 of this aquifer's freshwater was exhausted.

United States (USA)
The Ogallala Aquifer in the central USA, which is one of the world's largest aquifers, is rapidly depleting in some areas due to increased urban use and continued agricultural use. Underneath parts of eight states, this vast aquifer is composed mostly of Ice Age petrified water. In the drier regions of the aquifer, annual replenishment is believed to be only approximately 10% of yearly water extractions. According to a United States Geological Survey (USGS) study, about 30% of all depletion in the twentieth century occurred between 2001 and 2008 [7]. Major consumers of aquifer water in the USA are oil and coal exploration and agricultural irrigation. Total groundwater losses in the USA increased in the late 1940s and have remained nearly linear throughout this century. Groundwater depletion will unfavourably affect the long-term sustainability of groundwater resources to meet the country's water demands. In Central Texas, the Edwards Aquifer is an example of a large, long-lived carbonate aquifer. About two million people, who traditionally depend on this carbonate aquifer for quality water, still have access today, thanks to massive recharge from nearby streams, rivers, and lakes. Population growth in recharge areas poses the greatest threat to this resource.

Canada
The Basement Water Sand (BWS) aquifer is a discrete sand mass at the foot of the McMurray Formation in the Athabasca oil sand area of northeastern Alberta, Canada. They are trapped beneath water-resistant bitumen-saturated sand that is used to obtain well bitumen for the creation of synthetic crude oil because they are saturated with water. They are deep-lying, salinity replenished by the underlying Devonian formations, and shallow-lying, non-salinity charged by surface waters. Whether bitumen is obtained by open pit mining or by field techniques such as

steam-assisted gravity drainage (SAGD), BWS typically presents challenges and is subject to wastewater discharge at some sites.

South America

One of the largest aquifer networks in the world and a major source of freshwater, the Guarani Aquifer lies beneath Argentina, Brazil, Paraguay, and Uruguay. It has a surface area of 1.2 million km^2, a volume of approximately 40,000 km^3, a thickness of 50–800 m, and a determined depth of approximately 1800 m and is named after the Guarani tribe.

India

India consumes more water from aquifers than the USA and China combined, making it the largest aquifer user of the world. Today, groundwater is the principal source of water for the majority of Indians who depend on it for both domestic and irrigation practices. The extensive extraction of this invaluable resource has contributed to its alarming decline, while groundwater sparked the Green Revolution that made India a country with safe food supplies. Groundwater will become even more substantial as a result of climate change, which will make precipitation patterns more unpredictable. Falling groundwater tables now pose a danger to 63% of India's districts. Smallholder farmers are particularly vulnerable in areas where water tables are below 8 m and poverty rates are alarmingly high at 9–10%. At least 25% of India's agriculture will be at risk if present trends continue. In 2020, the World Bank began supporting a government programme in over 9000 water-scarce village councils in seven states such as Haryana, Uttar Pradesh, Rajasthan, Gujrat, Maharashtra, Madhya Pradesh, and Karnataka in India. These states account for about 25% of the areas with contaminated groundwater. The region contains a variety of hydrogeological systems, including shallow hard rock aquifers in the Indian Peninsula and deep sedimentary aquifers in the Indian Gangetic Plain.

1.5 Summary

Water chemistry encompasses the chemistry and quality of water in complex environments. Water chemistry studies characterize water quality by measuring the concentrations of parameters and comparing them to potable water and irrigation water standards. We collected groundwater samples and analysed their physicochemical properties to understand the chemistry of the water. Hydrogeology, on the other hand, deals with how water enters the soil layer, how it moves underground, and how groundwater interacts with adjacent soils/sediments and rocks. These processes take place in aquifer cellars. Aquifers are sedimentary and rock/mineral masses that contain groundwater. Groundwater is a term used to describe precipitation that seeps beyond the surface and into the ground and collects in the open space below the surface. There are two general kinds of aquifers: confined and unconfined.

Rock weathering, evaporation, ion exchange, and carbonate mineral dissolution were recognized as other significant hydrogeochemical courses affecting the chemistry of groundwater in the study zone.

References and Further Study

1. USGS. *Chemical characteristics of water in the aquifer system.* Chapter 6. United States Geological Survey. https://pubs.usgs.gov
2. Miao, Q., Li, X., Xu, Y., et al. (2021). Chemical characteristics of groundwater and source identification in a coastal city. *PLoS One, 16*(8), e0256360. https://doi.org/10.1371/journal.pone.0256360
3. Garrels, R. M., & Mackenzie, F. T. (1967). Chapter 10: Origin of the chemical composition of some springs and lakes. In W. Stumm (Ed.), *Equilibrium concepts in natural water systems* (p. 222). American Chemical Society. https://doi.org/10.1021/ba-1967-0067.ch010
4. Adams, S., Titus, R., Pietersen, K., Tredoux, G., & Harris, C. (2001). Hydrochemical characteristics of aquifers near Sutherland in the Western Karoo, South Africa. *Journal of Hydrology, 241*(1–2), 91–103. https://doi.org/10.1016/S0022-1694(00)00370-X
5. Alberto, W. D., María del Pilar, D., María Valeria, A, et al. (2001). Pattern recognition techniques for the evaluation of spatial and temporal variations in water quality – A case study. *Water Research, 35*(12), 2881–2894. https://doi.org/10.1016/S0043-1354(00)00592-3
6. Hartmann, A., Goldscheider, N., Wagener, T., et al. (2014). Karst water resources in a changing world: Review of hydrological modeling approaches. *Reviews of Geophysics, 52*, 1–25. https://doi.org/10.1002/2013RG000443
7. USGS. (2018). *Groundwater decline and depletion.* Water Science School, United States Geological Survey.

Chapter 2
Groundwater: Sources, Functions, and Quality

Abbreviations

BBS	Bangladesh Bureau of Statistics
DOC	Dissolved organic carbon
FAO	Food and Agriculture Organization, United Nations (UN)
IWMI	International Water Management Institute
NAQUA	National Aquaculture Group (Saudi Arabia)
OXFUM	Oxford Committee for Famine Relief, founded in the UK in 1942
STP	Sewage treatment plant
UNICEF	United Nations International Children's Emergency Fund
USD	United States dollar
US-EPA	United States Environmental Protection Agency
WHO	World Health Organization

M. S. Islam, *Hydrogeochemical Evaluation and Groundwater Quality*,
https://doi.org/10.1007/978-3-031-44304-6_2

2.1 Groundwater: *An Unlimited Renewable Resource*

Groundwater is one of the greatest valued and unlimited renewable natural resources, although perhaps one never sees it or even realizes it is there. It is freshwater situated in the underground pore space of soil and rocks. It is also water that moves inside aquifer underneath the water table (Fig. 2.1). Occasionally, it is useful to differentiate among groundwater that is strictly connected with surface water and bottomless groundwater (called fossil water) in an aquifer.

A big volume of water is present in the ground beneath our feet, and the public all over the world makes countless use of it. Around 30% of all readily accessible freshwater in the world is aquifer water. A unit of rock or an unconsolidated deposit is named an aquifer when it can produce a serviceable volume of water (see Chap. 1). The depth at which soil pore spaces or fractures and voids in rock become entirely saturated with water is said to be the water table. Groundwater is restored from surface sources of the water body; it may naturally discharge from the water table at leaks and springs and can form wetlands. It is also frequently extracted for industrial, municipal, and agricultural use by hand and engine wells. Classically, it is assumed to be water flowing through the shallow aquifer, but in the practical sense, it can also comprise soil moisture, frozen soil, steady water with very low penetrability bedrock, and deep geothermal water. Groundwater is supposed to deliver lubrication that can perhaps affect the drive of faults. Much of Earth's matter probably comprises some water, which may be mixed with other fluids in some cases.

Groundwater is very often cheaper, very suitable, and less susceptible to contamination than surface water. Surface water has a great chance of polluting it and causing water shortages. Thus, it is generally used for public water supplies. For instance, groundwater delivers the chief source of functioning water storage in the USA, and highly populated countries such as Bangladesh, India, and China annually withdraw a large amount of groundwater. In Bangladesh, about 14 crore rural people used raw groundwater for every purpose. Except for some riverine countries, for

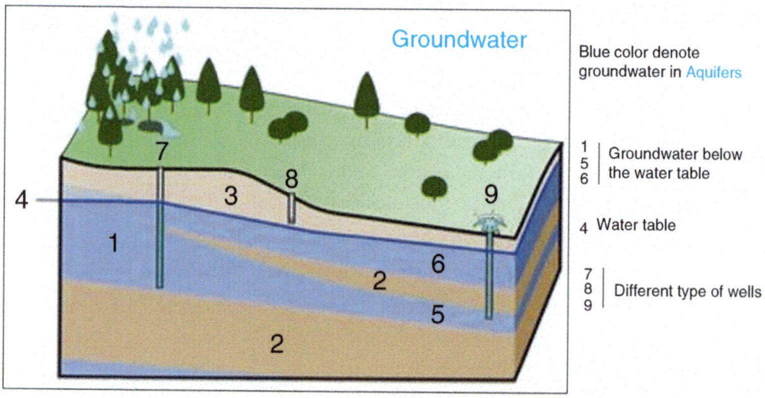

Fig. 2.1 Groundwater profiles beneath the Earth's surface

example, aquifer reservoirs contain far more water than all surface reservoirs and lakes in the USA, including the Great Lakes. Various municipal water supplies result exclusively from subsurface water. Worldwide, over 3 billion individuals depend on it as their prime water source. The utilization of groundwater has been correlated with environmental subjects. Perhaps, contaminated groundwater is less observable and more problematic to clean than contamination in surface water bodies. Groundwater contamination most often results from the indecorous disposal of household and industrial effluents on the land surface. Moreover, groundwater is vulnerable to seawater intrusion in seaside zones and can cause the land to subside when extracted untenably, leading to sinking cities (e.g. Florida, Bangkok, Indian shore) and loss in elevation (e.g. the central basin of California). These matters are made more complex by the rise of sea levels and other changes caused by global climate change.

2.2 Groundwater in the Bedrock

Groundwater is an important portion of the natural water cycle, and some portion of the rainfall that lands on the surface penetrates the subsurface. But it is difficult to imagine water underground. The portion that remains descends through the soil up until it reaches bedrock that is saturated by groundwater recharge (Fig. 2.2). Water in the saturated groundwater zone moves gradually and may ultimately discharge into oceans, rivers, streams, and lakes.

Groundwater is refilled by rain and, dependent on the local climate and lithological condition, is unequally dispersed in both quality and quantity. When

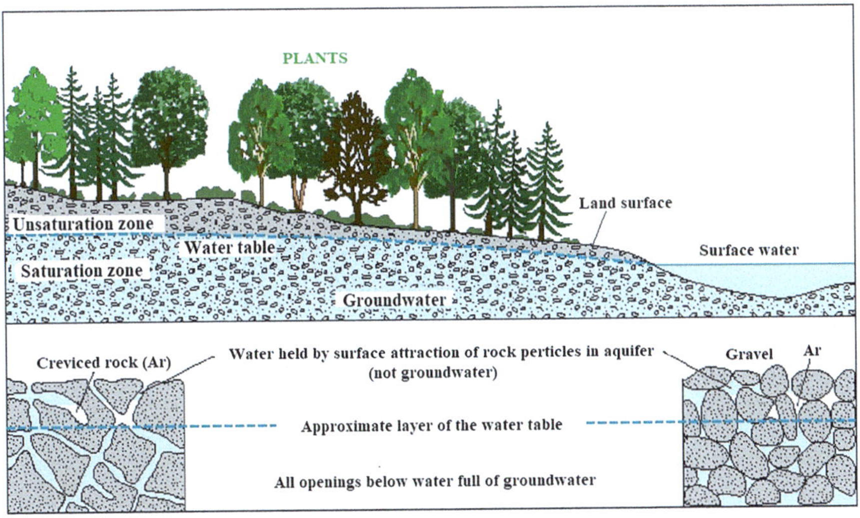

Fig. 2.2 Groundwater beneath the surface

precipitation occurs, some part of the water evaporates, some is transpired by plants, some streams are transported overland and collected in watercourses, and some portion penetrates the pores of the soil and rock minerals. The water that passes in the soil substitutes for water that has evaporated or is used by plants. Among the soil surface and the aquifer water is an area called the unsaturated area. In this area, there typically is at least a slight water, generally in smaller pores of the rock and soil; the greater openings generally contain air in place of water. After substantial rainfall, the zone may be almost saturated; after the widespread dry curse, it may be almost dry. Some part of the water remains in the unsaturated area by the surface attraction of rock particles, and it will not move toward or arrive at a water well (Fig. 2.2). After the water supplies for soil and plants are fulfilled, any extra water will penetrate to the water table at the upper portion of the zone under which the pores in rocks are saturated. Underneath the water table, all the vacuum in the rocks is full of water that transfers over the aquifer to springs, streams, or wells from which water is extracted. Normal replenishment of the aquifer at depth is a slow procedure because groundwater transfers gradually among the unsaturated zone and the aquifer.

Water flow in aquifer basements is extremely reliant on the penetrability of the aquifer substance. Porous substance contains consistent cracks or spaces that are large enough to allow water to transfer easily. In some porous matter, groundwater may flow quite a few meters in a day; in other places, it flows only a few cm in a century. Groundwater travels very gradually through comparatively resistant constituents such as clay and shale.

2.3 The Role of Groundwater in the Water Cycle

The involvement of groundwater in the water cycle is shown in Fig. 2.3. Groundwater can be viewed in the same terms as surface water: input, output, and storage. Natural entry into groundwater occurs through seepage from surface water. The natural outlets of groundwater are springs, which percolate into the sea. Due to the slow rotation speed, groundwater storing is usually much higher (by volume) than

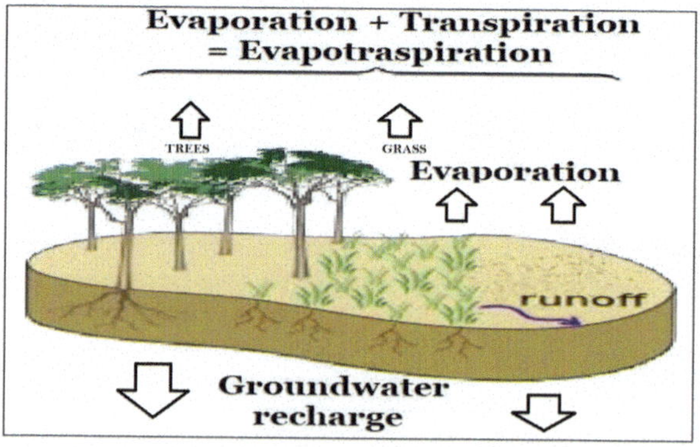

Fig. 2.3 Groundwater in the water cycle

the inflow to surface water. This difference makes it easier for humans to use groundwater for long periods without serious consequences. However, in the long term, the average leaching rate from a groundwater source caps the average water feeding from that source. It can be a long-term 'reservoirs' of the natural water cycle (with residence times ranging from days to thousands of years), as opposed to short-term reservoirs of water like the air and fresh surface water (with residence times ranging from minutes to years). It can take a long time for deep groundwater (which is fairly far from surface recharge) to complete its natural cycle.

2.4 Groundwater Recharge

Groundwater recharge is a part of the hydrologic cycle; water movement occurs downwards through water leaks from surface water sources (depression-focused recharge) or precipitation infiltrates (diffuse recharge) from the water table to the saturated zone. In these two processes, water arrives through an aquifer and encompasses water flow in the vadose zone. Depression-focused recharge is very important in arid regions. This type of recharge also deeply affects pollutant transport into groundwater. This is of great apprehension in areas with karst geological formations because water can ultimately dissolve tunnels to aquifers, or if not detached streams. Water recharge can help to transport additional salts or nutrients that gather in the root zone to deeper soil layers or into aquifer systems. Plant roots increase water saturation into groundwater dropping water runoff. Flooding for the moment increases riverbed penetrability by moving clay deposition downstream, which increases the aquifer recharge rate. The most common procedures to estimate recharge rates are isotopic tracers, chloride mass balance, the water balance method, the soil physics method, the groundwater level fluctuation method, and the estimation of baseflow to rivers.

Groundwater recharge can be performed through three processes:

- Natural recharge
- Artificial recharge
- Recharge through wetlands

Natural groundwater recharge occurs naturally by rainfall (rain feed) and to a lesser extent by surface water (rivers, ponds, lakes, etc.). Recharge may be hampered to a certain degree by human actions, as well as development, paving, or water logging. These actions can result in damage to topsoil, resulting in decreased water penetration, boosted surface runoff, and lessened recharge. Utilization of groundwater, specifically for irrigation events, may also reduce the water tables. Aquifer water recharge through natural processes is a significant course for justifiable groundwater management since the volume rate distracted from the aquifer in the long period should be less than or equal to the degree of volume that is recharged.

Artificial groundwater recharge occurs through man-made actions such as the storage of water by rainwater harvesting; making artificial lakes, ponds, reservoirs, etc.; and other surface water sources routed to the subsurface. This recharging process is becoming gradually important in India and the USA, wherever the misuse of groundwater for irrigation has led to aquifer resources becoming depleted. According to the recommendation of the International Water Management Institute in 2007, the Indian government to be paid 580 million USD in 2020 to fund dug-well recharge programmes in hundred districts within seven provinces where the water kept in hard rock aquifers had been overexploited [1].

Wetland support preserves the level of the water table and employs control on the hydraulic head. It delivers forces for groundwater recharge and discharge to other water sources as well. The degree of groundwater recharge through a wetland is reliant on the ratio of perimeter to volume, soil, vegetation, site, and water table rise. This type of recharge happens through the mineral soils that are found mostly around the boundaries of wetlands. The soil under maximum wetlands is quite water-resistant. A high perimeter-to-volume ratio, such as in small wetlands, means that the surface part concluded in which water can penetrate the aquifer is high. Groundwater recharge is characteristic in small size wetlands such as savanna potholes, which can donate suggestively to the recharge of local groundwater resources. Several investigators have revealed groundwater recharge of up to 20% of wetland volume per time of year [2].

2.5 Groundwater Availability

Groundwater makes up approximately 30% of the global freshwater demand, which is 0.75% of the total water volume, including oceans and ice. Almost all of the world's liquid freshwater is groundwater. Global groundwater storing is roughly equivalent to the total amount of freshwater deposited in the ice pack, including the north and south poles. This makes it a significant natural storage that can act as a buffer against surface water shortages, such as through times of drought.

The amount of groundwater in the aquifer is measured by measuring the water level in resident wells and examining the geological record from well bores to determine the extent, depth, and thickness of water-bearing sediment and rock. We can estimate. Before investing in production wells, test wells can be drilled to measure the depth at which water occurs, and soil, rock, and water samples can be collected for laboratory analysis. Pumping tests can be performed on test wells to characterize the aquifer flow.

Aquifer features vary depending on the geology and structure of the aquifer and the terrain in which the aquifer occurs. Generally, more productive aquifers occur in sedimentary strata. By contrast, weathered and cracked crystalline rocks provide small amounts of groundwater in numerous environmental conditions. The most productive groundwater sources include loose or poorly consolidated alluvial material that accumulates as valley-filling sediments in large river valleys and geologically subsiding tectonic basins.

Another estimation quantified that groundwater accounts for approximately 35% of all water extractions globally and surface water accounts for the other two-thirds. A similar approximation was published in 2021, which showed that 'groundwater is estimated to supply between a quarter and a third of the world's annual freshwater withdrawals to meet agricultural, industrial, and domestic demands'.

Global freshwater withdrawals are thought to have been about 600 km^3 per year in 1900 but increased to 3880 km^3 per year in 2017. This increase was particularly high in the period 1950–1980 (about 3% per annum), partly due to the rising rate of population growth and partly due to the rapid increase in groundwater exploitation, especially for irrigation. The growth rate is about 1% per year (as of 2022), which corresponds to the present population growth [3].

Estimates indicate that global groundwater loss is 100–300 km^3 per year. This decline is mostly due to the increase of irrigated agriculture in dry areas. The Asia-Pacific region is the world's major pumper of groundwater, containing seven of the top ten pumping countries (India, Pakistan, Bangladesh, China, Iran, Indonesia, and Turkey). These countries alone account for approximately 60% of the world's total groundwater withdrawals [4].

2.6 Groundwater Pumping

Groundwater occurs in the saturated rock and soil beneath the soil sub-layer. The public can drill simple hand wells into the aquifer and collect water if it is shallow and penetrable enough to permit water to move through it at a sufficiently quick rate. The depth of the groundwater table can naturally change over time owing to changes in weather patterns and rainfall events, streamflow change, geologic variation, and changes in anthropogenic activities. The pumping of well water can have a great deal of impact on water levels below ground, particularly in the locality of the well. If water is withdrawn from the aquifer at a quicker rate than it is refilled, whichever by penetration from the rainwater or streams, then the water table can become lower; this situation is called the 'cone of depression' around the well site. Depending on the geologic and hydrogeologic circumstances of the aquifer, the influence on the water table level can be short-lived or last for decades, and it can drop an insignificant amount or many hundreds of feet. Overwater mining can lower the water table so much that the wells no longer supply water and they can go dry. Recently, Bangladesh and India have faced this problem. In the dry season (April–June), due to lowering the water table, around 80% of hand tubes will become inactive. Overall, the performance of water pumping depends on the depth of the water table, which varies around the world. The deep or light green zone of the map of Fig. 2.4 is in very excellent conditions for water withdrawal, but the opposite situation occurs in the red zones.

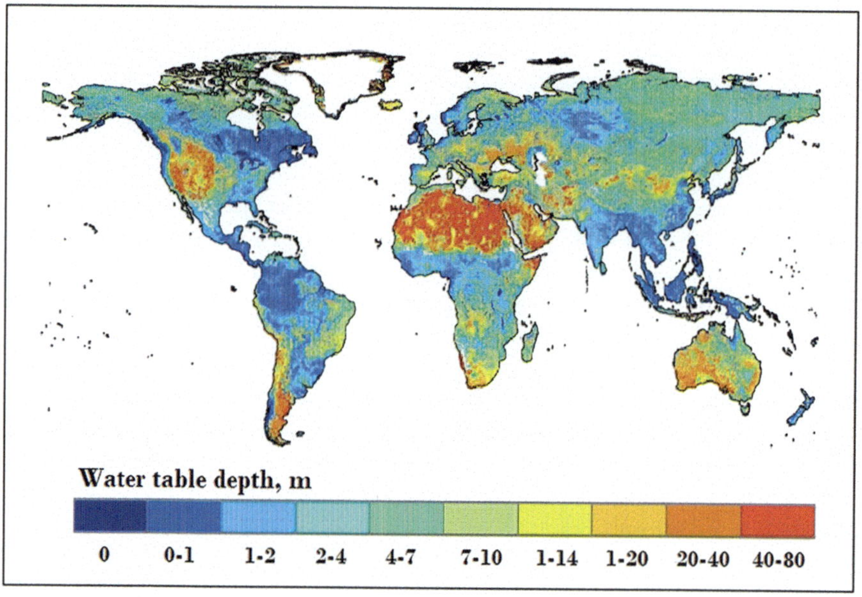

Fig. 2.4 The average depth pattern of the global groundwater table. The white areas indicate the regions where a stable shallow groundwater table may not exist. (Modified from Zeng et al. [5])

Figure 2.2 shows how the soil beneath the water table (blue line) is saturated with liquid water. Unsaturated zones (grey areas) overhead the water table still contain water (plant roots thrive in these areas) but are not completely saturated with water. Occasionally, the porous rock formations in the ground can tilt. Both below and above the porous layer, there may be a limiting layer of less porous rock, with the rock adjacent to the aquifer limiting the pressure of the porous rock and its water. When a well is drilled into this pressurized aquifer, the internal pressure is sufficient to push water out of the well to the surface, possibly completely out of the well without the aid of a pump. This is called a self-flushing well. Water pressure in a self-flowing well can increase relatively dramatically from day to day.

A connection does not essentially occur among the water-bearing capacity of rock and the depth at which they originate. A compact granite rock that will produce little or no water to a well may be uncovered at the land surface. On the other hand, a porous sandstone may situate hundreds of meters underneath the land surface and may produce hundreds of litters per minute of water. Rocks that harvest freshwater have been found at depths of more than 2000 m, and salty water has to originate from oil wells at depths of more than 10,000 m. Averagely, nevertheless, the penetrability and porosity of rocks decrease as their depth below the land surface increases; the cracks and pores in rocks at great depths are closed or greatly abridged in size because of the weight of the covering rock.

2.7 Groundwater Chemistry

Basically, groundwater is a very dilute solution of some natural components that are usually mixed at the aquifer level through a complex mechanism with several chemical changes. It contains many elements and free ions, both naturally occurring and man-made. Some are present in big concentrations, while others occur in trace quantities. Some are also chemically dissolved in the water, whereas larger constituent parts are generally suspended. Normally, it is convenient to divide dissolved ingredients into major components (main cations and anions) and trace metals. The level of dissolved ingredients is usually stated in mg/L for major components such as Na, Ca, Mg, HCO_3^-, and Cl^- and μg/L for trace elements such as As, Hg, Cd, and Co. Some rare and radioactive elements are expressed in ng/L (nanogram/litre). The main dissolved constituents of groundwaters are characteristically present at levels in the range of a few mg/L to more than a few hundred mg/L.

All dissolved components in groundwater consequently interact with the atmosphere, the surficial environment, soil, and bedrock. Generally, groundwater chemistry is mostly regulated by four factors: rechargeable surface water chemistry, man-made contamination, soil processes, and aquifer mineralogy (Fig. 2.5). The level and characteristics of dissolved ions differ in the age and depth of the aquifer. Groundwater contains much higher levels of most ingredients than surface water, and deep groundwater that has been in contact with rock formation for a long time tends to have elevated levels than shallow and young water. Younger groundwater is usually associated with land use pollution, whereas problems involving older waters are more likely to result from geogenic evolution processes.

Groundwater chemistry is impacted by many factors, such as the quality of penetrated water, the geographical characteristics of aquifer rocks/minerals, the partial pressure of subsoil carbon dioxide (pCO_2), the decay of organic matter, and numerous anthropogenic issues [6]. The chemical features of groundwater are a function of the dealings between water and the mineralogical construction in the aquifer system. Hydrogeochemical processes occurring inside this system are dissolution/weathering, precipitation, percolation, and ion exchange. These phenomena occur with the groundwater movement direction and rely on the residence time that

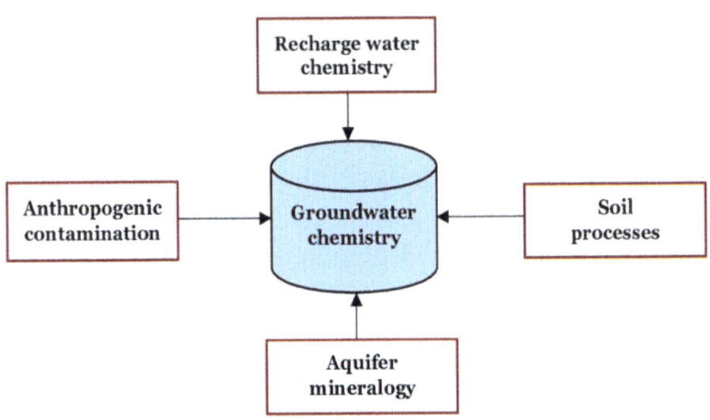

Fig. 2.5 Factors involved in groundwater chemistry

regulates the chemical arrangement of the groundwater. Aquifer water degradation occurred due to geological variation, overexploitation, trace metal contamination, local soil properties, and heavy agrochemical deposition on the topsoil. Water–soil and groundwater–rock interactions suggestively affect the aquifer water chemistry, and this mechanism defers with the place and period.

2.8 Groundwater Quality and Pollution

Groundwater is the chief valuable natural resource for all creature survival worldwide, while it is invisible. If its quality is deteriorating at an alarming rate owing to the shifting environment and severe human actions, it poses widespread health risks to the community who consume it and take baths using that water. Groundwater pollution occurs when unwanted harmful materials such as chemicals or microbes contaminate aquifer water bodies, degrading water suitability and causing toxicity to natural creatures or the environment. The pollution procedure of this resource is very complex and occasionally incredible to clean up. Polluted groundwater can lead to negative effects on animals, the environment, ecology, and human beings. Primarily, groundwater is the chief source of drinkable water for most people and animals worldwide. Once groundwater is polluted with toxic chemicals and microbes, humans and animals drink harmful materials through drinking water and then suffer physiological problems such as cholera, typhoid, diarrhoea, amoebae, and even cancer. Besides, the plants and vegetation that depend on groundwater are likely to dry up and lose their growth after using contaminated water. Therefore, the loss of plantations leads to an inequity in the ecosystem or biodiversity. Similarly, contaminated groundwater may seep into rivers, ponds, and lakes and lead to the loss of marine aquatic life (mostly fish), which is harmful to the environment and human beings. Finally, when groundwater is polluted with reactive chemicals, it may result in several chemical reactions that abolish the soil environment around the area. The significances of demolished soil include poor plant development and poor soil quality.

Groundwater is an important hidden resource because its quality and quantity vary with the lithological and hydrological environment. Usually, it is considered harmless to infectious contamination. However, due to the presence of inorganic pollutants from underlying minerals and rocks or anthropogenic activities, aquifers are not easily rehabilitated, exacerbating the effects of pollution. Hydrogeochemical processes include rock weathering, precipitation, charge/discharge, redox, ion exchange, mineral hydrolysis, the residence time of water mixing, etc. [6]. It can affect the compositional state of groundwater. Anthropogenic activities such as excessive groundwater depletion, chemical runoff, fertilizers and pesticides, industrial and transport waste, sewage seepage, and landfills also affect groundwater quality. Not only As pollution but also other heavy metals and several anions such as Co, Cd, Sb, Cr, Pb, Hg, F^-, NO_3^-, NO_2^-, SO_4^{2-}, PO_4^{3-}, etc., are very likely to contaminate groundwater by geogenic or anthropogenic activities [7]. Previous studies have focused on groundwater in South Asian countries, which contains excessive amounts of Ca, Mg, and Fe, which are not harmful metals. In Bangladesh, 37.258 billion tonnes of pesticides and 2.32 billion kg of chemical fertilizers are used on agricultural land each year [8]. The rest of this seeps through the topsoil and eventually reaches the aquifer, leading to water pollution. The chemical and bacteriological properties of groundwater are very important, as they determine whether it is appropriate for domestic, agricultural, and industrial use.

2.9 Assessment of Water Quality

Groundwater is the comparatively harmless source of potable and sustainable agricultural practices, particularly where surface water is unobtainable and polluted. The nature and method of water quality assessment are quite different from the utility of that water. The concentration ranges of some components in drinking water are very sensitive but not for irrigation. The irrigation water quality is mostly subject to the concentration and types of numerous dissolved solutes. Based on these salts, quite a few irrigation water quality parameters and indexing models were recognized to assess the suitability of water for sustaining the soil environment and improving crop harvest. Many of the areas are vastly irrigated zones of the world where groundwater is the only source of irrigation activities, but the quality of that water is not always justified. Some investigations have expected that groundwater degrades daily for irrigation and that degradation characteristically occurs due to regular geochemical dissimilarity, excess water withdrawal, enormous arsenic contamination, the interface of groundwater with saltwater, and agrochemical deposition on the soil surface. Maximum investigations have been conducted in coastal zones where salinity is a thoughtful problem. But the upper portions and river basin, calcium hardness, and excess trace metal load in groundwater are the main quality matters in many countries. So, it is vital to assess the water quality for groundwater irrigation practices. Several methods for the evaluation of irrigation water quality in different geomorphological areas will be explained in subsequent chapters.

The chemical standard for drinking water is quite different from that for irrigation water. Water quality standards and threshold limits for drinking water are generally guided by the WHO and the US-EPA, while many countries have national guidelines for their region. Numerous water quality indices, such as the heavy metal evaluation index (*HEI*), degree of contamination (C_d), and heavy metal pollution index (*HPI*), have been proposed for the valuation of water quality considering trace metal elements. These indices are used as supporting factors for the evaluation of the health risk assessment (*HRA*) of groundwater in an area. Risk measurement is well defined as a procedure used to estimation health effects that might result from exposure to carcinogenic and noncarcinogenic substances. It is classically performed in four steps: risk documentation, exposure evaluation, toxicity assessment, and risk categorization. The major routes of heavy metal intake in the human body are oral, dermal, and nasal through drinking water, foodstuff, and dust. Ingestion and dermal absorption are the common water intake paths.

Numerous indexing methods have been used to evaluate water suitability in a suitable and more comprehensible way. The water quality index (WQI) objects are assessed through a numerical single digit, calculated based on one system, which adapts all the dissimilar parameters and their values in the samples. This functioning method allows judgement of the quality of several water samples based on a single arithmetic value and not the water parameter values of each sample. Although there is no globally acknowledged amalgamated index method of water quality, some countries and zones have used or are using collective water quality datasets in the development of water quality indices. The maximum water quality index model is subject to the normalized data of water parameters. Frequent variations in water quality indices have been addressed in the literature over the past five decades. There is a vital need to establish a generally recognized WQI that is flexible enough to signify consumption or other purposes of water suitability for worldwide users.

A maximum water quality index was established for surface water, particularly for river water. An inadequate number of indexing methods are intended for groundwater only. Horton [9] was the first systematically projected index to assess water suitability by using the ten most often used water quality parameters, and this method was exclusively applied and acknowledged in Asian, European, and African states. In the middle of the 1990s, a new water quality index, the Canadian Council of Ministers of Environment Water Quality Index (CCEM-WQI), was developed and documented as the Canadian Water Quality Index (CWQI) in 2001 [10]. In this index method, the WQI was assessed based on the frequency of sampling variable quantity, failed variables, and deviation from the guideline values. Then, the CCME model was acknowledged as a suitable model for measuring the quality of drinking water globally through the wide utilization of the United Nations Environment Programme (UNEP). In addition, numerous indexing methods are traditionally used for the evaluation of water suitability.

2.10 Pathway of Groundwater Contamination

Figure 2.6 explains the total pathways of the groundwater contamination/pollution process. Groundwater pollutants come from two classes of sources: point sources and dispersed sources and nonpoint sources. Commonly, landfills, leaking gasoline storage and septic tanks, and unintentional spills are the main examples of point sources of pollution. Penetration from agricultural land treated with chemical pesticides and fertilizers is an example of a nonpoint source. Among the more substantial point sources are industrial waste dumping spots and municipal landfills. When either of these occurs in or near sand and gravel aquifers, the probability of extensive pollution is the greatest. Other point sources are separately less important, but they occur in large numbers across the world. Some of these hazardous and widespread sources of pollution are dense industrial organic liquids, leaks, septic tanks, and spills of petrochemical products. Dissimilar mechanisms affect the transport of contaminants, e.g. adsorption, diffusion, decay, and precipitation, in groundwater.

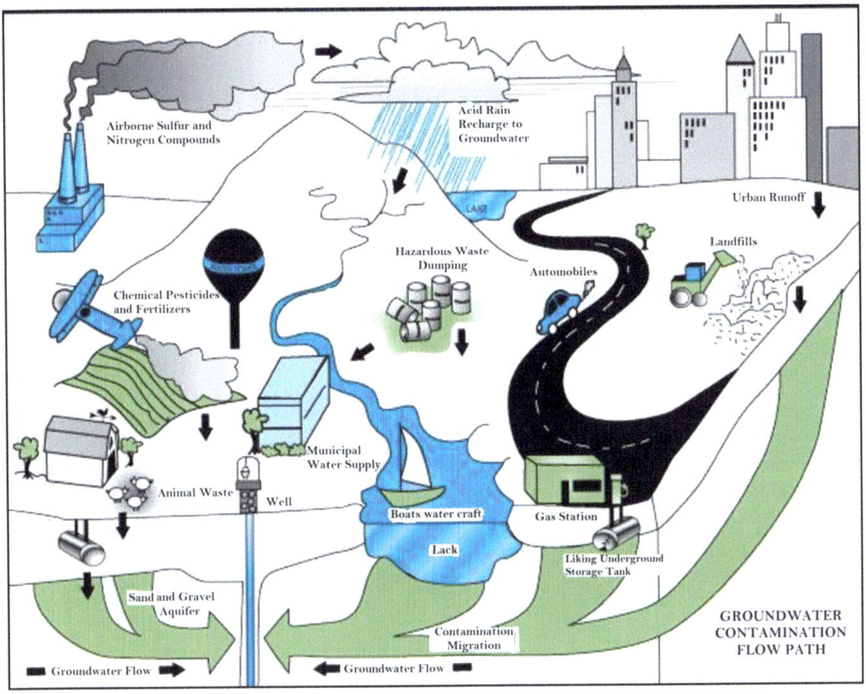

Fig. 2.6 Pathway of groundwater contamination. (Source: Zaporozec and Miller [11])

The collaboration of groundwater pollution with surface waters was analysed by hydrology transport methods.

Where the soil/land is sand or gravel type, the porosity of that soil is significantly high. In this case, the accumulated contaminants and harmful substances can easily penetrate the topsoil and ultimately reach groundwater. Some chemical species, such as nitrate, nitrite, phosphate, ammonium, potassium, and calcium, are added to groundwater through chemical manure leaching. Toxic elements such as lead, cadmium, zinc, chromium, copper, cobalt, and nickel are deposited in subsoil aquifers for industrial activities, municipal wastes, landfills, roadside wastes, metallurgical wastes, and agricultural activities. Due to overexploitation, hydrogeological changes with excess rock and mineral weathering are other causes of chemical contamination of groundwater. In addition, because of sea level rise, coastal aquifers are seriously vulnerable to saltwater intrusions.

2.11 Groundwater Contamination Sources

Groundwater typically appears clear and clean because the water obviously filters out suspended particulate material. But natural and human-induced chemicals and dissolved solutes can be found in groundwater. The sources of groundwater pollution in developing, densely populated, and agricultural countries are very common; those are geogenic, overwater mining, sewage, agrochemicals, sewage sludge, commercial and industrial leakages, waste disposal, on-site hygiene systems, landfill leachate, hydraulic rupturing, etc. Diverse physical aspects influence the transportation of contaminants, e.g. adsorption, precipitation, diffusion, and degeneration in groundwater. The contact of groundwater pollution with surface waters was analysed using hydrology transport models. Figure 2.7 shows the common sources of the groundwater contamination process. Categorically, these types of sources are as follows:

- Geogenic sources
- Agrochemicals and pesticides
- Industrial and commercial sources

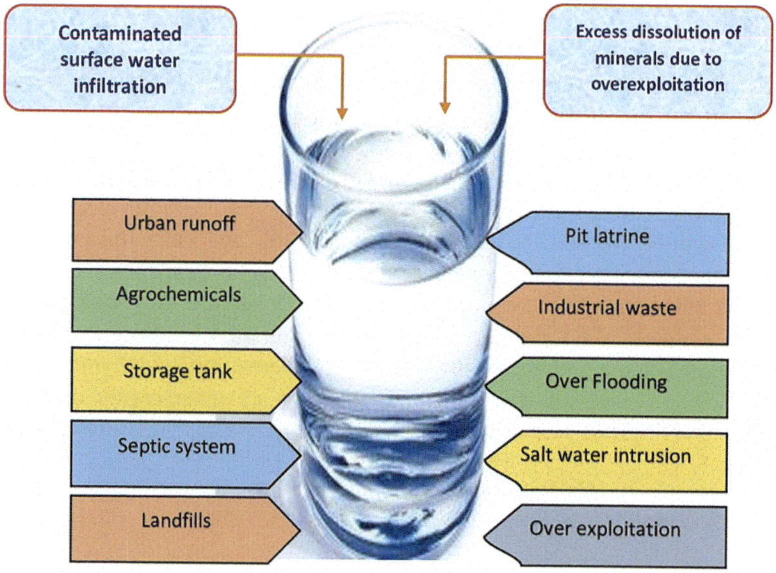

Fig. 2.7 Sources of groundwater contamination

2.11.1 Anthropogenic and Geogenic Processes

Geological processes are the main cause of groundwater pollution. This process occurs naturally due to geological processes. Arsenic and fluoride contamination occurs because aquifer sediments contain organic substance that creates anaerobic conditions in the aquifer. These conditions lead to the bacterial dissolution of iron oxide in the sediment, resulting in the release of arsenic, which is normally tightly bound to iron oxide in water. Thus, arsenic-bearing water is mainly enriched in Fe, but the incorporation of dissolved As and Fe is often obscured by secondary processes. Significantly high levels of fluoride in groundwater are naturally caused by Ca deficiency in the aquifer system. Excess levels of other parameters such as salinity, iron, manganese, chromium, uranium, and radon in groundwater can also be geological sources. Bangladesh is the 12th most densely populated and 9th most populous country in the world. As the population increases, the problem of waste disposal is increasing nationwide. Solid waste generation in Bangladesh is about 22.4 million tonnes per year, and in Dhaka City and its surrounding areas alone, 60,000 cubic meters of sewage is generated per day from 7000 large-scale industries without any treatment. In the Dhaka metropolitan area, 30,000 cubic meters of liquid domestic sewage per day is generated [12]. A tannery in Bangladesh produced huge amounts of effluent containing toxic chemicals and discharged it into neighbouring bodies of water without environmentally hazardous treatment. It is estimated that tanning waste discharged worldwide contains 300–400 million tonnes of trace metals, toxic sludges, solvents, alcohol, and other wastes that enter our waterways each year. Like other developing countries, Bangladesh suffers from inadequate wastewater treatment infrastructure, a lack of modern waste management, and systemic failure of the local wastewater treatment system. In addition to nutrients and pathogens, untreated wastewater can contain large amounts of toxic heavy metals and other inorganic and organic contaminants that can enter and leach into groundwater systems. Harmless wastewater from sewage treatment plants (ETPs) can also enter aquifers through leaching when discharged into local surface waters. Therefore, constituents not removed by simple sewage treatment plants (STPs) can also enter the aquifer system. This is because hospital waste, drug residues, and other trace contaminants in faeces and urine are only partially removed in local sewage treatment plants, with the remaining discharge to surface watersheds and groundwater. In agriculture, sewage sludge and diffused wastewater can also lead to faecal contamination of groundwater.

In African and Southeast Asian countries, over 60% of households with drinking water are contaminated with faeces. Pathogen contamination of groundwater can also occur through leaching and seepage into aquifers from unsanitary on-site mine cleanup systems, which is a major source of microbial contamination. Microbial outcomes and transport are highly complex, and the interactions between them are still unclear. Bangladeshi soils are sandy and very coarse-grained, so liquids can easily leak from septic tanks, pass through unsaturated soil zones, and eventually reach the water table. During their passage through the soil, microorganisms can die or adsorb soil particles, which mainly depends on the transit time between the septic tank and the well. Pathogens are generally killed within 50 days after crossing the soil layer. Pathogen removal levels vary widely with aquifer type, soil type, distance, and other environmental events, such as heavy rains and floods.

Landfill leachate, another source of groundwater contamination from sanitary landfills, can cause groundwater pollution. Toxic chemicals from leachate can enter groundwater through runoff and rainfall. New landfills must be lined with clay or other synthetics and leachates to protect the surrounding groundwater. However, overgrown landfills lack these measures and are often located near surface water and porous soils. A closed landfill can still pose a groundwater hazard if it is not covered with waterproofing solid to prevent the entry of harmful substances before closure.

2.11.2 *Chemical Pesticides and Fertilizers*

Agriculture plays an important role in the socioeconomic condition of the whole world. Globally, the ever-collective demand for food and agricultural resources has controlled an increase in agrochemical toxicity and environmental threats. In 1965, the worldwide consumption of nitrogen, phosphorus, and potash fertilizers was 46.3 million metric tonnes, but by 2019 this amount had increased to over 190 million tonnes [13]. Global chemical fertilizer use amounted to 136.8 kg per hectare of agricultural land in 2018, but this amount was almost double in 1976 when it stood at 70.9 kg per hectare. In Bangladesh, Brazil, China, and Indonesia, over 300 kg of chemical fertilizers was consumed per hectare of farmland in 2018 for the overproduction of crops (Table 2.1). Sulphate, nitrate, phosphate, and some semitoxic metals can also pass into groundwater via the overemployment of fertilizers, as well as compost spreading. A high degree of application of P- and N-containing chemical fertilizers in the country collectively with the high water

Table 2.1 Total amount consumed of chemical fertilizer and pesticides in several countries

	Chemical fertilizer consumption		Pesticide consumption	
Country	1961	2018	1990	2019
France	123.59	172.68	5.14	4.46
USA	41.33	128.77	2.14	2.54
Vietnam	16.04	415.28	–	–
Brazil	11.43	304.66	0.88	5.94
China	7.04	393.22	5.83	13.70
Pakistan	1.38	155.99	0.18	0.05
Bangladesh	2.61	318.47	0.13	1.72
India	2.17	175.02	0.44	0.36
Indonesia	7.55	236.44	0.08	0.03

Note: Figures are in kg per hectare

Data source: Our world in data: https://ourworldindata.org/grapher/fertilizer-use-in-kg-per-hectare-of-arable-land

solubility of nitrate and phosphate leads to increased runoff into surface water as well as discharge into groundwater, consequently creating groundwater pollution. Imperfect management practices in all types of chemical fertilizer use can introduce both nutrients (nitrate/phosphate) and pathogens into the groundwater. The extra use of animal dung or compost may also result in groundwater pollution with medicinal deposits resulting from veterinary medications.

Approximately 45% of crops are spoiled annually by pest infestation and insecticide attacks. Pesticides are toxic and deadly chemical materials used to kill, insecticides, other animals, unwanted plants and wides, or fungi that cause economic damage to crops and decorative plants but are also totally harmful to humans and the well-being of domestic animals. Pesticide use increases shortly to achieve food security for increasing demand. The estimated global use of pesticides in 2012 was about 2600 × 10^6 kg [10]. The global picture of pesticide use in 2019 is presented in Fig. 2.8. Pesticide intoxication of groundwater is a question of national deliberation because it is used for consumption purposes by around 95% of the nation's rural population and the maximum municipal water supply. Earlier in the 1970s, it was believed that soil represented a natural protecting filter that caught up pesticides from reaching unconfined groundwater aquifers. Presently, investigations have given away that this is a fully wrong conception. Where groundwater tables are in height and the soil is coarse, sandy, and loose, there is a greater chance that groundwater can be polluted by the leakage of pesticide residues through the soil. The Leaching Potential Index (LPI) of soil and sediment in most of the areas of this

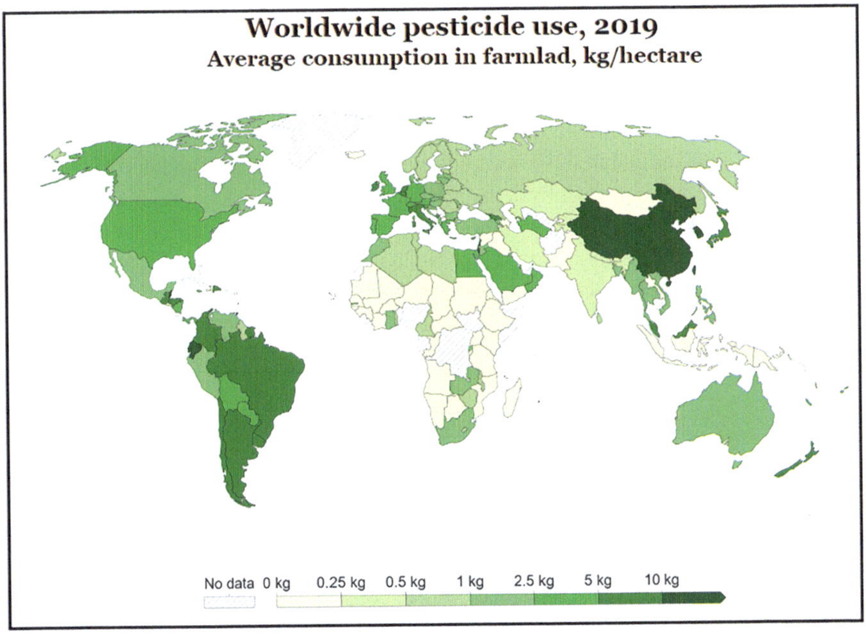

Fig. 2.8 Worldwide pesticide consumption in 2019. (Modified from FAO; OurWorldinData.Org)

type of zone is adequately high to leach pesticides into subsurface systems. Similar to most other chemical pollutants, pesticides can effortlessly reach water-containing layers from farming fields, causing the leaching of contaminated surface water, inappropriate dumping, unintended or accidental spills, seepages, and even the injection of waste materials into wells. Pesticide levels found in groundwater are generally low, and often the controlling human health-based limits exceeded are also very low. Organophosphorus and organochlorine pesticides appear to be one of a few hazardous, soluble, persistent, and mobile pesticides able to reach in the sources of drinking water. Organochlorines are very toxic to the human body at low doses. Overall, more pesticide substances are being detected as groundwater quality monitoring programmes have become more wide-ranging, though much less monitoring has been directed in less developed countries due to the high investigation costs.

Pesticides were first introduced in India and Bangladesh in 1952 to control pests in cultivation and improve productivity, and since then, the use of pesticides has been increasing day by day. Pesticide use increased rapidly in the early 1970s after the introduction of the modern, highly productive Boro rice variety. Use declined dramatically between 1973 and 1974, by which time the government had eliminated subsidies for pesticides, and in 1978, they were completely phased out. Pesticide use has increased again. Sales doubled in the late 1980s. Over the years, the use of pesticides in the agricultural sector has increased. Worldwide pesticide consumption in 2019 (modified by FAO; OurWorldinData.Org), and this is projected to continue for the next periods due to technological and socioeconomic development. Earlier in the past, 4500 pesticides and 37,258 million tonnes (MT) of pesticides were used in Bangladesh's agriculture, whereas in 2006 this figure was 31,522 MT. Knowledge and awareness of pesticide use are inadequate. Excessive and unscientific pesticide use can cause serious public health concerns, primarily through residues in food and water.

Pesticides are the number one threat to the aquatic environment not only in Bangladesh and India but also around the world. Some researchers agree that nanoscience can solve these problems through the use of nano-pesticides. N, P, B, K, Mn, Cu, Mo, Zn, Fe, Ni, and nanocarbon tubes show good efficiency as nano-fertilizers. Meanwhile, nano-insecticides and nano-designs, such as Zn, Mo, Ag, Cu, ZnO, SiO_2, etc., show excellent broad-spectrum effects for repelling insects and pests, in contrast to traditional insecticides.

2.11.3 Industrial and Commercial Sources

Weather processing and mining facilities are the main contributors to the presence of anthropogenic metals in groundwater. Oil and gasoline spills associated with underground pipelines, tanks, and gas stations can release benzenoids and other insoluble petroleum hydrocarbons that can rapidly penetrate aquifers. Many dangerous toxic chemicals used in the chemical industry, chlorinated hydrocarbon solvents, especially wood preservatives, can leach into industrial and low-permeability formations

Table 2.2 Common sources of groundwater pollution from industry

Source	Major related factors
Surface and ground storage reservoirs/tanks, discharging effluent pipe, and other transfer systems	Hidden leakage and insufficient bunding to retain main failures
Soak ways and liquid waste injection bores	Contamination because of unsuitable dumping practice
Industrial drains/collectors	Seepage for the reason of poor maintenance
Liquid runoff and process inlets	Leakage because of poor maintenance and construction
Wholesale chemical storing zones	Poor storage and management procedures and leakage
Catastrophic and accidental release	Plant fire, impact, explosion, and loss of substance to the ground
Process-waste (mainly solids) disposal sites	The leak of leachate through the failure of design and weak construction

and eventually deposit into soil layers. Other relevant sources of groundwater pollution include chemical spills from industrial operations, particularly from leather, textile, and chemical plants. The chemical spills occurring during transport, seepage from mining operations or urban runoff, illegal dumping of waste, deicing chemicals from airport, and road salt. Additionally, declining groundwater tables due to overexploitation and the burial of corpses and their subsequent degradation are the secondary causes of groundwater pollution. Since groundwater is part of the hydrologic cycle, it even causes air pollution. The most common groundwater pollution sources from industry are shown in Table 2.2.

2.12 Summary

Groundwater is water that flows from the surface of the Earth into cracks and crevices in soil and rocks. Almost all of the freshwater in the world is groundwater, making it an integral part of life on Earth. The empty subterranean space filled with groundwater is called the saturation zone and is the space below the water table. Water then flows down until it reaches an impassable rock formation, where it flows out of the ground through rocks and sedimentary layers called aquifers. It is an imperative source of drinking water and an important resource for agriculture, industry, and natural ecosystems. Although the groundwater supplies are threatened by overexploitation and pollutants, several agencies have argued that greater legal protection and oversight of this resource are needed.

References and Further Study

1. Groundwater Recharge. Source: Wikipedia (https://en.wikipedia.org/wiki/Groundwater_recharge) and International Water Management Institute (www.iwmi.cgiar.org).
2. Weller, M. W. (1981). *Freshwater marshes, ecology, and wildlife management.* University of Minnesota Press.
3. IBRD-IDA. (2023). *Annual freshwater withdrawals.* World Bank Data & Food and Agriculture Organization, AQUASTAT Data. https://data.worldbank.org/indicator/ER.H_2O.FWTL.K3
4. US-GS. (2018). *Groundwater decline and depletion.* Water Science School, The United States Geological Survey.
5. Zeng, Y., Xie, Z., Liu, S., et al. (2018). Global land surfacemodeling including lateral groundwater flow. *Journal of Advances in Modeling Earth Systems, 10*, 1882–1900. https://doi.org/10.1029/2018MS001304
6. Islam, M. S., & Mostafa, M. G. (2022). Evaluation of hydrogeochemical processes in groundwater using geochemical approaches and geostatistical models in the upper Bengal basin. *Geofluids, 2022*, 9591717. https://doi.org/10.1155/2022/9591717
7. Li, P., Karunanidhi, D., Subramani, T., et al. (2021). Sources and consequences of groundwater contamination. *Archives of Environmental Contamination and Toxicology, 80*, 1–10. https://doi.org/10.1007/s00244-020-00805-z
8. Faruq, A. N. (2018). Agriculture and pesticide consumption in Bangladesh (conference paper). In *Conference on effluent control and waste disposal in pesticide industry-with especial agenda on-integrated pest management with special emphasis on bio-pesticide component*, Bogor, Indonesia. https://www.researchgate.net/publication/327592585
9. Horton, R. K. (1965). An Index Number System for Rating Water Quality. *Journal of the Water Pollution Control Federation, 37*, 300–306.
10. CCME. (2001). *Canadian water quality guidelines for the protection of aquatic life.* Canadian Environmental Quality Guidelines CCME Water Quality Index 1.0 Technical Report. Ottawa.
11. Zaporozec, A., & Miller, J. C. (2000). Groundwater Pollution. UNESCO-PHI, Paris, 1–24.
12. NSSB. (2020). *Solid waste management in Dhaka City – A review on the present status and possible solutions.* Nature Study Society of Bangladesh (NSSB).
13. Pretty, J., & Bharucha, Z. P. (2015). Integrated pest management for sustainable intensification of agriculture in Asia and Africa. *Insects, 6*(1), 152–182. https://doi.org/10.3390/insects6010152

Chapter 3
Water Analysis

Abbreviation

AAS	Atomic adsorption spectroscopy
DO	Dissolved oxygen
DOC	Dissolved organic carbon
EC	Electrical conductivity
Eh	Redox potential
NTU	Nephelometric turbidity units
TDS	Total dissolved solids
TH	Total hardness

M. S. Islam, *Hydrogeochemical Evaluation and Groundwater Quality*,
https://doi.org/10.1007/978-3-031-44304-6_3

3.1 Outline of the Methods

The investigation focused on the geochemical description of groundwater, interpretation of the water quality data, and evaluation of the water suitability for drinking and irrigation purposes. Samples were collected from the upper Ganges River basin (the Padma River, Kushtia District) of Bangladesh in three periods, viz. pre-monsoon (PRM), monsoon (MON), and post-monsoon (POM), for 2019–2020 and 2020–2021. About 40 water samples ($n = 40$) were collected to analyse different physicochemical parameters. Numerous sophisticated techniques, including computer software and statistical approaches, were used to assess the geochemistry and water quality in the study areas.

Water analysis is carried out through quantitative and qualitative experimental works. Sample collection, preparation, and characterization; statistical analysis of the dataset; and assessment of geochemical processes, drinking water quality, and irrigation water suitability were the main parts of this book. A model of the flowchart of the methods is represented in Fig. 3.1.

Some water parameters should be measured on the spots, and most of the physicochemical parameters are analysed at the well-equipped laboratory. All experiments must be performed very carefully with three replicates using the standard methods of analysis. Any study characterized the samples using several analytical methods, including titrimetric, gravimetric, colorimetric, and spectrophotometric methods. The detailed materials and methods used in the present book for the implementation of the experimental works are discussed in this chapter.

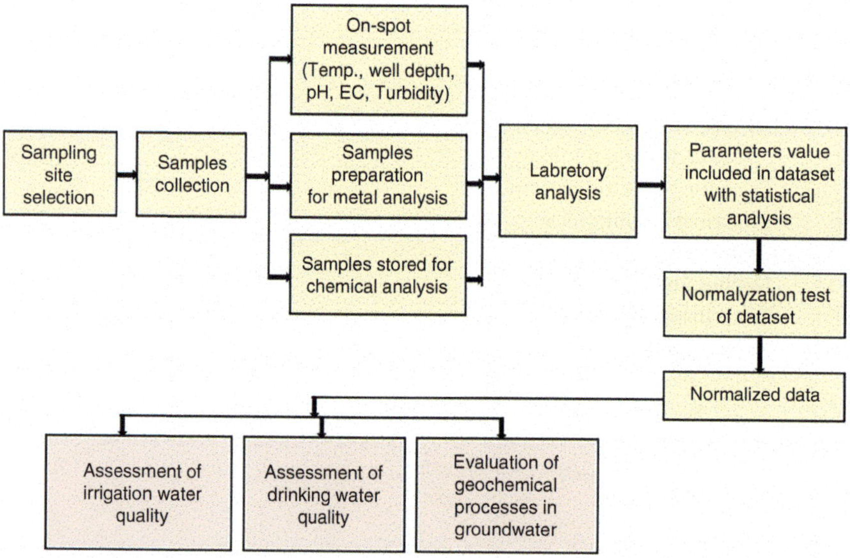

Fig. 3.1 Flowchart of methods to evaluate the geochemical processes and water quality

3.2 Study Area

The study areas included the geographically investigated area, part of Kushtia District, which is situated at 23°41′ and 24°11′ north latitude and 89°22′ east longitude. The gross study area is 1620.5 sq. km and is circumscribed by the Padma River (rename of the Ganges River) and other three-branched small rivers, making a large deltaic plain in the world (Fig. 3.2). The population of this area is approximately 2.5 million, and most inhabitants are engaged in agricultural actions [1]. The soil characteristics in the investigated zone are very good and heavily fertile, as the area placed in the valley of the Padma River, and the physiographic condition belongs to the upper Ganges floodplain. The soil surface characterizes unvarying geomorphology that appears to level the landscape with an elevation of about 9 m above sea level, but in some areas it contains low depressions and fairly higher ridges [2]. The investigating area is enclosed by a subtropical highly humid climate with a hot and rainy monsoon and a marked dry period in the winter. The maximum

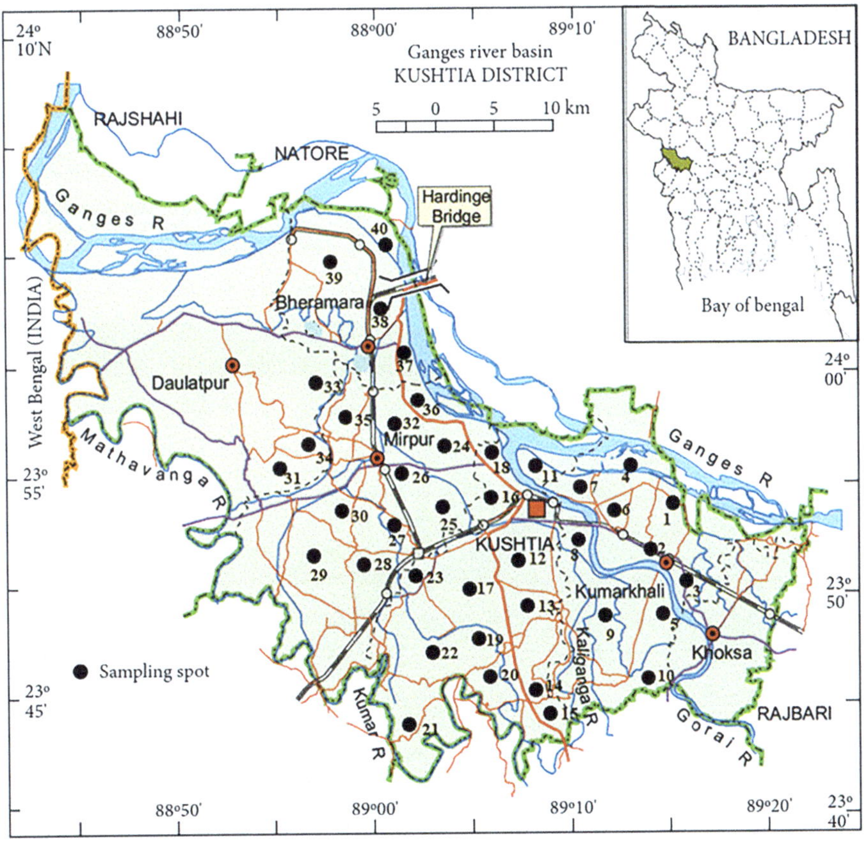

Fig. 3.2 Map of the study area and sampling points

temperature is mostly detected in May–June and the lowest in December–January. The zone received a total rainfall of 1165 mm/y. About 90% of extracted groundwater is utilized for irrigation events, and the remaining groundwater is consumed as domestic water. Recently, some large and medium–large industrial plants have grown in the municipality area, and plants consume groundwater for every industrial use. So, groundwater suitability should be evaluated as the most effective in terms of agricultural purposes in the study area.

3.2.1 *Geological Formations*

The study region is bounded by the Padma River (Ganges River), and the other three-branched rivers created a large deltaic sedimental basin (Figs. 3.2 and 3.3). The sampling sites with S1–S10 are positioned in a relative lowland area with a low topographic gradient. The primary aquifer of the study area contains unconsolidated alluvial sediments that are spread over the surface by penetrable sand, silt, fine gravel, and clayey soil. The geology of the study area encompasses young alluvial sediment deposits, deltaic silt deposits, stream and floodplain deposits, calcareous sandstone, and conglomerate. According to subsurface hydrogeological information, it seems that major good aquifers in this zone are found between 25 and 140 m depth from the surface level. The thickness of the local aquifer fluctuates because of the belongings of basement rock depth and the transverse extent of the aquifer. The groundwater movement route in the study zone is naturally from east–north to west–south.

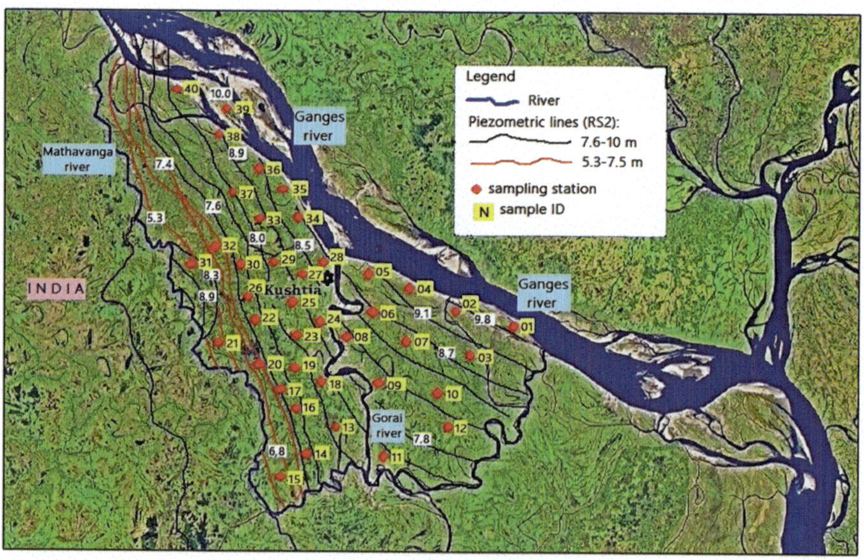

Fig. 3.3 Sampling location and piezometric lines (RS2) in the study area

3.2.2 Hydrogeological Settings

Geologically, the study area falls within the extreme front of the Ganges–Bengal basin which is located in the floodplain of the Padma River, classified by active low-relief Ganges floodplain sediments; it is crossed by rivers and canals and surrounded by river marshes and marshes. The surface lithology of the study area comprises of alluvial sands in the south of the country and deltaic silt and sands in the north. Alluvial sands are part of the active Ganges floodplain, overlying delta sands and slot deposits exposed further south. The lithology consists of coarse sand and fine gravel at the bottom of the study area and fine sand and silt at the top of the study zone. The study of BGS-DPHE [3] showed that the study area with a four-layer aquifer system including upper and lower shallow aquifers and upper and lower deep aquifer systems crosswise the Brahmaputra River just before its meeting with the Ganges. Figure 3.3 displays a geological cross section of the country as well as study area. The thickness of the aquifer depends on the depth of the bedrock and the lateral extent of the aquifer.

In the study area, the primary aquifer contains of loose river sediments covered with impermeable silt and clay. Subsurface geological information indicates that most of the aquifers in the area are between 20 and 150 metres deep, and the thickness of the aquifer differs with the depth of the bedrock and the lateral extent of the aquifer. The direction of groundwater flow within the study area is usually north to south (Fig. 3.4). During the monsoon season, rainfall and floods cause

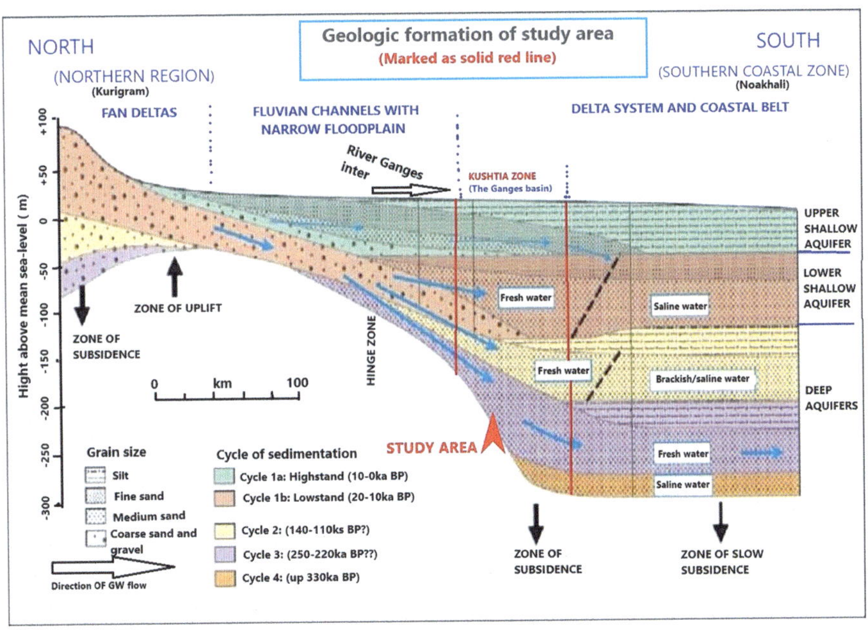

Fig. 3.4 Hydrogeologic formation of the study area. (Modified copy of BGS-DPHE [3])

groundwater recharge and raise the groundwater level. After the monsoon season ends, some of the water from rivers, streams, ponds, and lowlands is replenished. Piezometric groundwater levels drop during the dry season due to overwater mining for irrigation with low specific yields.

3.3 Sampling Strategy

The selection of sampling sites, collection, transportation, nature of water parameters, and analysis method of groundwater samples are key factors in the rationality of sampling and analysis strategy of this study. The sampling area was rural close to the city and involved the community throughout the sampling operations. The study was separated into two groupings of sampling stations such as groundwater for drinking and for irrigation. Possible sources and contamination are different in the two sampling groups. In both samplings, nonpoint sources (diffuse sources) were most commonly related to point sources (single sources). Pollution from nonpoint sources is the main cause of water contamination in Bangladesh, and the lack of identifiable sources is a challenge. In the study area, runoff and floodwaters that accumulate pollutants such as toxic sediments, surface toxins, nutrients, pesticides, fertilizers, inorganic contaminants, and animal and human excreta can flow into water bodies. Most of these substances leach underground and eventually reach aquifers. For other reasons, the overall water suitability of the study area deviates from the external sources mentioned above. For example, Table 3.1 shows sampling descriptions and possible sources of contamination for groups of sampling stations.

For physicochemical analysis of groundwater, it is better to select the three sampling periods, viz. pre-monsoon (PRM), monsoon (MON), and post-monsoon (POM), of a year than single sampling. Periodic and spatial variations in water data can help to reach a concrete decision on the determination of local geochemical facies and gross water quality. Groundwater should be collected from shallow and deep aquifers that differ from region to region. Samples can be collected randomly or selectively from the hand pump well, dug well, and engine well to cover the topographical extension of the study zone and the main geologic backgrounds.

3.4 Sampling of Groundwater

In general, the purpose of an aquifer monitoring programme and sample collection under that programme is to identify a representative portion of the aquifer and obtain field measurements at selected times and locations. These samples and subsequent analysis are used to identify sources of pollution, assess groundwater quality, develop water management plans, measure water balances and remediation strategies, and demonstrate environmental due diligence.

Table 3.1 Sampling descriptions and possible pollution sources of the sampling sites

Sampling field	Sample ID	Point sources	Nonpoint sources
Residential area Shallow well Relatively highland Mostly in river valley area	1, 2, 3, 4, 5, 6, 7, 8, 11, 18, 19, 23, 24, 26, 32, 35, 36, 37, 38, and 39.	On-site septic systems, leaky septic tank Landfill Livestock wastes	Atmospheric deposition and hydrologic modification Microorganisms and nutrients coming from pet animal washdowns, livestock, and damaged septic arrangements Drainage, seepage Flooding or overflow Land runoff Soil erosion and sediment runoff
Agricultural area Deep well Typically, in deltaic basin area Comparatively lowland	9, 10, 12, 13, 14, 15, 16, 17, 20, 21, 22, 25, 27, 28, 29, 30, 31, 33, 34, and 40.	Livestock wastes Landfill	Land runoff, flooding, waterlogging Chemical fertilizers and pesticides that penetrate the soil layer and make their way into an aquifer Soil erosion and washed from heavy agricultural farming Contaminants in rainwater Atmospheric deposition and hydromodification Organic wastes from farmlands

3.4.1 Sampling Methodology

Groundwater samples were collected in prewashed high-density polyethylene (HDP) plastic or glass bottles as stated by the standard process. After 3–5 min of pumping, the water samples were collected to avoid any debris. For trace element analysis, the samples were well preserved with analytical research grade HNO_3 and frozen at 4 °C for analysis.

Common purposes for applying groundwater sampling programmes may include investigating groundwater quality, detecting and assessing groundwater pollution or contamination, and providing information for groundwater resource management planning. To collect representative samples within a given aquifer, the methodology used must accurately extract samples whose composition replicates the spatial and periodic variations of the aquifer. When variations in the vertical or areal composition of groundwater are already known, specific methodologies are required to detect these variations.

Stagnant water in wells is open to the atmosphere, may change its physicochemical properties, and is not representative of aquifer water samples. Therefore, wells should be cleared before sampling by pumping out an amount of water equal to at least five times the internal volume of the well. If it is essential to flush the wells before sampling, do not use an airlift pump, as the mixture of atmospheric oxygen and groundwater can alter the DO of the samples.

As many parameters as feasible must be measured in the sampling field or once conceivable after the sample has been collected. It is most imperative for physico-chemical parameters such as pH, EC, turbidity, or DO. To precisely measure these parameters, a continuous quantity technique, such as a flow-through cell system, which minimizes contact between the samples and the air, should be employed. The use of such a flow-through cell system will also contribute to the determination of stable chemistry before sample collection.

3.4.2 Types of Sampling

(a) *Pump Sampling*

A pump sampler is shown in Fig. 3.5. The instrument is set up by lowering a pump into a borehole to draw enough groundwater to measure the chemical properties of the groundwater. Variable pump speed helps match extraction speed to wellbore refill speed, speeding up the sampling process. If the water chemistry is stable, the sample can be collected in a suitable container. Samples collected in this manner may contain a mixture of groundwater entering the wellbore from various depths through the open casing. Therefore, unless the aquifer conditions are known to be vertically uniform, this method can be expected to produce composite samples or samples of approximately average composition. Do not take a water sample until the pump has run long enough to

Fig. 3.5 Pump sampling instruments

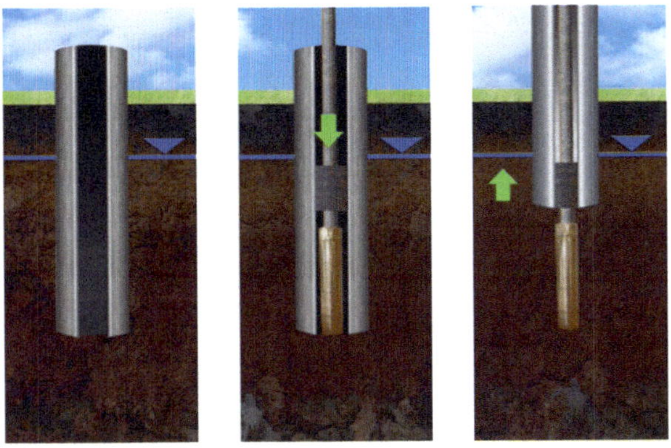

Fig. 3.6 Depth groundwater sampling

remove all stored water in the well. The amount of water that needs to be pumped prior to sampling can be calculated using normal pump capacity and borehole size. Whenever physical parameters are used, there is no need to take samples unless there is significant variation in the physical parameters. This means less than ±10% variation in pH, turbidity, EC, or DO or less than ±0.2 °C.

(b) *Depth Sampling*

In depth sampling, a specific sampling device is lowered into the wellbore and activated when the required depth is reached (Fig. 3.6). After collection, the water sample is transferred to a suitable container. This technique is only useful if the depth of water entry into the well is known. Depth samples should not be taken from the solid casing of the borehole, as water cannot be invented at that depth. Even samples taken from the open or screened portion of the well may not be representative of the depth chemistry due to natural or induced currents in the aquifer.

3.5 Groundwater Sampling Log

To confirm the dependability and interpretability of the analysed data, suitable documents should be combined into the monitoring programme that accounts for sample driven from collection to data inclusion and guarantees that analytical data are attributed to the correct site. A suitable chain of custody information for collected groundwater samples begins with the accomplishment of a sampling report. Sampling and well-purging information should contain the following items:

- Date and time of sampling
- Name of the site of the monitoring borehole and description of the piezometric line, coordinates, and any other related information
- Particulars of the borehole and aquifer (e.g. depth of pump, bore dimensions, casing state, screened interval, etc.)

Table 3.2 Groundwater sampling log

Sampling log						
Station name				Location		
Well no.			Sample ID		Date	Time
Well depth			Well type		Well diameter	
Pump type			Purging rate		Others	
Field data						
S. ID	Temp.	pH	Electric cond. µS/cm	DO	Turbidity	Comments
Signature/name of the sample collector:						

- Vertical water level
- Period and rate of pumping prior to sampling
- The appearance of the sample at the collection time (e.g. colour, odour, and clarity)
- Nature of pretreatment (if needed) and preservation procedure
- Groundwater quality parameters measured in the field (in situ)
- Name of the sample collector
- Any data that may control the results of analysis (e.g. generation of bubbles, debris, sediment, etc.)

After collecting the sample, it should be kept, handled, and transported in such a way as to avoid damage to sampling bottles or labels, minimize degradation of the sample, and avoid pollution of the sample. Upon transfer to the analytical laboratory, information concerning the time between sampling and analysis, storage and protection practice employed at the laboratory, and analytical procedure used should be documented. A simple groundwater sampling log is shown in Table 3.2.

3.6 Equipment for Water Analysis

Figure 3.7 shows a set of groundwater sampling equipment. The container to be used to collect samples must be properly cleaned and disinfected. Field kits should be calibrated along with the instruction manuals, and an adequate number of sampling pots should be prepared. The chemical components of interest in consequent

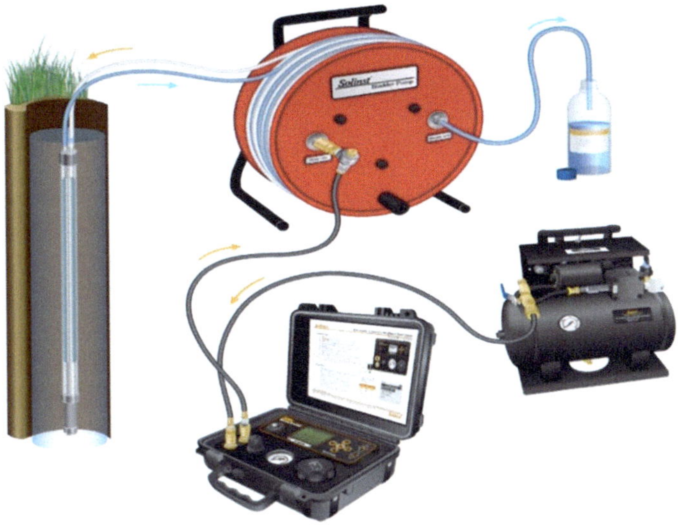

Fig. 3.7 Sampling equipment

laboratory analysis dictate the type of bottle to be used for sampling and sample storage, e.g. borosilicate glass or high-density polyvinyl plastic. An example of necessary equipment for the collection of groundwater samples is as follows:

- Sample labels, field sheets (Table 3.2), and chain of custody forms
- Bottle for in situ measurements, rather than a flow-through cell system or another pot that minimizes contact between the groundwater and air
- Pump or bailer suitable for the dimensions of the boreholes to be sampled
- Vessel/container for collecting the sample
- Nitrile hand gloves
- Sample bottles/tube
- Field parameter (physical) test kits or multimeter (thermometer, pH metre, EC metre, turbidity metre, DO kits, etc.)
- Filtration equipment
- Portable small refrigerator
- Suitable preservative, e.g. conc. Nitric acid (HNO_3)
- Personal protecting kit and first aid box

3.7 Field Parameters

Some water parameter values change with time variation, e.g. turbidity will be reduced, but DO will be increased with sampling time. Thus, these types of parameters should be measured in situ and are called field parameters. Such parameters are:

- Temperature (°C)
- pH (unitless)
- Electrical conductivity, EC (μS/cm)
- Turbidity (NTU)
- Dissolved oxygen, DO (mg/L)
- Alkalinity/acidity (mg/L)
- Reduction/oxidation potential, Eh (mV)

3.8 Sample Filtration, Preservation, Transport, and Storage

Groundwater sample filtration at the site can extend the period in which precise results can be obtained from the analysis. For example, levels of NH_3 or NO_3^- can decrease with an increase in the time in water sampling. Filtration extends the period that a typical result can be returned from a sample following collection. Besides, filtration is needed to eliminate colloidal or particulate substances, which delivers additional facies for adsorption. There are two chief methods utilized in the filtration of water samples: (a) vacuum filtration systems and (b) pressure systems employing gravitational pressure.

The suitable sample container/bottle, preservation technique, and holding time of the groundwater sample are described in Table 3.3. Though sample preservation will limit degradation, it is suggested that dispatch to the laboratory for chemical analysis be conducted as soon as possible.

3.9 Analytical Procedure

A total of 40 sampling stations in the Ganges River basin in the middle-western portion of Bangladesh (Figs. 3.2 and 3.3) were selected for this study during the pre-monsoon (PRM), monsoon (MON), and post-monsoon (POM) periods in 2019–2020 and 2020–2021. Groundwater samples were collected arbitrarily from the selected hand tube well, shallow engine pump, and semi-deep wells, and their depths ranged from 22 to 110 m. Samples were collected in prewashed high-density polyethylene (HDP) plastic containers as stated by the standard process (Table 3.3). After 3–5 min of pumping, the water samples were collected to avoid any debris. For trace metal analysis, the samples were conserved with AR grade HNO_3 and reserved at 4 °C for further analysis.

Due to the evaluation of hydrogeochemical processes and assessment of water utility, a total of 30–32 physicochemical parameters, i.e. temperature, pH, electrical conductivity (EC), total dissolved solids (TDS), turbidity, dissolved oxygen (DO), dissolved organic carbon (DOC), total hardness (TH), Cl^-, NO_3^-, NO_2^-, SO_4^{2-},

Table 3.3 Sample container types and preservation (US-EPA [4])

Determinant	Type of container	Preservation procedure
pH	Glass or HDPE plastic	Fill the container to eliminate air and refrigerate it
Trace metals	Acid-washed plastic or glass	Add HNO_3 at pH 1–2 and refrigerate. Must be filtrated the sample before acidification
NH_3	Glass or HDPE plastic	Refrigerate
Major cations, e.g. Ca^{2+}, Mg^{2+}, etc.	HDP plastic	Acidification is not needed
NO_3^-, CO_3^{2-}, HCO_3^-	Glass or HDPE plastic	Filter on site and freeze
Cl^-	Glass or HDPE plastic	No need
CN^-	Glass or HDPE plastic	If no nosy compound is present, then add NaOH to a pH larger than 12
NO_2^-	Glass or HDPE plastic	Refrigerate
PO_4^{3-} (total P)	Glass or HDPE plastic	For dissolved concentration determination, filter on spot and refrigerate. For total P determination, freeze
Total N	Glass or HDPE plastic	Refrigerate
SO_4^{2-}	Glass or HDPE plastic	Refrigerate
Hydrocarbons	Glass, solvent washed	Do not prerinse and fill the sample bottle. Acidify with H_2SO_4 or HCl to pH 1–2 and freeze

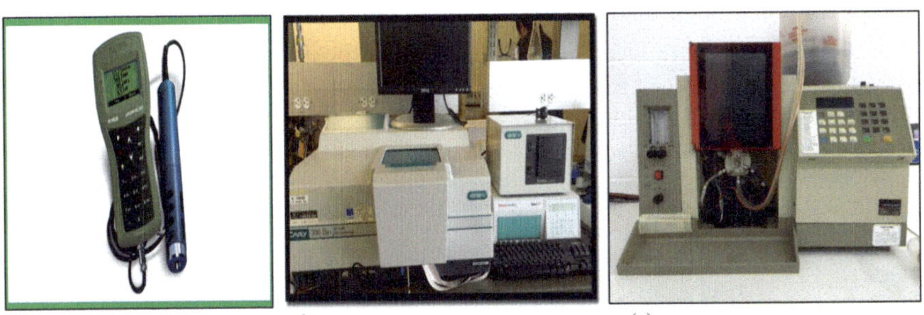

(a) (b) (c)

Fig. 3.8 Various types of water quality analyser (**a**) potable digital multimeter (**b**) UV–visible spectrophotometer (**c**) AAS

PO_4^{3-}, HCO_3^-, Na^+, K^+, Ca^{2+}, Mg^{2+}, B, Fe, Mn, Cr, Co, Pb, Ni, Si, Cd, As, Cu, Zn, etc., should be measured. The pH, EC, and TDS were estimated in situ by a portable multimeter (Fig. 3.8). Calcium (Ca^{2+}), magnesium (Mg^{2+}), bicarbonate (HCO_3^-), carbonate (CO_3^{2-}), ammonia (NH_3), and total hardness (TH) were measured by the titrimetric method [4]. Sodium (Na^+) and potassium (K^+) were analysed using a flame photometer. A UV spectrophotometer (Fig. 3.8) was used for the determination of chloride (Cl^-), nitrate (NO_3^-), nitrite (NO_2^-), sulphate (SO_4^{2-}), and phosphate (PO_4^{3-}) using the respective standard solutions. Trace metals, viz. boron (B), iron (Fe), manganese (Mn), cobalt (Co), cadmium (Cd), chromium (Cr), nickel (Ni), lead (Pb), arsenic (As), copper (Cu), and zinc (Zn), were analysed by the well-accepted method through an atomic absorption spectrophotometer (AAS).

Quality control for all metal analyses was performed according to individual instruction manuals, and method precision was ≥95% with a confidence interval (*CI*) of correlation coefficient $r = \sim 1$ for each calibration curve. Each method was recalibrated after running ten samples, and all quantitative analyses were performed in triplicate to ensure accuracy. Additional evidence for the correctness of the data was provided by the fitting formula (Eq. 3.1):

$$\text{Charge balance error, CBE} = \frac{\sum M_c|N_c| - \sum M_a|N_a|}{\sum M_c|N_c| + \sum M_a|N_a|} \times 100 \qquad (3.1)$$

where M_c and N_c are the molar concentration and charge number of the cation; likewise, M_a and N_a are the same for the anion. All calculated charge balance errors should be within ±5%. Also, $TDS_{measured}$ and $TDS_{calculated}$ ratios were computed for the use of quality-control measures. The calculated ratio fluctuates from 1 to 1.3, which shows the accuracy of the analytical data [4]. Probable water parameters are measured by the following procedures.

3.10 Analysis of Physicochemical Water Parameters

Physicochemical characteristics mention things that involve the features of both physical and chemical properties, meaning that they are reliant on, or formed by, the mutual actions of physical and chemical attributes. Except for some physical properties (e.g. test, colour, odour, etc.), all the water variables are included in this class of water parameters. In this subsection, the detailed analytical methods and detected water parameters are discussed.

3.10.1 Field Parameters

The temperature, pH, EC, turbidity, and DO were estimated by a thermometer, pH metre, EC metre, turbidity metre, and DO metre, respectively, at the sampling station just after collecting groundwater samples. Besides, these parameters may be measured by a portable digital multimeter (Fig. 3.8).

pH: The pH metre should be calibrated according to the manufacturer's requirements using a two-point or three-point calibration with buffer solutions of known concentration. The choice of calibration standard to use depends on the expected properties of the water being tested. Highly acidic or alkaline water requires calibration with a standard suitable for the expected range.

Temperature: The temperature of the water sample usually corresponds to the average temperature of the area where the monitoring wells are located. As the sample is taken, the value changes towards ambient temperature, so it should be recorded as soon as a stable reading is gotten after sampling.

Electrical conductivity (EC): The EC of natural waters is a count of the chemical ion loading of the aqueous medium. As with pH measurements, EC metre calibration should be performed using standards of known concentration appropriate for the expected EC range of the water being tested.

Turbidity: Turbidity is a measure of the amount of suspended solids in a water sample. Turbidity can be measured with a turbidimeter or multimeter.

Dissolved oxygen (DO): The amount of dissolved oxygen in a sample varies with depth, temperature, and biological needs. DO measurements are most accurate when the probe is placed in a closed flow cell that eliminates contact between air and water.

Redox potential (Eh): Eh measures the oxidation or reduction potential of aqueous systems. Field gauge return values in millivolts. Negative values indicate reducing conditions, and positive values indicate oxidizing conditions.

Alkalinity/acidity: Alkalinity/acidity can come from natural sources such as dissolved carbonate minerals and humic acids. Alkaline or acidic field tests can be performed using commercially available test strips or indicator solutions.

3.10.2 Lab Parameters

3.10.2.1 Total Dissolved Solids (TDS)

(a) The TDS in the water sample can be computed from this reading using formula (3.2):

$$\mathrm{TDS} = (0.548 \times \mathrm{EC}) + \left(2.2 \times 10^{-6} \times \mathrm{EC}^2\right) \tag{3.2}$$

(b) Initially, a 150 ml Pyrex glass beaker was taken and dried in an oven at 105 °C for 24 h. Then, it was allowed to cool and weighed accurately. One hundred millilitres of water sample was filtered by oven-dried Whatman 42 filter paper into the above glass beaker and evaporated to dry at 105 °C for 24 h. Then, the beaker was allowed to cool and weighed. The heating and cooling process was repeated until a constant weight was obtained. Then, TDS is calculated using Eq. (3.3):

$$\text{Total dissolved solids}\left(\text{TDS}, \frac{\text{mg}}{\text{L}}\right) = \frac{(X - Y) \times 1{,}000{,}000}{V} \quad (3.3)$$

where X = final weight of beaker (after filtration) in g; Y = initial weight of beaker (before filtration) in g; V = volume of a water sample taken = 100 ml.

3.10.2.2 Total Hardness (TH)

The TH of the water samples was measured by the EDTA titration method at pH 10. One hundred millilitres of a water sample was placed in a conical flask with 1 ml of ammonia buffer with pH 10 and two drops of Eriochrome Black T indicator. The solution was titrated with standard EDTA solution until the colour changed from wine red to blue. Equation (3.4) is used for the calculation of total hardness:

$$\text{Total hardness}, \frac{\text{mg}}{\text{L}} = \frac{\text{ml of EDTA} \times 1000}{\text{ml of sample}} \quad (3.4)$$

3.10.2.3 Calcium Ion (Ca^{2+})

Reagents: (i) 0.01 (M) EDTA solution 37.225 g EDTA +1000 ml distilled water; (ii) 1 N NaOH: 40 g NaOH +1000 ml distilled water; and (iii) murexide indicator: 150 mg murexide +100 ml ethylene glycol.

Procedure: A 25 ml sample was titrated with 0.01 M EDTA solution by adding 1 ml 1 N NaOH solution and one drop of murexide indicator; the endpoint changed from pink to purple:

Calculation:

$$\text{Ca hardness as mg/L CaCO}_3 = \frac{X \times Y \times 1000}{V}$$

where X = volume of EDT; Y = mg of $CaCO_3$ per 1 ml EDTA; V = ml of sample; and Ca present in sample (mg/L) = calcium hardness × 0.4.

3.10.2.4 Magnesium Ion (Mg^{2+})

Reagents: (i) 0.01 (M) EDTA; (ii) 1 (N) NaOH; (iii) murexide indicator; and (iv) EBT indicator.

Procedure: (a) A 25 ml water sample was taken to determine the total hardness (TH) using 0.01 (M) EDTA and EBT indicators. (b) Calcium hardness was determined as described above by using 0.01 M EDTA and murexide as indicators. (c) The amount of Mg was calculated based on two titter values:

Calculation:

$$\text{Mg hardness}, \frac{\text{mg}}{\text{L}} = (\text{total hardness} - \text{Ca hardness})$$

$$\text{Mg present in water sample}, \frac{\text{mg}}{\text{L}} = \text{Mg hardness} \times 0.244$$

3.10.2.5 Sodium (Na^+)

Preparation of sample for trace metal analysis: A 100 ml water sample was placed into a beaker, and 2 ml of conc. HNO_3 and 3 ml of conc. HCl were added. This mixed solution was heated at 90–95 °C on a hot plate until the volume of water got 10–15 ml. Then, it was detached from the hot plate and allowed to cool. Then, the samples were filtered with a 0.45 μm pore membrane filter. Lastly, the volume was made up to 100 ml by adding distilled water. This process was followed to measure the trace metals in the samples.

The Na^+ in the groundwater samples was measured by the atomic absorption spectrophotometric (AAS) method at 330.2 nm with 0.7 nm silt.

Reagent: *1000 mg/L standard solution*: The solution of Na^+ ions was prepared by accurately taking 2.54 g of analytically pure and dry 99% sodium chloride into a 1000 ml volumetric flask, adding distilled water gradually, and shaking well. Lastly, it was made up to the mark by adding distilled water.

Suppressing agents (0.1% potassium solution): 5.2 g of KCl was dissolved in a 1000 ml volumetric flask and gradually brought up to the mark with distilled water. After then, 100 ml of 1% K^+ ion solution was added and diluted to 1000 ml with the same water to a final concentration of 0.1% K. Two drops of 0.1% K solution were added to every standard and sample to decrease the proportion of relative standard deviation (RSD).

Calibration curve: The amounts of 1, 2, 5, 8, and 10 ml Na^+ ion-containing standard solutions were taken in five different 100 ml volumetric flasks, and distilled water was slowly added and carefully shaken. The absorbance of the samples was measured by AAS. A calibration curve was made by plotting the absorbance compared to the concentration (mg/L) of Na.

Calculation:

$$\text{Sodium}\left(\text{Na}^+\ \text{ion}, \frac{\text{mg}}{\text{L}}\right) = \text{concentration of Na from calibration curve} \times \text{dilution factor}$$

3.10.2.6 Potassium (K^+)

The K^+ in the water samples was determined by the AAS method at 766.5 nm with 0.7 nm silt.

Reagent: *1000 mg/L standard solution*: The standard solution of Na^+ was made by accurately taking 1.910 g of analytical grade 99% KCl in a 1000 ml volumetric flask, adding distilled water gradually, and shaking well. Finally, it was made up to the mark by adding the same water.

0.1% potassium ion solution: 1 g of KCl was dissolved into a 100 ml volumetric flask and made up to the mark by adding distilled water slowly. Two drops of 0.1% K solution were added to every standard and sample to decrease the fraction of relative standard deviation (RSD).

Calibration curve: The proportions of 1, 2, 3, 4, and 5 ml K^+ ion-containing standard solutions were taken in five different 100 ml volumetric flasks, and distilled water was slowly added and carefully shaken. The absorbance of the water samples was measured by AAS. A calibration curve was made by plotting against the absorbance and concentration (mg/L).

Calculation:

$$\text{Potassium}\left(K^+\ \text{ion}, \frac{\text{mg}}{\text{L}}\right) = \text{concentration of K from calibration curve} \times \text{dilution factor}$$

3.10.2.7 Iron (Fe)

Apparatus: AAS wavelength at 248.3 nm with 0.7 nm silt.

Reagent: The standard solution of iron ions of 1000 mg/L was prepared by taking accurately 4.98 g of analytical grade (99% pure and dry) $FeSO_4.\ 7H_2O$ in a 1000 ml volumetric flask, and then distilled water was slowly added and shaken well made up to the mark with same water.

Calibration curve: The proportions of 4-, 8-, 12-, and 16-ml Fe ion-containing standard solutions were taken in four different 100 ml volumetric flasks were gradually added to distilled water, carefully shaken, and lastly made up to the mark with distilled water. The absorbance of the water samples was measured by AAS. A calibration curve was made by plotting the absorbance compared to the concentration.

Calculation:

$$\text{Iron}\left(\text{Fe}, \frac{\text{mg}}{\text{L}}\right) = \text{concentration of Fe from calibration curve} \times \text{dilution factor}$$

3.10.2.8 Manganese (Mn)

Method: AAS method at 279.48 nm wavelength with 0.2 nm silt.

Reagent: A 1000 mg/L standard solution of Mn was prepared by accurately taking 3.07 g of analytical grade (99% pure) $MnSO_4.10H_2O$ in a 1000 ml volumetric flask, adding distilled water gradually, and shaking well. Lastly, it was made up to the mark with the same water.

Suppressing agent: The aqueous solution of 2% Ca was made by accurately taking 2.0 g of $CaCO_3$ in a 100 ml volumetric flask and adding 3 ml of 1 (M) HCl solution to this flask. Then, it was dissolved with distilled water gradually, made up to the mark with distilled water slowly, and shaken well. At that time, 10 ml of 2% $CaCO_3$ solution was taken and diluted to 100 ml with distilled water to a final concentration of 0.2% $CaCO_3$. This solution was used for maintenance in every standard and sample to reduce the percentage of relative standard deviation (RSD).

Standard curve: The proportions of 0.5-, 1-, 2-, 3-, and 5-ml Mn ion-containing standard solutions were placed in five separate 100 ml volumetric flasks, and distilled water was slowly added, carefully shaken, and finally made up to the mark with distilled water. The absorbance of the water samples was measured by AAS. A calibration curve was made by plotting the absorbance against the concentration (mg/L).

Calculation:

$$\text{Manganese}\left(\text{Mn}, \frac{\text{mg}}{\text{L}}\right) = \text{concentration of Mn from calibration curve} \times \text{dilution factor}$$

3.10.2.9 Lead (Pb)

Method: AAS method at 283.3 nm wavelength with 0.7 nm silt.

Reagent: A 1000 mg/L standard solution of lead ions was ready by accurately taking 1.599 g of analytical grade 99% $Pb(NO_3)_2$ in a 1000 ml volumetric flask, adding distilled water gradually, and shaking well. Finally, it was made up to the mark by adding distilled water.

Calibration curve: The proportions of 1, 2, 3, 4, and 5 ml Pb ion-containing standard solutions were taken in five different 100 ml volumetric flasks, and distilled water was slowly added and carefully shaken and lastly brought up to the mark with distilled water. The absorbance of the water samples was determined by AAS. A calibration curve was made by plotting the concentration (mg/L) against the absorbance.

Calculation:

$$\text{Lead}\left(\text{Pb}, \frac{\text{mg}}{\text{L}}\right) = \text{concentration of Pb from calibration curve} \times \text{dilution factor}$$

3.10.2.10 Chromium (Cr)

Method: AAS method at 248.3 nm wavelength with 0.7 nm silt.

Reagents: *(i) 1000 mg/L standard solution*: The standard solution for Cr ions was prepared by accurately taking 2.8290 g of analytical grade potassium dichromate ($K_2Cr_2O_7$) into a 1000 ml volumetric flask. Then, 5 ml of H_2SO_4 was added, and distilled water was slowly added, shaken well, and lastly made up to the mark. *(ii) Chromium standard solution corresponding to 10 mg/L of chromium:* 10 ml of the Cr stock solution was taken into a 1000 ml volumetric flask by pipette. Twenty millilitres of nitric acid was added, brought up to the mark with water, and mixed well. *(iii) Chromium standard solution corresponding to 0.4 mg/L chromium:* 20 ml of the Cr standard solution was taken into a 500 ml volumetric flask by pipette. Ten millilitres of HNO_3 was added and brought up to the mark with distilled water and mixed well. This solution was prepared on the day of use.

Standard curve: The proportions of 0.5-, 1-, 2-, and 3-ml Cr ion-containing standard solutions were taken in four separate 100 ml volumetric flasks and slowly added to distilled water and carefully shaken and lastly made up to the mark with distilled water. The absorbance of the water samples was measured by AAS. A calibration curve was made by plotting the absorbance against the concentration (mg/L).

Calculation:

$$\text{Chromium}\left(\text{Cr}, \frac{\text{mg}}{\text{L}}\right) = \text{concentration of Cr from calibration curve} \times \text{dilution factor}$$

3.10.2.11 Cadmium (Cd)

Reagent: A 1000 mg/L standard solution of Cd ions was prepared by accurately taking 2.282 g of analytical grade $3CdSO_4.8H_2O$ in a 1000 ml volumetric flask, and distilled water was slowly added and shaken well. Finally, it was made up to the mark by adding distilled water.

Calibration curve: The proportions of 0.1-, 0.2-, 0.4-, and 0.6-ml Cd ion-containing standard solutions were taken in four different 100 ml volumetric flasks, and distilled water was slowly added and carefully shaken and finally made up to the mark with distilled water. The absorbance of the water samples was measured by AAS at 228.80 nm wavelength with 0.7 nm silt. A calibration curve was made by plotting the absorbance against the concentration (mg/L).

Calculation:

$$\text{Cadmium}\left(\text{Cd}, \frac{\text{mg}}{\text{L}}\right) = \text{concentration of Cd from calibration curve} \times \text{dilution factor}$$

3.10.2.12 Arsenic (As)

The As ion in groundwater samples was determined using the AAS method by graphite furnace at 193.7 nm wavelength with 0.7 mm silt. The standard solutions of As ions were prepared by taking 1.3201 g of analytical grade As_2O_3 in a 100 ml volumetric flask, slowly adding distilled water, shaking well, and lastly making up to the mark with the same water.

Calibration curve: The proportions of 1-, 2-, 3-, 4-, and 5-ml arsenic ion-containing standard solutions were taken in four different 100 ml volumetric flasks, and distilled water was added and shaken and lastly made up to the mark with distilled water. The absorbance of the samples was measured by AAS.

Calculation:

$$\text{Toal inorganic As}\left(\frac{\text{mg}}{\text{L}}\right) = \text{concentration of As from calibration curve} \times 1000$$

3.10.2.13 Copper (Cu)

Apparatus: AAS wavelength at 324.75 nm with 0.7 nm silt.

Reagent: A 1000 mg/L standard solution of Cu ions was prepared by taking exactly 3.93 g of pure (99%) $CuSO_4.5H_2O$ in a 1000 ml flask, and distilled water was slowly added and shaken well. Lastly, it was made up to the mark with distilled water.

Standard curve: The proportions of 0.5, 1.0, and 2.0 ml Cu ion-containing standard solutions were taken in three separate 100 ml volumetric flasks, gradually added to distilled water, carefully shaken, and lastly brought up to the mark with distilled water. The absorbance of the water samples was determined by AAS. A standard curve was made by plotting the absorbance against the concentration (mg/L).

Calculation:

$$\text{Copper}\left(\text{Cu}, \frac{\text{mg}}{\text{L}}\right) = \text{concentration of Cu from standard curve} \times \text{dilution factor}$$

3.10.2.14 Zinc (Zn)

Apparatus: AAS wavelength at 213.86 nm with 0.7 nm silt.

Reagent: A 1000 mg/L standard solution of Zn^{2+} ions was prepared by accurately taking 4.98 g of analytically pure (99%) $ZnSO_4.7H_2O$ in a 1000 ml volumetric flask, adding distilled water gradually, and shaking well. Finally, it was made up to the mark with distilled water.

Standard curve: The volumes of 1-, 2-, 3-, and 4-ml Zn ion-containing standard solutions were taken in four different 100 ml volumetric flasks, distilled water was slowly added and carefully shaken, and lastly made up to the mark with distilled water. The absorbance of the water samples was determined by AAS. A standard curve was made by plotting the absorbance against the concentration (mg/L).

Calculation:

$$\text{Zn}\left(\text{Zn}, \frac{\text{mg}}{\text{L}}\right) = \text{concentration of Zn from standard curve} \times \text{dilution factor}$$

3.10.2.15 Chloride Ion (Cl^-)

Reagents: *Indicator preparation (5% K_2CrO_4):* 5 g of K_2CrO_4 was placed in a 100 ml volumetric flask and dissolved in 50 ml of distilled water. Then, 0.0141 N $AgNO_3$ solution was added dropwise to a K_2CrO_4-containing flask until the first enduring red precipitate was formed. The prepared solution was filtered, gradually diluted with distilled water, and finally made up to the mark.

Preparation of 0.0141 N $AgNO_3$: A total of 2.397 g of pure $AgNO_3$ salt was weighed out, shifted to a 1000 ml volumetric flask, and gradually brought up to the mark with distilled water. The resulting prepared solution was 0.0141(N). The solution was standardized by the NaCl solution. Analytical grade NaCl was dried overnight and cooled at normal temperature. NaCl (0.25 g) was weighed into flasks and dissolved in distilled water. Lastly, it was diluted to 100 ml with distilled water. To regulate the pH of the solution, little amounts of $NaHCO_3$ were added till the dizziness stopped. Around 2 ml of K_2CrO_4 solution was added, and the solution was titrated to the first enduring appearance red colour of K_2CrO_4.

Procedure: Initially, 100 ml of the water sample was placed in a 250 ml conical flask. One millilitre of K_2CrO_4 indicator was added to the flask and shaken gradually. Then, the sample was titrated with 0.0141(N) $AgNO_3$ solution until a brick red colour appeared, representing the endpoint. The volume of 0.0141(N) $AgNO_3$ solution used for titration was noted.

Calculation:

$$\text{Concentration of chloride, } (\mathrm{Cl}^-) = \frac{(X \times N)}{V} \times 35{,}450$$

where X = ml of 0.0141 N $AgNO_3$ solution used for titration from a burette; N = normality (0.0141 N) of $AgNO_3$ solution; V = volume (100 ml) of water sample.

3.10.2.16 Bicarbonate Ion (HCO_3^-)

Reagents: 0.1 N HCl solution; phenolphthalein indicator; and methyl orange indicator.

Procedure: A 100 ml water sample was placed into a 250 ml conical flask, and two drops of methyl orange indicator were added and shaken slowly. The water sample was then titrated with 0.1 N HCl until the colour changed to orange. The volume of the HCl solution in the burette was noted.

Calculation:

$$\text{Bicarbonate ion } (\mathrm{HCO_3}^-) = \frac{A \times N \times 1000 \times 50}{V}$$

where A = ml of HCl used for titration; N = concentration (0.1 N) of HCl solution; V = volume (100 ml) of water sample.

3.10.2.17 Sulphate Ion (SO_4^{2-})

The sulphate ion (SO_4^{2-}) in the samples was measured using a UV-spectrophotometric method at a wavelength of 420 nm, which is described below:

Reagent: *Buffer solution:* The buffer solution was prepared by accurately taking 30 g of $MgCl_2.6H_2O$, 5 g of sodium acetate (CH_3COONa), 0.111 g of sodium sulphate (Na_2SO_4), and 20 ml of glacial acetic acid (CH_3COOH) in a 1000 ml volumetric flask. Then, it was dissolved in distilled water and shaken slowly. Lastly, it was made up to the mark with the same water.

Barium chloride: A 20- to 30-mesh $BaCl_2$ solid crystal was used to make constant turbidity.

Standard sulphate solution: The standard sulphate solution was made by weighing it out accurately in 0.1479 g anhydrous Na_2SO_4, and it was placed into a 1000 ml volumetric flask. Then, it was dissolved in distilled water and shaken well. Lastly, it was made up to the mark with distilled water. It was made of 1 ml = 100 μg sulphate ion.

Procedure: *Formation of barium sulphate turbidity:* A 100 ml water sample was taken in a 250 ml flask, and a 20 ml buffer solution was added and mixed with a stirring apparatus. At this time, stirring a spoonful of $BaCl_2$ solid crystals was added, and the solution was stirred for 2 minutes.

Estimation of turbidity: After stirring, the solution was poured into the absorption cell, and turbidity was determined within 5 min.

Preparation of calibration curve: The concentrations of 2-, 4-, 6-, 8-, and 10-mg/L of Na_2SO_4 solutions were taken into four separate 1000 ml volumetric flasks, and distilled water was slowly added and carefully shaken, and finally made up to the mark with distilled water. The light absorbance of the water samples was measured using a UV spectrophotometer. A standard curve was prepared by plotting the absorbance against the concentration (mg/L).

Calculation:

$$\text{Sulphate}\left(SO_4{}^{2-},\frac{\text{mg}}{\text{L}}\right)=\text{concentration of } SO_4{}^{2-} \text{ from standard curve} \times \text{dilution factor}$$

3.10.2.18 Nitrate (NO_3^-)

The nitrate ion (NO_3^-) in the water sample was measured using the UV-spectrophotometric method at 220 and 275 nm wavelengths.

Reagents: *Nitrate-free water:* Double distilled water was used to prepare all used solutions and dilutions.

The stock solution of nitrate, solid potassium nitrate (KNO_3), was dried in an oven at 105 °C for 24 hours. Approximately 0.7218 g KNO_3 salt was weighed out and transferred to a 1000 ml volumetric flask and slowly brought up to the mark with distilled water. It was made of 1 ml = 100 μg NO_3-N solution.

Intermediate nitrate solution: A 100 ml stock nitrate solution was taken in a 1000 ml volumetric flask, and distilled water was added slowly and finally made up to the mark where 1 ml solution = 10 μg NO_3-N. Two millilitres of chloroform was used per 1 L of the solution to reserve it for at least 6 months.

1 (N) HCl solution: 83 ml of conc. HCl was measured and transferred into a 1000 ml volumetric flask and finally made up to the mark by gradually adding distilled water.

Procedure: *Treatment of sample:* 1 ml of 1 (N) HCl solution was mixed carefully in 50 ml of clear water sample and filtered.

Preparation of calibration curve: The volumes of 0.50-, 1.0-, 1.5-, and 2.0-ml intermediate nitrate solutions were taken in three separate 1000 ml volumetric flasks, distilled water was slowly added and carefully shaken, and lastly made up to the mark by adding distilled water. The absorbance of the water samples was determined by a UV spectrophotometer. A calibration curve was made by plotting the absorbance against the concentration (mg/L).

Spectrophotometric determination: The absorbance was read against double distilled water set to zero absorbance. A wavelength of 220 nm was used to obtain a NO_3^- reading, and a wavelength of 275 nm was used to estimate interference because of dissolved organic matter (DOM).

Calculation: The concentration of NO_3^- was measured using a UV spectrophotometer at two wavelengths, viz. 220 and 275 nm. The absorbance for samples and standard solutions was taken at two wavelengths, and the absorbance reading at 275 nm was deducted from the reading at 220 nm for each sample and standard solution.

Calculation:

$$\text{Nitrate}\ (\mathrm{NO_3}^-, \mathrm{mg/L}) = \frac{\mu\text{g of } \mathrm{NO_3}^- - N \text{ from calibration curve}}{\text{Volume (ml) of sample taken}}$$

$$\text{Nitrate}\ (\mathrm{NO_3}^-, \mathrm{mg/L}) = [\mathrm{NO_3}] \times 4.429$$

3.10.2.19 Phosphate (PO_4^{3-})

Reagent: *Ammonium metavanadate (NH_4VO_3) solution*: 2.5 g of NH_4VO_3 was placed in a 1000 ml volumetric flask in which 20 ml of conc. HNO_3 was added. The mixture was diluted after filtration with distilled water and brought to the mark.

Ammonium pentamolybdate solution: 5 g of $(NH_4)_6Mo_7O_{24}.4H_2O$ was placed in a 100 ml volumetric flask and diluted to the mark with distilled water.

Sulphuric acid (H_2SO_4) solution: 1:6 H_2SO_4 was prepared by adding distilled water and kept in a glass bottle.

Standard phosphate (PO_4^{3-}) solution: Accurately, 0.4391 g KH_2PO_4 was placed into a 100 ml volumetric flask, dissolved in distilled water through thorough shaking, and finally brought up to the mark with distilled water. Ten millilitres of this solution were transferred into a 100 ml volumetric flask and brought up to the mark with distilled water, in which 1 ml of solution is equivalent to 100 mg PO_4^{3-}.

Procedure: A proportion of 0.5, 1, 3, 5, and 7 ml intermediate PO_4^{3-} solutions were taken in five separate 100 ml volumetric flasks. Distilled water was added to the flasks slowly and carefully shaken and finally made up to the mark. A 2 ml standard sample, 5 ml $(NH_4)_6Mo_7O_{24}.4H_2O$ solution, 5 ml NH_4VO_3 solution, and 5 ml H_2SO_4 solution were placed into a 50 ml volumetric flask. The absorbance was taken against distilled water previously set at zero absorbance. A wavelength of 450 nm was used to obtain PO_4^{3-} ion absorbance. A calibration curve was made by plotting the absorbance against the concentration (mg/L).

Calculation:

$$\text{Phosphate}\left(PO_4^{3-}, \frac{\text{mg}}{\text{L}}\right) = \text{concentration of } PO_4^{3-} \text{ from standard curve} \\ \times \text{dilution factor}$$

3.11 Categorization of Studies

Thirty water quality parameters of 40 groundwater samples were measured in three seasons each year. The results of the water analysis are used as an instrument to (a) identify the processes and mechanisms influencing the groundwater chemistry (geochemical facies), (b) identify the origin of trace metals in groundwater, (c) evaluate drinking water quality, (d) assess irrigation water quality, and (e) assess industrial water quality in the study area. A fixed number of parameters are used for these different purposes. The detected parameters are divided into five categories for the above five purposes and are shown in Table 3.4.

3.12 Summary

Water quality analysers are used for the chemistry of monitoring processes, including water quality, process optimization, and control. Water quality parameters are of three types, physical, physicochemical, and biological, and are tested according to the desired water parameters. Water quality parameters often sampled include pH, conductivity, TDS, hardness, major cations, anions, trace metals, etc. Biological and other parameters are not significant for groundwater. Later, the water quality dataset will be used for geochemical investigation of local aquifers and evaluation of water suitability for drinking, irrigation, and industrial purposes.

Table 3.4 List of studies and parameters used in each investigation

Study	Subcategories of investigation	Used parameters	Chapter
(a) Evaluation of geochemical characteristics	(i) General geochemical processes	Temperature, well depth, pH, EC, TDS, Na, K, Ca, Mg, TH, Cl^-, SO_4^{2-}, PO_4^{3-}, NO_3^-, HCO_3^-, saturation index (SI), LSI, and $\log_{10}pCO_2$	4
(b) Origin identification of the trace metals	(ii) Sources and dissolution of trace metals	Well depth, pH, EC, TDS, TH, and trace metals	5
	(iii) Sources and dissolution of Fe and Mn	pH, EC, turbidity, Fe, Mn, SO_4^{2-}, NO_3^-, HCO_3^-, DO, DOC, and $\log_{10}pCO_2$	
(c) Assessment of drinking water quality	(i) Heavy metal pollution indices	B, Fe, Mn, Cr, Pb, Co, Ni, Cd, Cu, Zn, and As	6
	(ii) Canadian water quality index (CWQI)	Detected all water quality parameters	
	(iii) Weighted average water quality index (WWQI)	Detected all water quality parameters	
	(iv) Human health risk assessment (HRA)	B, Fe, Mn, Cr, Pb, Co, Ni, Cd, Cu, Zn, and As	
(d) Evaluation of irrigation water suitability	(i) Irrigation water quality parameters	Ca, Mg, Na, K, Cl^-, and HCO_3^-	7
	(ii) Simsek water quality index (SWQI)	pH, EC, TDS, SAR, Cl^-, NO_3^-, HCO_3^-, B, Fe, Mn, Cr, Pb, Co, As, Cd, Cu, and Zn	
	(iii) Meireles water quality index (MWQI)	SAR, EC, Na^+, Cl^-, and HCO_3^-	
	(iv) Canadian water quality index (CWQI)	SAR, EC, Na^+, Cl^-, and HCO_3^-	
	(v) Diagram methods	EC, TDS, SAR, Na%, Ca, Mg, Na, K, Cl^-, and HCO_3^-	
	(vi) integrated irrigation water quality index ($IIWQ_{index}$)	Detected all water quality parameters	
(e) Assessment of industrial water quality	The Langelier saturation index (LSI) Chloride–sulphate mass ratio (CSMR) Puckorius scaling (PSI) Ryznar stability (RSI) Aggressiveness (AI) Larson–Skold (LI) Corrosion (CI)	Temperature, pH, TDS, total hardness (TH), total alkalinity (T. Alk.), Ca^{2+}, Cl^-, HCO_3^-, CO_3^{2-}, SO_4^{2-}, NO_3^-, Fe, and Cu.	8

References and Further Study

1. BBS. (2020). *Bangladesh burrow of the statistics yearbook*. Ministry of Planning, People's Republic of Bangladesh.
2. DPHE/BGS/DFID. (2000). *Phase I: Groundwater studies of arsenic contamination in Bangladesh*. Department of Public Health and Engineering, Govt. of Bangladesh.
3. BGS-DPHE. (2001). *Arsenic contamination of groundwater in Bangladesh*. British Geological Survey and Department of Public Health Engineering, Vol. 2, Final Report, BGS Technical Report WC/00/19.
4. US-APHA. (2005). *Standard methods for the examination of the water and wastewater* (21st ed., p. 1134). APHA (American Public Health Association), AWWA, WPCF.

Chapter 4
Evaluation of Hydrogeochemical Processes

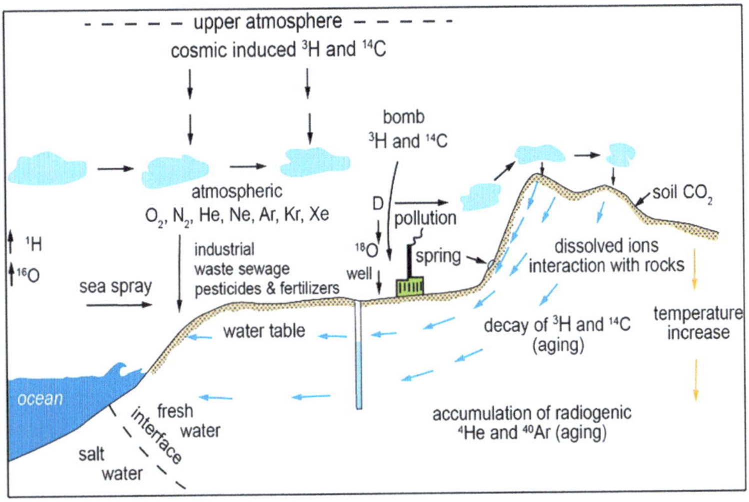

Abbreviations

BCI	Bicarbonate index
CAI	Chloro-alkaline index
FAO	Food and Agriculture Organization
HCA	Hierarchical cluster analysis
MON	Monsoon
PCA	Principal component analysis
pCO_2	Partial pressure of carbon dioxide
POM	Post-monsoon

M. S. Islam, *Hydrogeochemical Evaluation and Groundwater Quality*,
https://doi.org/10.1007/978-3-031-44304-6_4

PRM	Pre-monsoon
SI	Saturation index
TDS	Total dissolved solids
WHO	World Health Organization

4.1 Hydrochemical Characteristics and Water Chemistry

Groundwater chemistry is subject to issues such as the overall local geology, quality and quantity of leaching water, dissolution of the diverse rocks and minerals in aquifer basements, soil conditions, and contamination source potentiality. The dealings of these factors create a very complex water chemistry. The reactions between water and rock minerals impact total water chemistry and are valuable in understanding the source of the solutes in groundwater mixture.

To evaluate the geochemical characteristics, interpret the water quality data, and assess the water suitability for drinking (Chap. 6) and irrigation (Chap. 7), the original water data obtained from 40 monitoring sites around the upper Ganges River basin (Kushtia District) of Bangladesh were considered. Samples were collected for three periods, viz. pre-monsoon (PRM), monsoon (MON), and post-monsoon (POM) from each year of 2019–2020 and 2020–2021. These groundwater samples are calcite type (Ca–HCO_3), but other types, such as sodic (Na–Cl), dolomite (Mg–HCO_3), mixed (Ca–Mg–Na–Cl–HCO_3), etc., from different topographical areas, are taken into account for the same evaluation procedure. In this chapter, first, the investigated water quality parameter values/concentrations are included methodically. For the proper presentation of results, some statistical indicators and diagrams/pie charts were used. Ranges of parameter values with the environmental impacts of each detected water parameter are discussed in the subsequent subsections. Later, according to the findings of the water data, three assessments, viz. geochemical characterization of groundwater, sources and mobilization of trace metals (Chap. 5), and determination of water quality, were performed. Numerous multivariate statistical techniques, linear regression models, biplots, computer software, etc., are utilized to interpret the evaluation part.

Usually, analysed water parameters are divided into two categories: (a) geochemical parameters and (b) trace metals. The first classes are used to evaluate the geochemical processes and irrigation water quality. The second is used for the assessment of metal toxicity to the human body through the calculation of the degree of pollution and human health risk. The hydrochemical statistical analyses of physicochemical water parameters (except trace metals) of 40 groundwater samples during the three periods are revealed in Table 4.1. The average values of geochemical data of groundwater for the PRM, MON, and POM periods in 2 years (2020 and 2021) are presented separately in Appendices I–III. The results showed that the standard deviation (±SD) varied widely from the average value. A minimum of

Table 4.1 Summary of the analysed data of the groundwater in the pre-monsoon (PRM), monsoon (MON), and post-monsoon (POM) of 2019–2020 and 2020–2021 sampling campaign, with basic statistics

		Temp.	pH	EC	TDS	Ca	Mg	Na	K	TH	Cl	SO_4	PO_4	NO_3	HCO_3
	Pre-monsoon (PRM), $n = 40$														
Minimum	32.5	6.63	367.5	220.3	54.0	13.3	5.25	0.31	121.5	14.35	2.77	0.23	0.81	270.8	22
Maximum	39.0	8.04	1043.5	676.5	155.2	63.7	71.00	2.96	564.0	60.70	45.74	2.71	13.86	708.2	125
Mean	**35.5**	**7.03**	**670.0**	**413.4**	**96.11**	**32.73**	**14.27**	**1.27**	**362.8**	**31.18**	**16.42**	**0.99**	**4.11**	**419.6**	**50**
Stand. deviation	0.15	0.23	172.3	113.3	22.91	12.27	11.67	0.70	93.66	9.37	9.13	0.58	3.69	110.7	31.9
Variance	0.05	0.05	29689	12841	524.7	150.7	136.1	0.49	8771	87.88	83.41	0.34	13.6	12262	1017
Skewness	0.39	1.91	0.39	0.57	0.23	0.78	3.27	0.70	−0.07	0.87	1.22	1.20	1.06	1.25	0.96
	Monsoon (MON), $n = 40$														
Minimum	30.5	7.06	560.5	369.0	61.5	11.4	4.43	0.23	157.1	11.04	2.08	0.10	0.72	292.6	22
Maximum	37.5	8.41	1237.0	815.9	181.8	65.8	70.14	2.77	594.4	57.33	43.95	2.81	13.83	722.7	125
Mean	**32.0**	**7.43**	**867.5**	**558.2**	**110.4**	**31.08**	**12.72**	**1.11**	**396.1**	**28.52**	**15.73**	**0.91**	**4.18**	**441.9**	**50**
Stand. deviation	0.88	0.23	179.2	114.7	24.47	11.66	11.58	0.71	93.5	9.11	9.22	0.60	3.84	110.5	31.9
Variance	0.15	0.05	32118	13148	598.6	135.8	134	0.51	8749	83.03	84.99	0.35	14.74	12208	1017
Skewness	0.61	1.89	0.43	0.55	0.33	0.86	3.41	0.77	−0.04	0.80	1.14	1.30	1.10	1.22	0.96
	Post-monsoon (POM), $n = 40$														
Minimum	24.5	7.14	663.0	449.0	67.3	14.0	3.91	0.22	225.7	12.9	2.95	0.25	0.80	249.1	22
Maximum	32.0	8.91	1710.0	1108.1	184.3	52.5	50.25	2.56	613.3	41.8	41.71	2.14	18.34	815.6	125
Mean	**28.5**	**7.80**	**947.1**	**600.6**	**114.2**	**29.0**	**11.5**	**1.00**	**404.8**	**27.1**	**15.10**	**0.90**	**3.70**	**448.8**	**50**
Stand. deviation	0.61	0.40	220.0	156.0	26.2	9.40	8.54	0.64	93.8	7.60	9.24	0.49	3.66	124.5	31.9
Variance	0.09	0.16	48491	24346	688.5	88.3	72.9	0.41	8800	57.82	85.46	0.242	13.4	15489	1017
Skewness	0.71	0.07	0.93	−0.002	0.50	0.69	2.83	0.70	0.60	0.32	1.34	0.79	2.01	0.60	0.96

Note: (a) n = number of samples, depth = water depth, EC = electrical conductivity, TDS = total dissolved solids, TH = total hardness
(b) All parameter units are in mg/L except depth in m, EC in µS/cm, temperature in °C, and pH

10 variables out of a total of 15 were not homogenous, and the chemical composition of the samples was impacted by several influencing factors.

The analytical accuracy for the analysis of ions was measured by computing the standardized ionic (inorganic) charge balance error, which is represented in Eq. 4.1:

$$\text{Charge balance error, CBE} = \frac{\sum M_c|N_c| - \sum M_a|N_a|}{\sum M_c|N_c| + \sum M_a|N_a|} \times 100 \tag{4.1}$$

Here, M_c and N_c are the concentration (mole/L) and cationic charge, respectively; likewise, M_a and N_a are the molar concentration for the anions. Also, $TDS_{measured}$ and $TDS_{calculated}$ ratios are also used for quality-control measures of analysis. The calculated ratio of these both terms varies from 1 to 1.3 for all samples, which demonstrates the precision of the analytical data obtained.

The measured common ions such as Ca^{2+}, Mg^{2+}, Na^+, K^+, Cl^-, NO_3^-, SO_4^{2-}, HCO_3^-, and PO_4^{3-} of water samples are usually sufficient to give a charge balance because those ions carry 99% of charge in any natural freshwater. It was found that about 95% of the water samples showed a charge imbalance mainly in favour of positive charge excess, but some samples contrariwise displayed a negative charge deficit. In this case, the concentration of several heavy metals does not count in the calculation of CBE. The maximum charge balance that was calculated was 10.2%. An excess of positive charge higher than 5% approves with the database of the dissolved load in the summer period when a larger inequity appeared throughout the low water flow periods. This disproportion of negative charges could be correlated to the fact that no measure was made of organic substances, which is primarily produced by biological actions during dry (spring and summer) periods.

4.1.1 Physicochemical Water Parameters

Physicochemical parameters of groundwater samples, such as physical appearance, temperature, pumping well depth, pH, electrical conductivity (EC), total dissolved solids (TDS), water turbidity, total hardness (TH), dissolved oxygen (DO), etc., in water are not water quality indicators, but they are very significant and vital characteristics and can influence the growth of biota in water systems and thus influence the water quality. These parameters indicate the ionic loads in water. The parameters such as pH, EC, and TDS are the key water variable quantities that are essentially used to assess the geochemical procedures of groundwater. The values of the parameters in different sampling periods are stated in Table 4.1 and Fig. 4.1.

The results showed that the pH values were below the standard ranges, but the EC, TDS, and TH (mg/L) of the groundwater samples exceed the WHO and national guideline bar for both domestic and irrigation usages (vertical lines). All parameter values of POM are higher than those of PRM. In different coastal areas, groundwater carries excess dissolved sodium salts and obviously obtains higher values of EC and TDS. It was also found that the EC and ionic concentrations in some arid areas in the world show an imbalanced result.

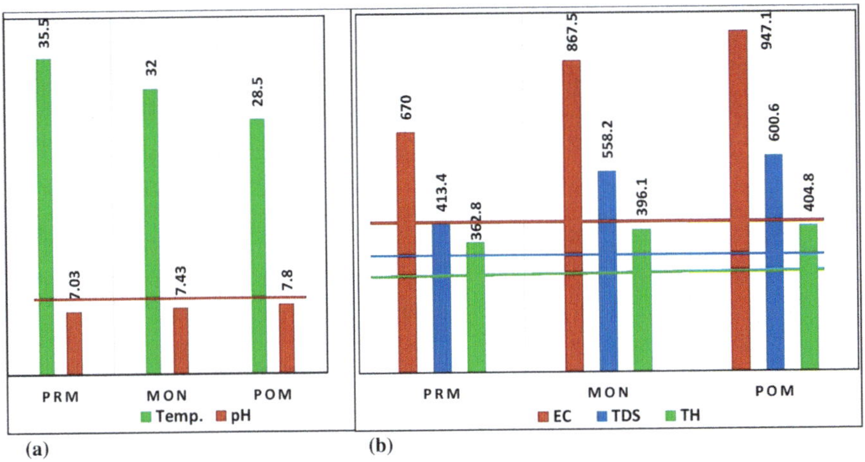

Fig. 4.1 Average values of (**a**) temp. (°C) and pH. (**b**) EC (μS/cm), TDS (mg/L), and TH (mg/L) in the different sampling seasons

Temperature

Temperature is a very important physical parameter that impacts water chemistry. In particular, it affects the dissolved gas concentration and equilibrium reactions in water. In tempered areas where the average air temperature is over 30 °C, the DO value is less than 2 mg/L, and the partial pressure of CO_2 in groundwater decreases. These two dissolved gases very much regulate the organic reactions and metal dissolution into the water phase from metal-laden rocks. In addition, several bicarbonate/carbonate equilibrium reactions occur in groundwater that may be influenced by water temperature.

pH

pH measures the acidity or alkalinity of water. A pH less than 7 indicates an acidic medium, and a pH greater than 7 indicates a basic medium. Although pH has no traditional influence on the water user, it is one of the very important and indispensable operative water quality parameters. The WHO and other native drinking water quality standards have suggested the suitable range of pH for drinking purposes from 7.5 to 8.5 (Table 4.2), i.e. the slightly alkaline nature of water is essential for good health. A pH value of water higher than 8.5 or below 6.5 can generate discolouration etching or scaling. The mean pH values were 7.03, 7.43, and 7.80 in the PRM, MON, and POM sampling period, respectively (Table 4.1). In the PRM season, the acidic character of water was mostly ascribable to natural biogeochemical activities, plant root respiration, and the discharge of organic acid from the decay of peripheral biological matter. The somewhat basic nature of samples in the MON and POM was due to the heavy mineralization and vast weathering of carbonate and bicarbonate materials during the recharge of the ground aquifer through filtration and penetration of surface water in rainy season. The elevated pH values with the higher HCO_3^- load (448.7 mg/L) in the POM seasons designated the free H^+ ion bond to the buffering

Table 4.2 Drinking and irrigation water quality standards

	Drinking water quality standard				Irrigational water quality standard		
Parameters	BDWS[a]	WHO[b]	US-EPA[c]	INDIA[d]	BIWS[e]	FAO[f]	US-EPA[g]
pH	6.5–8.5	7.5–8.5	6.5–8.5	7–8.5	7.5–8.5	6.5–8.4	7.5–8.0
EC (μs/cm)	–	–	–	–	750	350–500	–
TDS (mg/L)	1000	600	500	500	–	450–2000	500–1000
T. hardness (mg/L)	200–500	500	–	300	–	–	–
Na (mEq/L)	8.7	8.6	1.3–2.6	8.0	–	0–40	–
K (mEq/L)	0.3	–	–	–	–	0–0.05	–
Ca (mEq/L)	3.75	5.0	–	3.75	–	0–20.0	–
Mg (mEq/L)	2.5	12.5	–	2.5	–	0–5.0	–
Cl^- (mEq/L)	4.2–17	7.0	7.0	7.0	17.0	0–30	–
NO_3^- (mg/L)	10	50 (as N)	10 (as N)	45	–	0–10 (as N)	–
SO_4^{2-} (mg/L)	400	500	250	200	–	0–800	–
PO_4^{3-} (mg/L)	6.0	–	–	–	0.2	0–2(as P)	–

[a]Department of Public Health and Engineering, Bangladesh (2017)
[b]WHO-Drinking water standard, 4th ed. (2022)
[c]US-EPA-Drinking water standard (2018)
[d]Drinking water standard for India (IS10500, 2012)
[e]Bangladesh irrigation water standard (2009)
[f]FAO-Water quality for agriculture (1985)
[g]US-EPA-Guidelines for water reuse (2004)

agent HCO_3^-. Bicarbonate ions (HCO_3^-) regulate the basic and acidic characteristics of the water medium through equilibrium states. In Upper Egypt [1] and northern China's [2] arid area (rainfall less than 5 mm), the measured pH value was 8.1 (mean), whereas in the coastal areas of India [3], Nigeria [4], and Djibouti [5], this value was just over 7.

Electrical Conductivity (EC)
The total dissolved solids (TDS) in water systems regulate the electrical conductivity (EC), which measures the ionic concentration in water that permits it to transmit current. No standard value of EC was projected by the WHO or any country, but some water experts projected that drinking water typically accepted conductivity from 50 to 500 μS/cm and with mineralized water recorded values are over 500 μS/cm, whereas an EC value of 350 to 750 μS/cm is satisfactory for irrigation water. Water with elevated EC may have an undesirable taste, cause staining, and form a scale on the wall of the water containers and supply pipes. Several investigations have revealed that the groundwater in both shallow and deep aquifers has a high electrical conductivity in the southern coastal area of Bangladesh. Elevated levels of some

chemical constituents such as sodium (Na^+) and chloride (Cl^-) ions (over 3000 μS/cm) of shallow groundwater in the coastal belt of Bangladesh were found. For this reason, higher value of EC (over 2000 mg/L) as well as TDS (over 2000 mg/L) was gated in these areas' groundwater. The groundwater of other coasts in India, such as Tamil Nadu and Telangana, contains large amounts of sea salt, and the EC value was found to be over 5000 μS/cm. Several studies have illustrated that salinization is a serious environmental issue influencing soil and water qualities, agricultural production, and tree plantation as well as creating trouble in the natural ecosystem in the coastal region of the world. Table 4.1 shows that the EC values of the water ranged from 367.5 to 1043.5 μS/cm, 560.5 to 1237.0 μS/cm, and 663.0 to 1710.0 μS/cm in the PRM, MON, and POM, respectively. A higher level of EC was observed in the MON and POM seasons than in the PRM season, which may be due to the heavy mineralization of water during surface runoff and filtration in the rainy period. The great variances in EC values (±SD: 172.3 to 220.0) are characteristically attributed to geochemical actions such as filtration and recharge of rain or any surface water, evaporation, exchange of ions, and soil/sediment dissolution. The higher values of EC are attributable to the impact of the geology and impacts of anthropogenic events, which vary broadly in space. The dissolution or weathering of rocks/minerals with water thus results in the accessibility of more ions or electrolytes in the groundwater and leads to a higher EC value. In similar geological settings of Mexico [6] and Northern Ghana [7], the EC was less than 400 μS/cm, which indicated less mineralization of groundwater. A fantastic result was observed in the shallow groundwater of Omoku, Nigeria [8], in which the concentrations of ionic components, as well as EC (14.7–53.0 μS/cm), are very low, indicating that very poor mineralization occurs at the aquifer level.

Total Dissolved Solids (TDS)

Total dissolved solids (TDS) mean dissolved inorganic salts, minerals, and less amounts of organic substances in natural water. Its value has a significant relationship with the EC. A higher TDS indicates that the water is highly mineralized. The TDS ranges from 500 to 1000 mg/L, which is the allowable limit for potable uses (Table 4.2). In the coastal region of Bangladesh, some research showed that TDS levels exceeded the standard limit [9]. A higher value of TDS in potable water is not serious problem but may impact human health at low scale, i.e. people suffering from constipation, heart disease, and kidney function. The analysis results illustrated that the TDS values were gated to be in higher ranges of 413.4–670.5 mg/L (PRM), 369.0–815.9 mg/L (MON), and 449–1108 mg/L (POM) in the samples (Table 4.1). During the monsoon time, the surface runoff of rainwater passes in the sandy-type soil through percolation and infiltration and lastly reaches the aquifer with excess mineral concentrations. The results presented that the EC and TDS values were comparatively lower in the recharge area and higher in the discharge area in the study zone. Thus, when the water moved through the recharge to discharge region, it dissolved additional ingredients along its movement of journeys. The analysis results exhibited that higher EC and TDS values were found in the POM (just after the rainy period) sampling period related to the PRM due to the mineralization of the aquifer.

Another picture was found in northern Ghana [7], where the measured value of TDS in groundwater was just over 200 mg/L, which is very low, corresponding to the same geology of Bangladesh. The values of TDS are characteristically lowest at the place of infiltration, designated recharge regions, typically with TDS values like to those of the precipitation in the area, and maximum at the place of discharge, therefore after it has moved through the rock media and usually dissolved more salts along its route of travel. More materials are dissolved along the migration path. Therefore, more minerals dissolved by the water make more electrolytes/ions available in the groundwater system and correspondingly higher EC values. High EC values and extreme outliers are due to geological influences and human activities that vary greatly in space.

Total Hardness (TH)

Total hardness (TH) is caused mainly by geogenic processes in aquifers. Among the physicochemical parameters of groundwater, TH is very significant and considerable for domestic, industrial, and irrigation purposes. This parameter is not caused by a single component but by a diverse kind of dissolved multivalent metal ion, including the dominant Ca^{2+} and Mg^{2+} cations. In addition, other metals (ionic form), such as Ba, Al, Fe, Mn, Sr, and Zn, and the associated anions, such as bicarbonate (temporary hardness) and noncarbonate (permanent hardness), also contributed to the total hardness. Water containing $CaCO_3$ (equivalent value) at a level below 60 mg/L is typically considered soft water; 61–120 mg/L, moderately hard water; 121–180 mg/L, hard water; and more than 180 mg/L, very hard water. The study has revealed that the hardness value of groundwater samples of maximum zones of Bangladesh is higher than that of the standard limit of the WHO, and the waters were found to be hard to very hard [9]. Table 4.1 and Fig. 4.1b show that the samples contained very high TH values of 362.8, 396.1, and 404.8 mg/L in the three separate seasons. Thus, water hardness is a serious environmental concern in the sampling area. Generally, the TH value in coastal and arid areas' groundwater was found to be relatively low, but in upper zones, this value was very high. For example, the TH value in the groundwater of the south Gangetic plain (India part) [10] is much higher (303 mg/L), whereas in the Nigerian coast [4], this value is only 44.5 mg/L. Hard water has numerous effects on human health. Skin and hair are the most common parts of the human body that are directly affected by hard water. Additional precautions must be considered when the water is used for drinking and industrial purposes. Several studies have illustrated that hard water consumption is considered to be a substantial factor causing many diseases, including diabetes, gastroenteritis, and neural and cardiovascular disorders. The World Health Organization (WHO) has made the argument that the mineral content in hard drinking water on average can be helpful in circumstances where people are deficient in certain essential minerals. Not only for drinking purposes, hard water is also very harmful to irrigation (Chap. 7) and industrial water (Chap. 8). It can damage irrigation and industrial equipment due to scale formation. Soil properties may be changed through hard water, and this type of water directly affects plant growth.

4.1.2 Chemical Parameters

Physical parameter values mainly depend on chemical component loads in water. Table 4.1 and Fig. 4.2 represent the concentration and statistical data of different cations and anions in water samples in the sampling area. Most important and common cations include calcium (Ca^{++}), magnesium (Mg^{++}), sodium (Na^{+}), potassium (K^{+}), etc., and anions, including chloride (Cl^{-}), nitrate (NO_3^{-}), bicarbonate (HCO_3^{-}), sulphate (SO_4^{2-}), phosphate (PO_4^{3-}), etc., in groundwater are not serious pollutants; additionally, these anions are essential for human physiological functions in limited quantities, though an over-concentration of these ions can make the water harmful to any living biota. The levels of Na^{+}, K^{+}, and Cl^{-} in the groundwater of the sampling areas are very low, consistent with other coastal areas as well as other parts of Bangladesh. However, loads of Ca^{2+}, Mg^{2+}, and HCO_3^{-} were suggestively high and carried more than 90% of the total ions in aquifer water throughout the three sampling seasons (Fig. 4.2). These are the controlling ions of the collected samples, demonstrating that the water suitability was very poor. This might be due to the overexploitation of groundwater, ion exchange, lower river flow, excess dissolution of carbonate minerals, excessive soil erosion, and extra dissociation of carbonic acid (H_2CO_3). Usually, earth metal ions are derived from carbonate minerals (calcite, aragonite, and/or dolomite), where CO_2 is enhanced from anoxic and oxic degradation of organic substance. In groundwater, the aragonite/calcite ($CaCO_3$) and dolomite ($CaMg(CO_3)_2$) react with CO_2, resulting in an increase in the concentrations of Ca^{2+}, Mg^{2+}, and HCO_3^{-}. In the rainy season, dissolved CO_2 in rainwater can facilitate the rate of dissolution of carbonate minerals. Hence, the concentrations of Ca^{2+}, Mg^{2+}, and HCO_3^{-} in the POM sampling season were relatively higher than those in the PRM. The levels of other anions, such as SO_4^{2-}, NO_3^{-}, and PO_4^{3-},

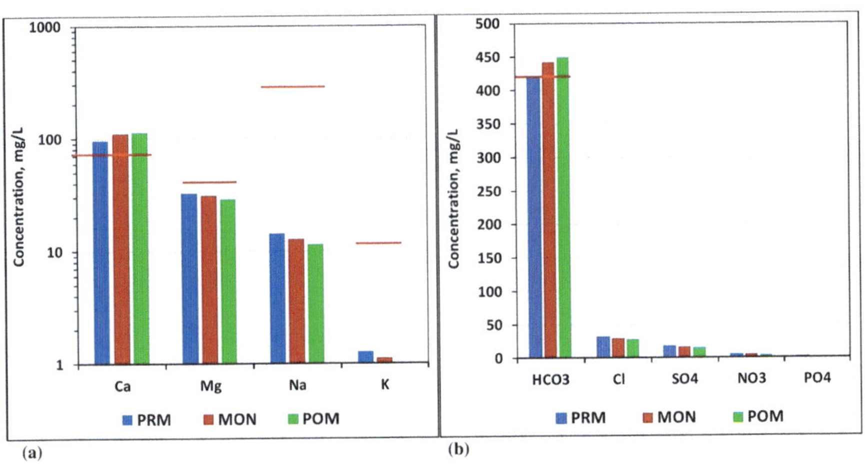

Fig. 4.2 Average levels of (**a**) major cations (mg/L) and (**b**) anions (mg/L) in the different sampling seasons

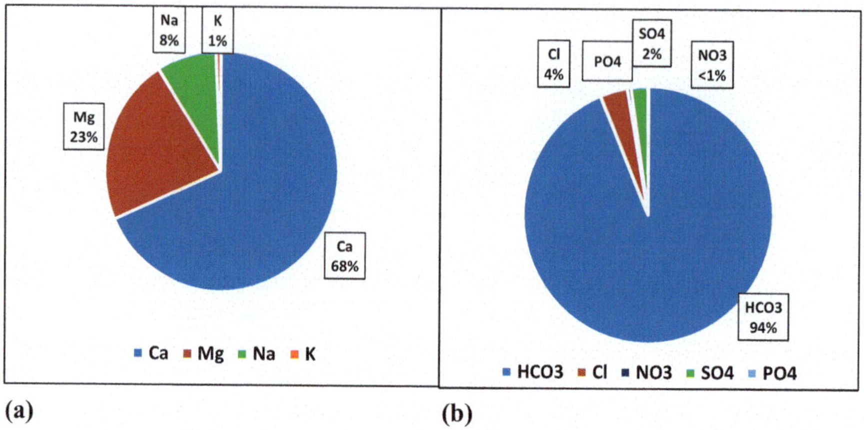

Fig. 4.3 Presence of major (**a**) cations and (**b**) anions of total concentration (mg/L) in groundwater of the study area

were found in comparatively lower ranges of concentration. On the other hand, in the seaside area, the levels of Na^+, K^+, and Cl^- totally control the water chemistry and geochemical processes. Due to sea level rise, the degree of saline water intrusion in coastal areas, soil, and aquifers is vastly affected by high sodicity. The concentration and environmental impacts of the analysed ions are discussed separately below:

(a) Cations

Calcium (Ca^{2+}), magnesium (Mg^{2+}), sodium (Na^+), and potassium (K^+) are the major cationic constituents in groundwater. The values of these ion concentrations are needed for the evaluation of geochemical facies and evaluation of irrigation water suitability. In addition, the hardness and scale formation nature of water depends mainly on the Ca^{2+} and Mg^{2+} concentrations. The average concentrations of those major cations are shown in Fig. 4.3. The present analytical results with other findings of a relevant study of those four abundant metals and their human health impacts are discussed below:

Sodium (Na^+)

Sodium is one of the most common and abounded inorganic constituents in aquifer waters and usually exists as chloride (Cl^-), bicarbonate (HCO_3^-), phosphate (PO_4^{3-}), nitrate (NO_3^-), and sulphate (SO_4^{2-}) salt. It is a crucial electrolyte that supports and preserves the balance of water or other fluids in the human body cell. Appropriate muscle and nerve functions, acid–base equilibrium in the blood, steady blood pressure level, and maintenance of the osmotic pressure of the plasma are also maintained by sodium ions. Deficient sodium ion in the human body fluid is also known as hyponatraemia, though the additional level of Na^+ raises blood pressure because it grasps excess liquid in the human body and upsurges the risk of stomach cancer, stroke, osteoporosis, heart failure, and kidney disease. In addition to being harmful to the human body, excess sodium in water may be very toxic to plants and crops. Due to sea level rises and increases in saltwater intrusion, a large portion of the coastal area becomes inefficacious to cultivation for sodium toxicity. Irrigation with water that has excess concentrations of Na^+ can harm the soil structure, making plant growth difficult. As said by the World Health Organization (WHO) and some native guidelines, the standard level of sodium in drinkable water is 200 mg/L or 8.7 mEq/L, and for irrigational use (FAO standard), it is equal to 40 mEq/L (Table 4.2). The samples of the study area (upland, far from the coast) contain a very low concentration of Na^+ compared to the maximum tolerable limits (Table 4.1). Other areas with the same geologic conditions, such as Talensi (Ghana) [11], Zamora (Mexico) [6], and Calabria (S. Italy) [12], as the discussible investigation area, have the same very low sodium concentration (>30 mg/L) in shallow groundwater, while some coastal groundwaters contain a large amount of this ion at over 1000 mg/L, which is completely unfit for domestic and other uses. In addition, it was found that some groundwater of arid areas such as the Ejina Basin (N. China) [2], Luxor (Upper Egypt) [1], and near New Delhi (India) [13] contains more than several hundred mg/L of sodium. Sodium-rich water makes the soil harder and is responsible for the corrosion of metal equipment.

Potassium (K^+)

Potassium (K^+) is usually found as chloride, bicarbonate, nitrate, and sulphate in groundwater, and its level is less than that of other vital cations in water. It is an essential constituent in the human nutrition, and along with Na^+, it maintains typical osmotic pressure in the body cell. In addition, it is a cofactor for numerous body enzymes and is vital for insulin ooze, muscle contraction, carbohydrate ingestion, creatinine phosphorylation, protein degradation, and nerve stimulation. Reducing Na^+ and cumulative K^+ in human food can help control hypertension and lower the risk of cardiovascular disease and death. Potassium-rich food can help insulin secretion in the human body, which controls the excess sugar in the blood of a diabetic patient. However, the additional amount of K^+ in the human body may be related to muscle weakness, depression, heartbeat disorder, etc. The potassium level in blood plasma is typically 3.5 to 5.3 mmol/L, and if blood contains 6 mmol/L, it can be risky, called *hyperkalaemia*, and typically necessitates immediate treatment. Presently, there are no definite guideline concentrations for K^+ in drinking water. The level of K^+ in all samples of the three sampling periods was found to be in safe ranges for human consumption (Table 4.1 and Figs. 4.2 and 4.3), but coastal groundwater contains freer K^+ than the upland groundwater.

Magnesium (Mg^{2+})

Mg^{2+} ions in groundwater are an important diet for biotics, control almost 300 biochemical reactions in the human body system, and support the production of protein and energy. It supports maintaining regular nerve and muscle systems, keeps the heartbeat steady, and helps to keep strong bones. It also helps regulate blood sugar levels in the human body. Standard Mg concentration in blood plasma is between 11.45 and 20.70 mg/L, and less than 11.45 mg/L is known as hypomagnesaemia. As stated by the WHO and BDWS, the usual concentrations for Mg^{2+} in drinking water should be 18 and 30 mg/L, respectively, and for irrigation activities, this value is 60 mg/L (Table 4.2). The analytical results for Mg^{2+} in water samples presented that the average level of this metal ion was within the safe range. Except some arid

(Rajasthan, India) and mountain range (Republic of Djibouti) areas, maximum groundwater (including the coastal belt) has Mg^{2+} within safe ranges in groundwater.

Calcium (Ca^{2+})

Calcium (Ca^{2+}) is an abundant and vital mineral in the Earth's crust, natural water, and the human body. This component plays a significant role in the human body's cell functioning. Excess calcium may cause of cancer, heart disease, hormone and fluid imbalance, neurodegenerative disease, muscle contraction, etc., along with the descent of the testis in the human body. Calcium is good for preventing osteoporosis in bones but bad for the brain, urinary tract, kidney, and arterial disease and solidity of bone restoration. The national and international guideline value for Ca^{2+} in drinking water is 75 to 100 mg/L. The maximum groundwater sample of the study area contains an excess of Ca^{2+}. Calcium concentration along with Mg^{2+} is primarily accountable for hardness, which is the major threat to household and industrial water in Bangladesh. Table 4.1 and Fig. 4.3 illustrate that the groundwater of the study area contains an excess amount of this metal ion. Over 100 mg/L Ca^{2+} (68% of all cations) is present in the samples of all sampling seasons (Fig. 4.4). Except for north-eastern Algeria (Ca^{2+}: 43.3–526.8 mg/L) [14] and Embu County, Kenya (Ca^{2+}: 184.6 mg/L) [15], the groundwater contains very little Ca in other African countries, viz. Nigeria (Ca^{2+}: 2–18 mg/L) [4], South Africa (Ca^{2+}: 19.65 mg/L) [16], and Egypt (Ca^{2+}: 34 mg/L) [1]. The Ca load in shallow groundwater of South Asian countries such as Bangladesh, India, Pakistan, and Nepal is somewhat higher (>60 mg/L) than that in other Asian countries.

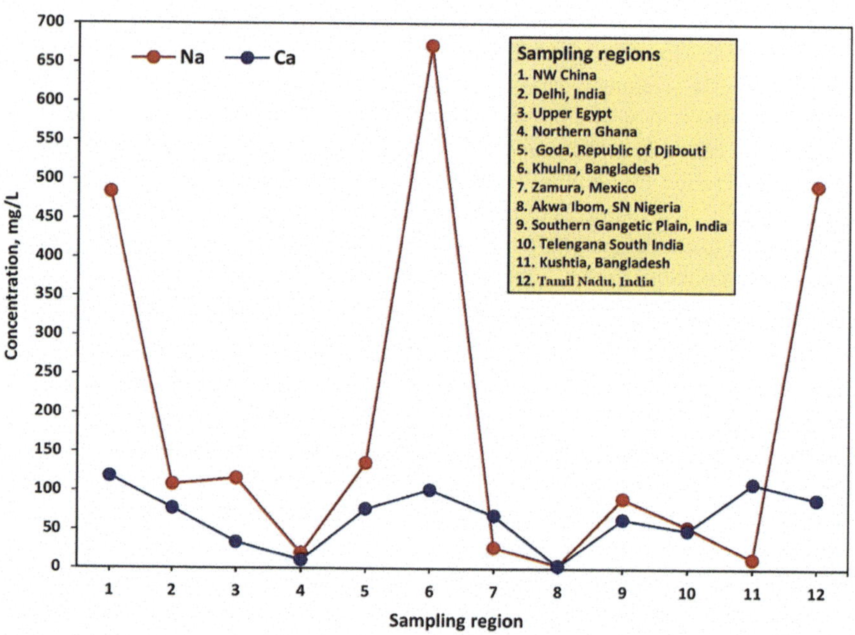

Fig. 4.4 Comparable loads of Na^+ and Ca^{2+} in groundwater of different sampling regions

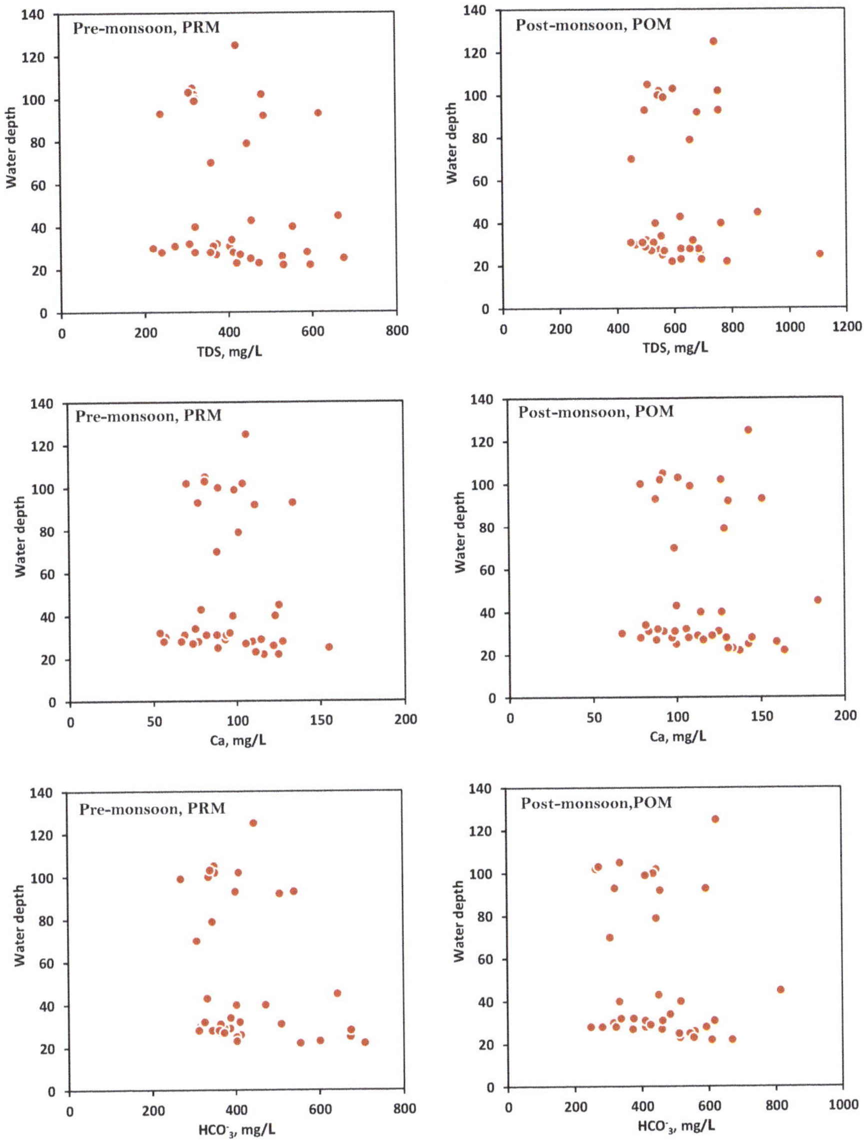

Fig. 4.5 Biplots of groundwater depth vs. (**a**) EC (μS/cm), (**b**) TDS (mg/L), (**c**) Ca (mg/L), and (**d**) HCO_3^- (mg/L)

It was observed that maximum coastal groundwater contains less Ca^{2+} relative to Na^+ concentration. In the aquifer base, excess Na^+ can replace Ca^{2+} through a natural ion exchange process. Some water data are represented in Fig. 4.5, where it is clear that the concentrations of Na^+ and Ca^{2+} are proportional to each other, but Ca^{2+} does not increase enough compared to Na^+.

(b) Anions

Chloride, bicarbonate, nitrate, sulphate, and phosphate are the major anionic constituents in groundwater. The concentration of these ions is needed for the evaluation of geochemical processes and the assessment of irrigation water quality. The average concentrations of those major anions are shown in Table 4.1 and Fig. 4.2b. The analytical results of these five anions and their human health impacts are discussed below:

Chloride (Cl^-)

Chloride is very common and essential anion in natural freshwater and acts as a comparatively minor pollutant in drinking water. Every mineral and rock contains some Cl^- as $CaCl_2$, $MgCl_2$, NaCl, and KCl, and it is vastly mobile in ground aquifers. A mature human body contains approximately 80 gm of Cl^- ions, and it maintains the metabolic actions in the human and plant body, human cellular activities, and fluid balance in the body. Chloride has no harmful activities in the human body except in cases of cardiac problems, but an additional quantity of Cl^- as NaCl may cause human hypertension and can disintegrate the water supply metal pipes. However, Cl^- is not a traditional pollutant, but because of the unpleasant taste and odour of drinking water, the BDWS, WHO, and US-EPA have set the maximum guideline value of secondary level of Cl^- in drinking water that ranges between 150 and 250 mg/L (Table 4.2). Excess Cl^- is a very toxic component to crops and several plants, and it should not contain an elevated level of Cl^- in irrigation water. Several studies have shown that the shallow groundwater of southern seaside zones in Bangladesh and others contains a very high concentration of Cl^- at several hundreds or thousands of mg/L, which is very far from standard levels. For example, the groundwater of the Khulna, Bagerhat, and Satkhira Districts of Bangladesh carries over 5000 mg/L Cl^-, which makes them completely unfit for use [9]. But the present samples contain a very low concentration (~30 mg/L) of chloride compared to the suitable level. This kind of water is not harmful for irrigation and industrial purposes.

Bicarbonate (HCO_3^-)

Bicarbonate ions remain as Ca, Mg, and Na bicarbonate, and these are very common and naturally occurring abundant component in the ground aquifer system. It plays curiously essential roles in several biological actions. There are no regional or international guideline levels for HCO_3^- in domestic water, while it is connected to the total hardness of the water. Numerous studies showed that the HCO_3^- level in the groundwater of Bangladesh wide-ranging from 60.4 to 700.5 mg/L [9]. The average concentration of HCO_3^- in the groundwater was found to be 419.6 to 448.8 mg/L in different sampling periods of the sampling area (Table 4.1), which is somewhat higher than in similar geographic locations. Generally, the source of HCO_3^- is geogenic, and its level is subject to the degree of carbonate mineral dissolution, saturation index value, rate of penetrated air CO_2 into the aquifer, etc. However, the seasonal variation in this ion is very considerable; for example, the concentrations of HCO_3^- in the pre-monsoon and post-monsoon periods were measured at 88.57 and 362.44 mg/L, respectively, in Delhi (India) [13]. In the monsoon period, CO_2 in rainwater can help carbonate mineral dissolution. For this reason, it may be assumed that the concentration of HCO_3^- in the pre-monsoon was relatively higher than that in the post-monsoon period.

Nitrate (NO_3^-)
Nitrate is the common form of nitrogen in natural water, and high levels in drinking water may cause methemoglobinemia which is also called blue baby syndrome. The key sources of NO_3^- in water comprise inorganic nitrogenous fertilizers, wastewater treatment plants, and septic tanks. The concentration of NO_3^- in groundwater samples is presented in Table 4.1, and the results presented that the concentration of NO_3^- was below the guideline limit of the WHO. The groundwater of some arid and semiarid areas, such as Telangana, India (69 mg/L) [17], Halabja Saidsadiq Basin, Iraq (11.7 mg/L) [18], some arid areas of Egypt (56.5 mg/L) [1], and Karnataka state of India (45.5 mg/L) [19], contains an excessive quantity of nitrate, although the concentration of this ion in other regions of the world, including coastal areas, is within safe ranges.

Sulphate (SO_4^{2-})
Sulphate is a general anion of groundwater that originates from the dissolution and weathering of Na, K, Ca, and Mg rocks and some kinds of surfactants such as sodium dodecyl sulphate. There is no initial health-based guideline value for SO_4^{2-} in drinking water. But the US-EPA set a secondary recommendation value for SO_4^{2-} of 250 mg/L because water carrying higher levels may create an aggressive taste that makes it unsuitable for household use and increases the risk of dehydration from diarrhoea. If the concentration of SO_4^{2-} in drinking water was over 600 mg/L, laxative effects were reported. Tables 4.1 and 4.2 show that the level of SO_4^{2-} in all the water samples of the study area is far below the standard limit.

Phosphate (PO_4^{3-})
Phosphate comes from phosphorus-containing inorganic fertilizers and detergents in the surface water body. Phosphate (PO_4^{3-}) has no guideline concentration for drinking purposes like SO_4^{2-}. Though too much PO_4^{3-} can cause health problems, such as osteoporosis and kidney damage, it can accelerate eutrophication in inland water. The levels of phosphate in the groundwater samples of the study area were found to be within the acceptable range (Tables 4.1 and 4.2).

4.1.2.1 Overall Groundwater Chemistry

Statistical studies of the physical and chemical parameters of 40 groundwater samples during pre-monsoon (PRM), monsoon (MON), and post-monsoon (POM) seasons for the 2020 and 2021 campaigns are presented in Table 4.1. This shows that the standard deviations (±SD) deviate significantly from the mean and median for at least 10 of the 15 total variables. Additionally, it was seen that the values were not uniform.

Table 4.1 shows that the major anions appeared in the order HCO_3^- >> Cl^- > SO_4^{2-} > NO_3^- > PO_4^{3-}, while the main cations appeared in the order Ca^{2+} > Mg^{2+} > Na^+ > K^+. The water parameters such as pH, EC, and TDS are the regulating variables for evaluating geochemical processes. During the PRM period, the acidic character of groundwater was mostly due to natural biogeochemical processes, plant root respiration, and the leakage of organic acids from decay of external biological matter. The weakly alkaline nature of groundwater during the POM period was responsible for cation mineralization and carbonate/bicarbonate dissolution. Higher pH values with increasing HCO_3^- concentration (448.8 mg/L) during the POM season designated that free H^+ is bound to the buffer HCO_3^-. Bicarbonates control the acidity and alkalinity of aqueous solutions through equilibrium conditions. The EC values of the water samples range between 366 to 1035 and 662 to 1708 μS/cm for PRM and POM, respectively. The elevated EC values in the POM season related to the PRM season are likely due to dissolution processes during surface runoff and infiltration during the wet season. Huge differences in EC values (±SD: 172.5 and 206.1 for PRM and POM, respectively) are mainly attributed to geochemical activity such as ion exchange, rainwater seepage, evaporation, sediment dissolution, etc. The TDS value of water is usually EC-dependent. Throughout the rainy season (MON), surface runoff of rainwater enters the soil through infiltration and eventually into aquifers with higher concentrations of minerals. The results show that the EC and TDS values are comparatively low in the charge zone and high in the discharge zone. As the water progresses through the replenishment zone to the discharge zone, additional ingredients are dissolved along its movement route. Therefore, more ions or electrolytes were added to the groundwater, resulting in elevated EC and TDS values. As a result of the analysis, higher EC and TDS values were observed during the POM period (immediately after the rainy monsoon) than during the PRM season due to the mineralization of groundwater. Increased EC values are due to the influence of geological and man-made events.

Levels of Na^+, K^+, and Cl^- in groundwater in the study area are lower than those in the south and other regions of the country. The concentrations of Ca^{2+}, Mg^{2+}, and HCO_3^- were very high, with over 95% of the total ions in the groundwater throughout the sampling period. These were the predominant ions in the collected samples, representing that the water quality was very poor. This may be due to overmining of groundwater, cation exchange, excessive weathering of carbonate rocks, reduced river flow, soil erosion, and carbonic acid dissolution. Earth metal ions are generally derived from carbonate minerals (calcite, aragonite, and/or dolomite) and enriched by inhaled CO_2 from the decomposition of oxygen and oxygen-free organic matter. Calcite/aragonite ($CaCO_3$) and dolomite ($CaMg(CO_3)_2$) react with CO_2 to increase Ca^{2+}, Mg^{2+}, and HCO_3^- concentrations in groundwater. During the rainy season, CO_2 in rainwater can accelerate the dissolution of carbonate minerals. Thus, Ca^{2+}, Mg^{2+}, and HCO_3^- concentrations in the POM were higher than during the PRM period. In addition, Ca-rich calcareous concrete nodules developed in alluvial deposits (the main deposits in the study area). Concentrations of other anions, such as SO_4^{2-}, NO_3^-, and PO_4^{3-}, were found in relatively low concentration ranges.

Normally distributed data were used for multivariate statistical analysis. Data for most chemical parameters had a positive bias (Table 4.1). In statistical and probability theory, skewness is a measure of the asymmetry of a probability distribution about the mean of a random variable. If it is between 0.5 and 1, it designates that the data are moderately skewed. Further statistical tests, such as goodness tests, are needed to see which statistical method defines the observations. A univariate non-parametric test (Mann–Whitney U test) evaluates the null hypothesis that both samples come from the same population, and the test bases its calculations on the rank of the data. In this study, this test showed substantial differences between PRM and POM sampling seasons for several water variables. In addition, both seasonal and random variations were analysed and presented. EC, TDS, Ca^{2+}, and Cl^- values were not significantly different, but $p < 1$ for the remaining variables in both rounds of sampling.

4.1.3 Water Depth and Mineralization Potentiality

Mineralization potentiality in aquifers mostly depends on the depth of sampling. Water data vary with shallow, semi-deep, and deep aquifers, which has been proven by several investigations. The present groundwater samples were collected from the first shallow aquifer, which was up to 100 m beneath the surface layer, less than 100 years old, and constantly recharged by river streams and rainwater. Samples were collected randomly from the designated shallow and semi-deep hand tube wells and engine well, and their depths from the surface ranged from 22 to 125 m.

The total dissolved solids (TDS) in water are a significant parameter that helps to understand the salinity in groundwater and the possibility of saltwater intrusion. TDS constitutes inorganic salts or minerals and less amounts of organic materials that are dissolved in water, and it depends on the water's conductivity. Water with a high range of TDS values indicates that the water is tremendously mineralized. On the other hand, the Ca and HCO_3^- in groundwater samples are the major components that regulate the total mineralization process in the aquifer. Figure 4.5 shows the dependence of water depth on the EC, TDS, Ca, and HCO_3^- values. It was found that these four (4) parameters of groundwater strongly depend on water depth. All the indicators showed lower values in relatively deeper aquifers and higher values in shallow water. Therefore, deeper water is less mineralized than shallower water. This finding is not only a fact; the same results have been found in several studies.

4.1.4 *Normalization Test of the Dataset*

In the application of statistics, normality tests are used to regulate if a dataset is well modelled by a normal distribution and to calculate how possible it is for a random variable fundamental to the dataset to be normally distributed. Normally distributed data were used for multivariate statistical analysis. Data for most chemical parameters had a positive bias (Table 4.1). In statistics and probability theory, skewness is a measure of the asymmetry of a probability distribution about the mean of a random variable. If this is between 0.5 and 1, it indicates that the data are moderately skewed. Further statistical tests, such as goodness tests, are needed to see which statistical model describes the observations. A univariate nonparametric test (Mann–Whitney U test) evaluates the null hypothesis that both samples come from the same population, and the test bases its calculations on the rank of the data. In this study, this test showed significant differences between the PRM and POM sampling periods for several water variables (Fig. 4.6). In addition, both seasonal and random variations were analysed and presented. The values of EC, TDS, Ca^{2+}, and Cl^{-} do not show substantial differences, whereas significant differences were observed for the rest of the variables in both sampling periods at $p < 0{:}05$. Any water data could be tested by this process and may come to decision for the acceptability of that data.

The datasets were correspondingly subjected to normality tests; meanwhile, some of the statistical methods, such as cluster analysis, have a normal distribution. On the other hand, datasets that were not normally distributed were log-transformed and standardized to their *z*-score values which presented in Eq. 4.2 as follows:

$$z = \frac{(x - \mu)}{s} \tag{4.2}$$

where z, x, μ, and s are the z score, sample value, average value, and standard deviation (±), respectively. If the z score is 0, the data point's score is alike to the average score. A z score of 1 would show a value that is one standard deviation from the average value. It may be positive or negative. The positive value of z score representing that the score is above the mean value and a negative score signifying that it is below the mean value.

4.2 Evaluation of Geochemical Processes: *Statistical Approaches*

Several statistical approaches, i.e. Pearson's correlation matrix, principal component analysis (PCA), hierarchical cluster analysis (HCA), numerous bivariate models, linear regression, Gibbs and Piper plots, etc., were utilized to select the classification and solute source of the groundwater systems. All statistical calculations were accomplished using IBM-Excel Worksheet and IBM-SPSS v-26. A review of the

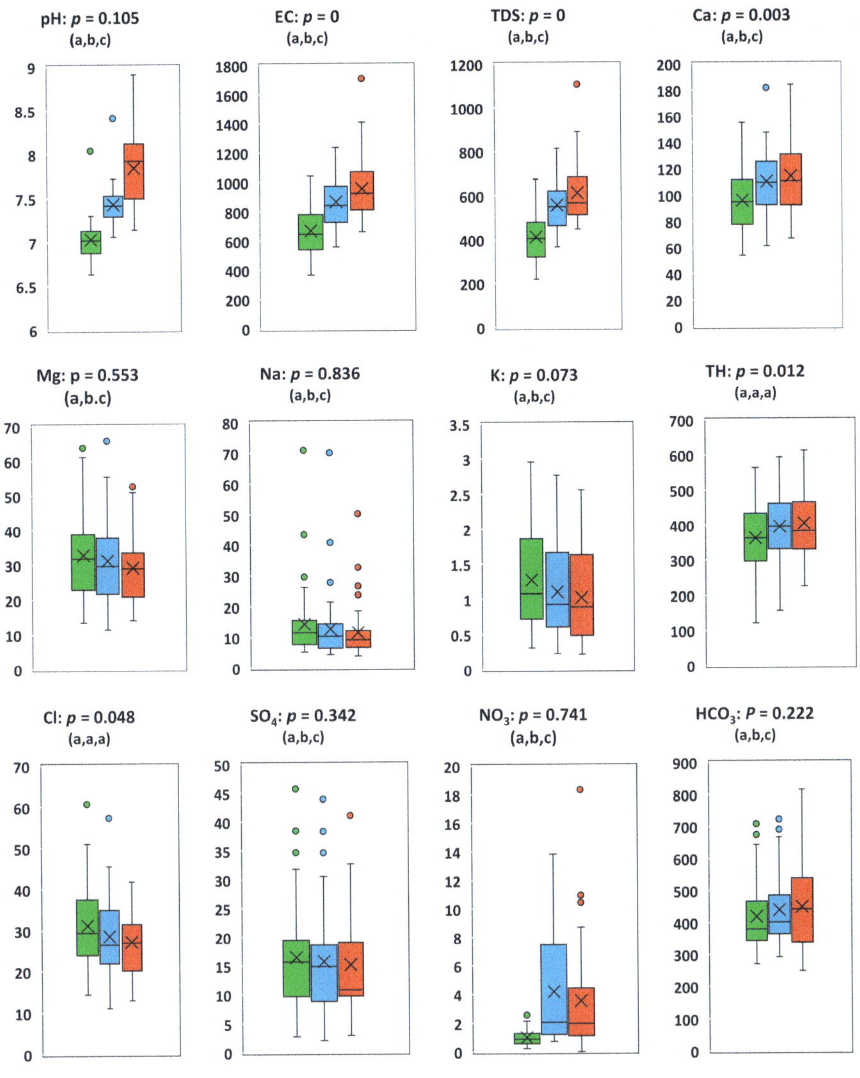

Fig. 4.6 Boxplot diagrams of the upper and lower quartiles (box); median (black line inside the box); 1.5× interquartile range, IQR (cross mark within the box); and outliers (circles) for nominated variables in the three sampling periods PRM (green), MON (sky blue), and POM (red). As stated by the Mann–Whitney U test, the letters ‘a’, ‘b’, and ‘c’ in each panel indicate provocatively different data distributions at $p < 0.05$ (two-tailed)

data revealed that the data were globally skewed. Thus, the data were log-transformed to more strictly resemble normally distributed data. All water variables were then standardized by calculating standard values (z-scores). This way, the statistical analysis gives equal weight to each variable. The strength of linear correlation, or degree of association, between two variables was evaluated by the

Pearson correlation coefficient r. After considering the two variables simultaneously, we used multiple linear regression analysis to assess their interdependence. In multiple regression, the coefficient of determination, the R^2 value, is easier to describe as a measure of association than the correlation coefficient, r. This is because R^2 corresponds to the part of the total variability of the dependent variable that can be influential by the independent variable.

Principal component analysis and factor analysis are different types of analysis. These analyses are ordered statistical processes, the initial method used to recognize the geochemical resolution or weathering that explains the mineralization of aquifers. These are identifying factors such as anthropogenic and geogenic processes that impact groundwater chemistry. The factors were independent, and a varimax rotation was used to identify the factors showing the highest (normal) variability. The variability is called the eigenvalue. Factors are assumed to have eigenvalues greater than 1, limiting the number of factors used in PC analysis. Factor analysis does not provide information about the interaction of factors. Hierarchical cluster analysis (HCA) is a valuable tool for organizing samples within classified collections. Q-mode and R-mode HCA of groundwater samples were presented in the outlined sample group along with the content of hydrochemical parameters to deliver useful statistics from PCA. HCA was calculated using the Ward cluster method, and the squared Euclidean distance was measured to determine the distance between clusters with similar parameter values. This study focused on selected water quality variables and used Ward's linkage regulation in the HCA. This regulation states that the shorter the distance between two data points, the greater the resemblance between them. A posterior analysis of variance (ANOVA) test with Bonferroni adjustment for multiple comparisons of means considering the 99% confidence level was performed to assess the effect of differences between water groups with similar chemical properties. In addition, Piper, Gibbs, and Chadha plots and some bivariate plots are produced to support other hydrogeochemical evaluation processes. To assess rock source identification, the Langelier saturation index (LSI) and Ca^{2+} and Mg^{2+} mass balances (in mEq/L) were calculated for all samples. Instead, using the geochemical computer programme PHREEQC v.3.0, the hydrogeochemical interactions and stability between rocks and water were determined by measuring the saturation index (SI) and their combined catalogue been studied. The method used the average chemical compositions of defined groundwater clusters/areas obtained from geostatistical analyses showing the evolution from the recharge zone (early) to the runoff zone (final). Anthropogenic origins were also considered in the statistical treatment, in addition to geological processes.

4.2.1 Multivariate Statistical and Spatial Approach

Throughout the earlier decades, attention to the hydrogeochemistry of aquifer water has increased, as established by numerous geochemical studies, which are progressively becoming a firm part of local and regional hydrogeological investigations. Earlier studies employed graphical demonstrations (Schoeller, Piper, Chadha, Stiff,

Gibbs, etc.) in visually describing differences in key ion chemistry of groundwater and categorizing water compositions into identifiable groups that are characteristic of analogous genetic history. Geo-environmental datasets and multivariate statistical techniques were successfully applied as tools in the study of groundwater geochemistry. The application of multivariate statistics to datasets facilitated the presentation of unseen structures in the datasets and helped resolve key geo-environmental problems at several scales. This geochemical data procedure operated on the concept that each aquifer zone has its single groundwater suitability signature based on the chemical makeup of the sediment that includes it. Moreover, thermodynamic calculations with the measured pCO_2 using PHREEQC-3v computer software were used to categorize the definite rocks that controlled the water equilibrium reactions in the solid phase. It is against this background that this study is pursuing the application of the multivariate statistical procedure with some biplots and programmes as a tool for a whole geochemical assessment of groundwater that enables the unveiling of hidden structures in the datasets and supports in defining the factors accountable for the nonconformity of groundwater quality. Several statistical and graphical techniques were used to evaluate the complete geochemical characteristics of groundwater.

Multivariate analysis is a statistical approach to multiple dependent variables as a consequence of one outcome. It is a very important technique to identify the correlation and association between water variables. In addition, these techniques provide the classification of components with several different cultures. The study used Pearson's correlation matrix, principal component analysis (with robust PCA), and hierarchical cluster analysis (R- and Q-mode) as the multivariate statistical analysis. In addition, numerous biplots, diagrams, regression models, and programmes are used to explore the geochemistry of groundwater.

4.2.1.1 Pearson's Correlation Matrix

Pearson's correlation matrix (PCM) for analysing groundwater geochemical variables in the PRM, MON, and POM periods is presented in Table 4.3 (a, b, and c). Among the variables, the correlation of EC with TDS, Ca^{2+}, Mg^{2+}, TH, SO_4^{2-}, and HCO_3^- of the samples was strongly associated ($r > 0.5, p = 0.01$, at 95% confidence interval, *CI*) all of the three sampling periods. The elevated value of EC is mostly caused by the divalent cation of earth metals and higher level of HCO_3^- but not by univalent ions, viz. Na^+, K^+, Cl^-, and NO_3^-. Total hardness (TH) was significantly correlated with Ca^{2+}, Mg^{2+}, HCO_3^-, and SO_4^{2-}, representing its Ca–Mg–HCO_3-type temporary hardness (the SO_4^{2-} level was very low). Therefore, Ca^{2+} and Mg^{2+} seem to be the main contributors to TH, stemming from the dissolution of calcite-type minerals by CO_2-loaded precipitation. pH illustrates a weak adverse correlation with carbonate-baring components such as Ca^{2+}, Mg^{2+}, and HCO_3^-. It was an insignificant link with the calco-carbonic equilibrium, where the pH impacts the weathering of carbonate-type minerals/rocks. A moderate positive relationship was found among Ca^{2+} and Mg^{2+}, suggesting that the maximum ions were engaged in

Table 4.3 Pearson's correlation matrix (PCM) of water parameters in the (a) PRM, (b) MON, and (c) POM sampling seasons in the study zone

	pH	EC	TDS	Ca^{2+}	Mg^{2+}	Na^+	K^+	TH	Cl^-	HCO_3^-	SO_4^{2-}	NO_3^-	PO_4^{3-}	Dep.
	(a) Pre-monsoon (PRM)													
pH	1													
EC	−0.12	1												
TDS	−0.22	0.96**	1											
Ca^{2+}	−0.15	0.78**	0.77**	1										
Mg^{2+}	0.03	0.50**	0.46*	0.45*	1									
Na^+	−0.27	0.43*	0.45*	0.25	0.08	1								
K^+	−0.11	0.55**	0.54**	0.44*	0.41*	0.62**	1							
TH	−0.1	0.58**	0.55**	0.65**	0.52**	−0.03	0.15	1						
Cl^-	0.04	0.39*	0.31	0.30	0.30	0.12	0.22	0.30	1					
HCO_3^-	−0.1	0.76**	0.78**	0.80**	0.46*	0.09	0.34	0.68**	0.17	1				
SO_4^{2-}	−0.25	0.76**	0.80**	0.66**	0.35*	0.32	0.47*	0.52**	0.18	0.61**	1			
NO_3^-	−0.16	0.29	0.34*	0.18	−0.05	0.31	0.17	0.22	0.28	0.10	0.38*	1		
PO_4^{3-}	0.11	0.13	0.02	0.06	0.05	−0.05	0.22	0.15	0.27	−0.05	0.11	0.03	1	
Depth	−0.24	−0.12	−0.15	−0.03	0.07	0.06	−0.07	0.12	−0.21	−0.20	−0.01	−0.19	−0.05	1
	pH	EC	TDS	Ca^{2+}	Mg^{2+}	Na^+	K^+	TH	Cl^-	SO_4^{2-}	PO_4^{3-}	NO_3^-	HCO_3^-	Depth
	(b) Monsoon (MON)													
pH	1													
EC	−0.13	1												
TDS	−0.23	0.96**	1											
Ca^{2+}	−0.20	0.76**	0.75**	1										
Mg^{2+}	0.03	0.50**	0.44*	0.43*	1									
Na^+	−0.27	0.47*	0.47*	0.26	0.04	1								
K^+	−0.10	0.61**	0.57**	0.46*	0.40*	0.62**	1							
TH	−0.14	0.56**	0.55**	0.67**	0.57**	−0.05	0.17	1						

Cl^-	0.03	0.36*	0.27	0.32	0.35*	0.09	0.23	0.30	1					
SO_4^{2-}	−0.27	**0.77****	**0.80****	**0.66****	0.36*	0.33	0.48*	**0.54****	0.16	1				
PO_4^{3-}	0.10	0.13	0.02	0.11	0.13	−0.04	0.24	0.18	0.31	0.11	1			
NO_3^-	−0.16	0.26	0.31	0.20	−0.06	0.28	0.10	0.20	0.25	0.32	0.03	1		
HCO_3^-	−0.25	**0.73****	**0.73****	**0.70****	0.38*	0.18	0.41*	**0.61****	0.32	**0.59****	−0.01	0.05	1	
Depth	−0.22	−0.15	−0.17	−0.03	0.10	0.06	−0.05	0.13	−0.20	−0.02	−0.03	−0.19	−0.20	1
	pH	**EC**	**TDS**	**Ca^{2+}**	**Mg^{2+}**	**Na^+**	**K^+**	**TH**	**Cl^-**	**HCO_3^-**	**SO_4^{2-}**	**NO_3^-**	**PO_4^{3-}**	**Dep.**
	(c) Post-monsoon (POM)													
pH	1													
EC	−0.41*	1												
TDS	−0.20	**0.85****	1											
Ca^{2+}	−0.39*	**0.64****	**0.50****	1										
Mg^{2+}	−0.38*	0.33*	0.17	**0.54****	1									
Na^+	−0.37*	0.37*	0.21	0.43*	0.13	1								
K^+	−0.47*	0.32	0.10	0.45*	0.37*	**0.73****	1							
TH	−0.44*	**0.59****	0.41*	**0.94****	**0.80****	0.36*	0.48*	1						
Cl^-	0.03	0.33*	0.33*	0.36*	0.09	0.25	0.34*	0.29	1					
HCO_3^-	−0.38*	**0.52****	0.42*	**0.81****	0.45*	0.28	0.36*	**0.76****	0.27	1				
SO_4^{2-}	−0.47*	**0.60****	**0.57****	**0.54****	0.34*	0.28	0.28	**0.53****	0.15	**0.55****	1			
NO_3^-	−0.04	**0.50****	**0.51****	0.10	−0.21	0.22	−0.01	−0.02	0.20	0.12	0.27	1		
PO_4^{3-}	−0.10	0.05	−0.10	0.01	−0.06	−0.02	0.09	−0.02	0.13	−0.02	0.13	0.09	1	
Depth	−0.11	0.04	−0.11	−0.05	0.19	−0.01	0.05	0.04	−0.29	−0.17	0.02	−0.25	−0.06	1

*Correlation is considerable at the $p = 0.05$ level with a 95% confidence interval, *CI* (2-tailed)

**Correlation is considerable at the $p = 0.01$ level with 95% confidence interval, *CI* (2-tailed, bold)

numerous physicochemical reactions (ion exchange and redox) in the groundwater. On the other hand, Na^+, K^+, and Cl^- showed a weak association with Ca^{2+}, Mg^{2+}, HCO_3^-, and SO_4^{2-}. This was credited to the consequence of discharge and weathering of quick soluble Triassic or carbonate/lime compound. This type of weathering was buoyed by the Langelier saturation index (LSI). The computed index exposed that above 80% of samples had a positive LSI value (MON, 0.22; MON, 0.60; and POM, 0.91) for the sampling periods. Therefore, the aquifer water was supersaturated with carbonate minerals (e.g. calcite, aragonite, and dolomite). NO_3^- presented a negative and very weak correlation with other parameters, which designated that it might come from outside sources, i.e. from nitrogen-bearing inorganic fertilizers in the sampling area. All the water parameters were not associated with the depth of the water table because all samples were collected from the same upper shallow aquifer. In the investigated area, the deep aquifers started from over 150 m beneath the surface layer. The matrix table may help to explore the information for assessing the hydrogeochemistry of the study areas.

Some dissimilarities were observed in Table 4.3 (a, b, and c). In the pre-monsoon (PRM) and monsoon (MON) sampling periods, EC was strongly positively correlated with K^+ and Mg^{2+} but not with POM. On the other hand, TDS is strongly positively correlated with K^+ and TH in the PRM and MON sampling seasons but not in the POM. NO_3^- has a strong correlation with both EC and TDS in the MON and POM periods. This is because of the extra nitrate leaching from inorganic fertilizers in the heavy rainy season. Another observation is very clear: maximum parameters are negative or very weakly correlated with water depth in all sampling seasons. Therefore, heavy mineralization occurred in relatively shallow water rather than in semi-deep water.

A different picture was observed in the correlation matrix of coastal groundwater, where the dominant ions are Na^+ and Cl^- and have a very high value of EC. The matrix table for the coastal groundwater parameters of Khulna (Bangladesh) [20] and Goda (Djibouti) [5] showed that the EC and TDS are strongly positively correlated with Na^+ and Cl^- ($r > 0.9$) but not with earth metal ions and HCO_3^- ($r < 0.2$). On the other hand, Na^+ and Cl^- are highly correlated, but both are very weakly correlated with Ca^{2+}, Mg^{2+}, and HCO_3^-. So, coastal groundwater quality and the origin of solutes are completely different compared to upland groundwater (Ca^{2+}, Mg^{2+}, and HCO_3^- rich). In arid regions such as Upper Egypt [1], Karnataka (India) [19], and southern Punjab (Pakistan) [21], no groundwater quality parameter is strongly correlated with other parameters. In this case, it is not easy to assess the geochemical processes of groundwater.

4.2.1.2 Principal Component Analysis (PCA)

Principal component analysis (PCA) is an influential and flexible technique that has many uses and works for most statistical cases. This is a data reduction technique that helps analyse multivariate datasets to clarify large datasets and visualize correlations between variables and factors. It is hoped that we can limit the number of factors that cause dataset discrepancies. PCA was accomplished utilizing a correlation matrix, which brings the measurements onto a common scale, and the focal components were pulled out based on eigenvalues equal to or greater than 1 with the PCs being sorted in lessening order of variance, such that the axis of highest variance becomes the first principal component (PC1) and the axis of second highest variance becomes the second principal component (PC2). In general, the factors that control groundwater geochemistry in the region show some degree of correlation. A varimax rotation was used to confirm that the factors were not correlated with each other and that the variables were not significantly correlated with multiple factors. Varimax rotations produce orthogonal factor rotations, so the resulting factors are uncorrelated and understandable. Based on the above classification, the final factor model resulted in five factors. Parameters with high commonality are usually those that contribute suggestively to the factors.

An interelement correlation was determined for the 40 samples and 14 different water variables for the three sampling periods (Table 4.4 a, b, c and Fig. 4.7a–c). The results presented a total variance of 74.506%, 78.185%, and 74.49% in the pre-monsoon (PRM), monsoon (MON), and post-monsoon (POM) sampling rounds, respectively, with an eigenvalue greater than 1, as determined by five PCs of R-mode. Around 60% of the total variance in all sampling periods is shown in the first three loadings (PC1, PC2, and PC3). The positive and negative figures in the PCA table make clear that the water samples were influenced or uninfluenced by the presence of extracted loads on a precise component. The parameters such as EC, TDS, Ca^{2+}, SO_4^{2-}, and HCO_3^- displayed a strong connotation (bold type) with PC1 in the PRM and MON, but four (4) components, viz. EC, Ca^{2+}, TH, and HCO_3^-, exhibited the same loading as PC1 in the POM sampling season. A strong and moderately strong loading with Ca^{2+}, Mg^{2+}, SO_4^{2-}, and HCO_3^- suggested rock–water interaction with ion exchange in both periods. Elevated loading factors for Ca^{2+} and HCO_3^- correspond with events such as carbonate rock weathering that could be exposed to increased levels of Ca^{2+}, HCO_3^-, and Mg^{2+}. The first component is strongly associated with EC, whereas the role of the second to the fifth component in this variable is very insignificant. The strong correlation between EC, TDS, and TH showed the presence of enormous ionic components, mostly Ca^{2+}, Mg^{2+}, HCO_3^-, and SO_4^{2-}, which were collected by aquifer rock–water interactions and man-made pollution sources, like chemical fertilizers from agricultural runoff. Other ions were not significantly associated with EC in any component number (PC2 to PC5); therefore, they do not contribute to EC. However, robust PCA exhibited that only NO_3^- was strongly and moderately loaded for PC4 in the PRM and PC2 in the POM sampling period (Fig. 4.7a, c), respectively, which indicated that this parameter penetrates into ground layer from outside sources. However, HCO_3^- has a negative loading in the same groups. The contradictory evolution of NO_3^- and HCO_3^- could replicate the influence of acid–base equilibrium circumstances on water chemistry. Similar to NO_3^-, unaccompanied PO_4^{3-} was strongly loaded for PC4 in the PRM, and this variable comes from human-caused sources. FC2 was characteristically related to chemical composts.

Table 4.4 Principal component (five components extracted) loadings of the analysed parameters in the groundwater samples (sorted by size)

Parameter	*PC1*	*PC2*	*PC3*	*PC4*	*PC5*
	(a) Pre-monsoon (PRM)				
(1) EC	**0.938**	0.042	0.076	−0.052	0.101
(2) TDS	**0.937**	−0.047	0.042	−0.180	0.093
(3) Ca^{2+}	**0.865**	0.110	−0.189	−0.047	−0.020
(4) SO_4^{2-}	**0.828**	−0.141	−0.059	−0.062	−0.141
(5) HCO_3^-	**0.812**	0.271	−0.243	−0.269	0.071
(6) TH	0.692	0.367	−0.307	0.035	−0.260
(7) K^+	0.627	−0.303	0.290	0.337	0.391
(8) Mg^{2+}	0.580	0.309	−0.196	0.295	0.305
(9) Na^+	0.431	−0.672	0.333	0.065	0.319
(10) pH	−0.224	0.624	0.293	0.039	0.322
(11) Depth	−0.091	−0.314	−0.589	0.439	−0.012
(12) Cl^-	0.405	0.292	0.485	0.213	−0.233
(13) PO_4^{3-}	0.130	0.220	0.341	**0.745**	−0.322
(14) NO_3^-	0.362	−0.263	0.555	−0.291	−0.505
%Variance	***38.000***	***11.108***	***10.195***	***7.986***	***7.217***
%Cumulative	***38.000***	***49.108***	***59.303***	***67.289***	***74.506***
Eigenvalue	***5.700***	***1.666***	***1.529***	***1.198***	***1.083***
	(b) Monsoon (MON)				
EC	**0.940**	−0.036	0.077	−0.010	−0.145
TDS	**0.931**	−0.143	0.034	−0.120	−0.143
Ca^{2+}	**0.849**	0.089	−0.137	−0.109	−0.003
SO_4^{2-}	**0.823**	−0.136	−0.085	−0.087	0.054
HCO_3^-	**0.795**	0.104	−0.128	−0.267	−0.208
TH	0.692	0.403	−0.354	−0.171	0.245
K^+	0.661	−0.217	0.201	0.526	−0.228
Mg^{2+}	0.570	0.448	−0.244	0.266	−0.113
Cl^-	0.432	0.407	0.402	0.082	0.306
Na^+	0.455	−0.679	0.220	0.369	−0.048
pH	−0.258	0.514	0.379	0.097	−0.418
Depth	−0.099	−0.133	**−0.742**	0.450	0.312
PO_4^{3-}	0.166	0.414	0.311	0.511	0.395
NO_3^-	0.320	−0.273	0.428	−0.322	0.578
%Variance	***40.146***	***11.598***	***10.209***	***8.588***	***7.644***
%Cumulative	***40.146***	***51.745***	***61.953***	***70.541***	***78.185***
Eigenvalue	***5.620***	***1.624***	***1.429***	***1.202***	***1.070***
	(c) Post-monsoon (POM)				
(1) Ca^{2+}	**0.891**	−0.118	−0.063	0.227	−0.071
(2) TH	**0.882**	−0.308	−0.136	0.248	−0.055
(3) EC	**0.815**	0.344	−0.194	−0.196	0.071
(4) HCO_3^-	**0.786**	−0.051	−0.117	0.310	−0.075
(5) SO_4^{2-}	0.724	0.125	−0.210	−0.235	−0.184

(continued)

Table 4.4 (continued)

Parameter	*PC1*	*PC2*	*PC3*	*PC4*	*PC5*
(6) TDS	0.652	0.539	−0.336	−0.082	0.154
(7) Mg^{2+}	0.596	−0.543	−0.221	0.212	−0.010
(8) pH	−0.577	0.276	−0.100	0.388	0.014
(9) NO_3^-	0.274	**0.787**	0.012	−0.241	0.098
(10) K^+	0.609	−0.290	0.616	−0.156	0.122
(11) Na^+	0.551	−0.026	0.601	−0.235	0.385
(12) Depth	−0.030	−0.498	−0.278	−0.555	0.105
(13) Cl^-	0.410	0.335	0.401	0.427	−0.066
(14) PO_4^{3-}	0.045	0.105	0.344	−0.182	−0.825
%Variance	***36.482***	***13.213***	***9.329***	***8.398***	***7.066***
%Cumulative	***36.482***	***49.696***	***59.025***	***67.423***	***74.489***
Eigenvalue	***5.472***	***1.982***	***1.399***	***1.260***	***1.060***

Note: Bold figures designate strong loading with the component number (over 0.75)

The major component of nitrogen-bearing fertilizers, such as ammonium, was quickly oxidized to NO_3^-. The nitrification method produced an additional amount of H^+ in aquifer groundwater oxic conditions. Therefore, it resulted in a lowering value of pH. Conversely, in groundwater, the higher loading of HCO_3^- was connected to the dissolved CO_2, which was created from the decay of soil organic matter and plant root respiration. The connotation of these two parameters reflects the influence of inorganic fertilizers on groundwater pollution and can be termed the contamination factor. The following reactions can describe that process, where an increase in NO_3^- and HCO_3^- levels occurs in the aqueous medium:

$$NH_4^+(\text{nitrogenous fertilizers}) + O_2(\text{oxic condition}) \xrightarrow{\text{Nitrification}} NO_{3\,(aq)}^- + H_{(aq)}^+ + H_2O$$

$$CO_2(\text{leached and root respiration}) + H_2O \rightleftharpoons H_2CO_3$$

$$H_2CO_3 + H_2O \rightleftharpoons H_2CO_3^- + H_3O_{(aq)}^-$$

In addition, the water depth showed a strong association (negative) with PC3 in the monsoon (MON) season and a moderate negative association with the same principal component. Moreover, depth had a moderate negative association with PC4 of the POM samples. Therefore, well depth is a vital factor that may regulate the mineralization processes of groundwater.

Varimax-rotated PCA loading matrices concerning aquifers give a primary idea of water classes, types of rock weathering, groundwater pollution sources, etc. To explain the geochemical processes, numerous researchers essentially use this technique. Some case studies in which PCA was used for the detection of the degree of association among water parameters that provided the assumption on water class are stated in Table 4.5.

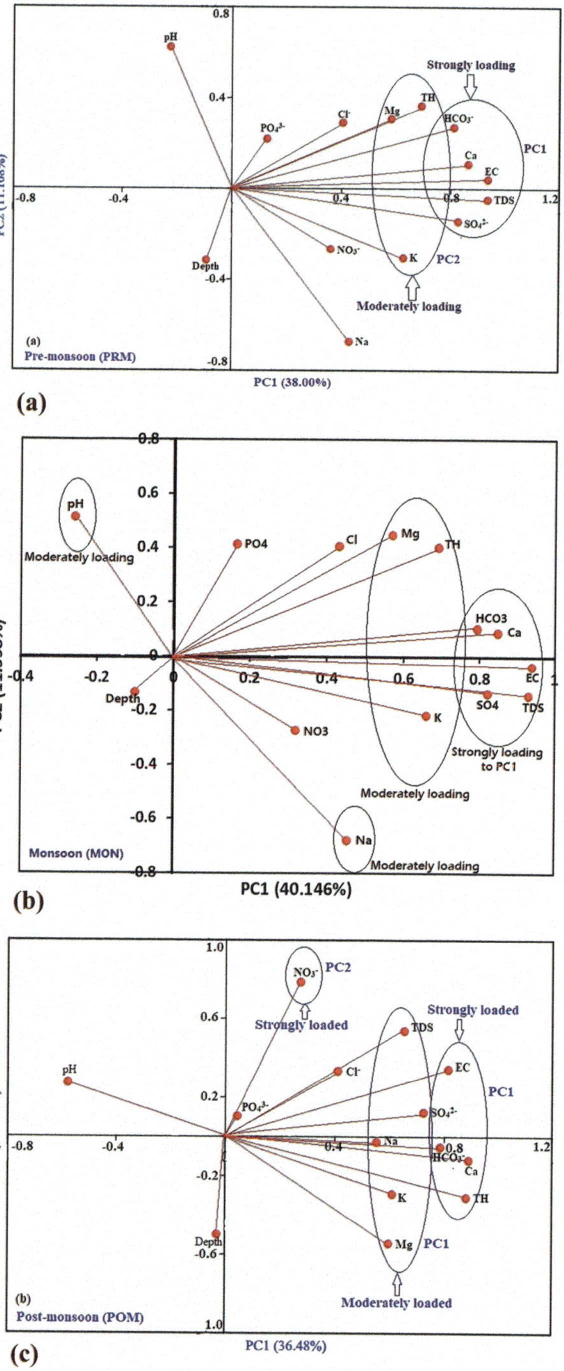

Fig. 4.7 Biplot (axes PC1 and PC2) of robust principal component analysis for (**a**) pre-monsoon (PRM), (**b**) monsoon (MON), and post-monsoon (POM) sampling seasons

Table 4.5 Application of PCA for groundwater in different geographical areas

Sampling site	PC1[a]	PC2[a]	Water class	Solute sources	References
Southern Gangetic Plain, Bihar, India	TDS, K^+, Ca^{2+}, Mg^{2+}, SO_4^{2-}, NO_3^-, and HCO_3^-	Na^+, Cl^-	Ca–Mg–HCO_3, Ca–Na–HCO_3	Silicate and carbonate weathering, mineral dissolution, anthropogenic activities	Sethi et al. [10]
Coastal alluvial aquifer of Akwa Ibom, Southeastern Nigeria	Ca^{2+}, Mg^{2+}	Cl^-, HCO_3^-	Ca–Mg–Cl–HCO_3, Na–K–Cl–HCO_3	Silicate weathering, ion exchange reactions, anthropogenic activities	Inim et al. [4]
Luxor governorate, Upper Egypt	EC, TDS, Na^+, Ca^{2+}, SO_4^{2-}, and Cl^-	pH	Ca–Na–HCO_3, Ca–Mg–Cl (mixed)	Calcite, aragonite, dolomite, anhydrite, gypsum, and halite dissolution	Alfy et al. [1]
Goda Mountains Range, Republic of Djibouti (Horn of Africa)	EC, Na^+, Cl^-, and SO_4^{2-}	HCO_3^-, NO_3^-	Ca–Mg–HCO_3	Rhyolite weathering, volcanic rock mineral dissolution	Ahmed et al. [5]
Talensi District, Northern Ghana	pH, EC, K^+, Ca^{2+}, Mg^{2+}, Cl^-, and HCO_3^-	SO_4^{2-}, F^-	Ca–Mg–Cl–SO_4, Na–K–Cl–SO_4, Mg–Ca–HCO_3	Silicate and carbonate mineral weathering, reverse ion exchange, anthropogenic activities	Chegbeleh et al. [11]
Kerala, India	Na^+ and Cl^-	F^-, HCO_3^-	Na–Cl, Ca–Na–HCO_3, mixed Ca–Mg–Cl	Ion exchange, reverse ion exchange, carbonate dissolution, and silicate weathering	Nandakumaran and Balakrishnan, [22]
Ghotki District, Sindh Province, Pakistan	EC, TDS, Na^+, K^+, Ca^{2+}, Mg^{2+}, Cl^-, HCO_3^-, and SO_4^{2-}	F^-, NO_3^-	Ca–Mg–Cl, Ca–Cl	Calcite and dolomite dissolution, human activities	Ghani et al. [23]

[a]Water parameters are strongly loaded with the component number (over 0.7)

4.2.1.3 Hierarchical Cluster Analysis (CA)

Hierarchical cluster analysis was accomplished for 40 groundwater samples with 14 geochemical parameters to classify the group of sampling spots (clusters) displaying similar water characteristics. In agglomerative schedule cluster analysis, based on sample ID, the most comparable variables are sited in one cluster and connected to a closely related cluster(s). Furthermore, clusters with fewer comparative connections are linked to form one large cluster. The hydrogeochemical parameters of samples in the three sampling rounds displayed six (6), six (6), and eight (8) smallest single clusters in the PRM, MON, and POM periods, respectively, and two (2) major cluster groups (cluster I and cluster II) in the above seasons based on a dendrogram using Ward's method (Fig. 4.8). On the other hand, based on water variables, two large cluster groups (cluster I and cluster II) are obtained from hierarchical cluster analysis. These clusters are verified according to the water table, land use, and hydrochemical processes. Hydrochemical processes are grouped into recharge, mixing with river water, and irrigation return flow. In the cluster figures, the phenon line is drawn at a linkage distance of approximately 3.5 in R-mode cluster analysis. Table 4.6 and Fig. 4.8 show a detailed description of both clusters.

Parameters that fit in the same cluster are likely to have been invented from the same rock or mineral source. Similar to PCA, cluster analysis places samples (variables) into groups based on the same characteristics and connotations with each other. In agglomerative schedule cluster analysis, the most identical variables are sited in one cluster and connected to a closely accompanying cluster(s) and further from clusters with less relative, all of which are connected to form one large cluster. The dendrogram demonstrates close connotations between EC, TDS, HCO_3^-, and TH in cluster I and K^+, PO_4^{3-}, pH, NO_3^-, Na^+, SO_4^{2-}, Mg^{2+}, Cl^-, Ca^{2+}, and well depth in cluster II (Table 4.6 and Fig. 4.8: A2, B2, C2). The first cluster (cluster I) displays a similar correlation between TH and TDS and proposes the dominion of groundwater by precipitation and contact with air CO_2. Additionally, cluster I comprises EC and HCO_3^-, which are related with each other, and the elevated level of EC is largely caused by HCO_3^- concentration and is confirmed through a several bivariate tests. Cluster II selected the probable influences of pollutants of infiltrating precipitation and/or recharge, possibly from agricultural input fertilizers and related anthropogenic actions. Chemical fertilizers, such as NKP, urea, and TSP fertilizer, stimulate groundwater PO_4^{3-}, K^+, and NO_3^- content, since these chemical fertilizers are composed mostly of such chemicals, though the weathering of K-rich feldspar rock is linked with the release of K^+ and other associated ions in water. Ca^{2+}, Mg^{2+}, Na^+, and Cl^- represent a groundwater system conquered by water–rock interactions, probably impacted by acidic groundwater conditions because of the lower pH (<7) of the most samples in the study areas (mostly in the PRM period). This association is categorized largely by silicate and carbonate mineral (calcite and dolomite) weathering, which releases such ions in the groundwater. However, what type of minerals is dissolved in the groundwater would be understood by Piper, Gibbs, and numerous bivariate diagrams. Likewise, SO_4^{2-} could also be introduced from the oxidation of pyrite rocks or the weathering of sulphate minerals, particularly in recharge areas where the bedrock of the aquifer is exposed to such situations.

The maximum water samples were collected from the river valley of the Ganges (the Padma) and statistically proved that the groundwater quality is impacted by external causes such as river morphology. Figure 4.9 shows the satellite image of the Padma River (downstream part of the Ganges River) in 1988 and 2018. This different image displays that it has been growing in size, converting in shape, and

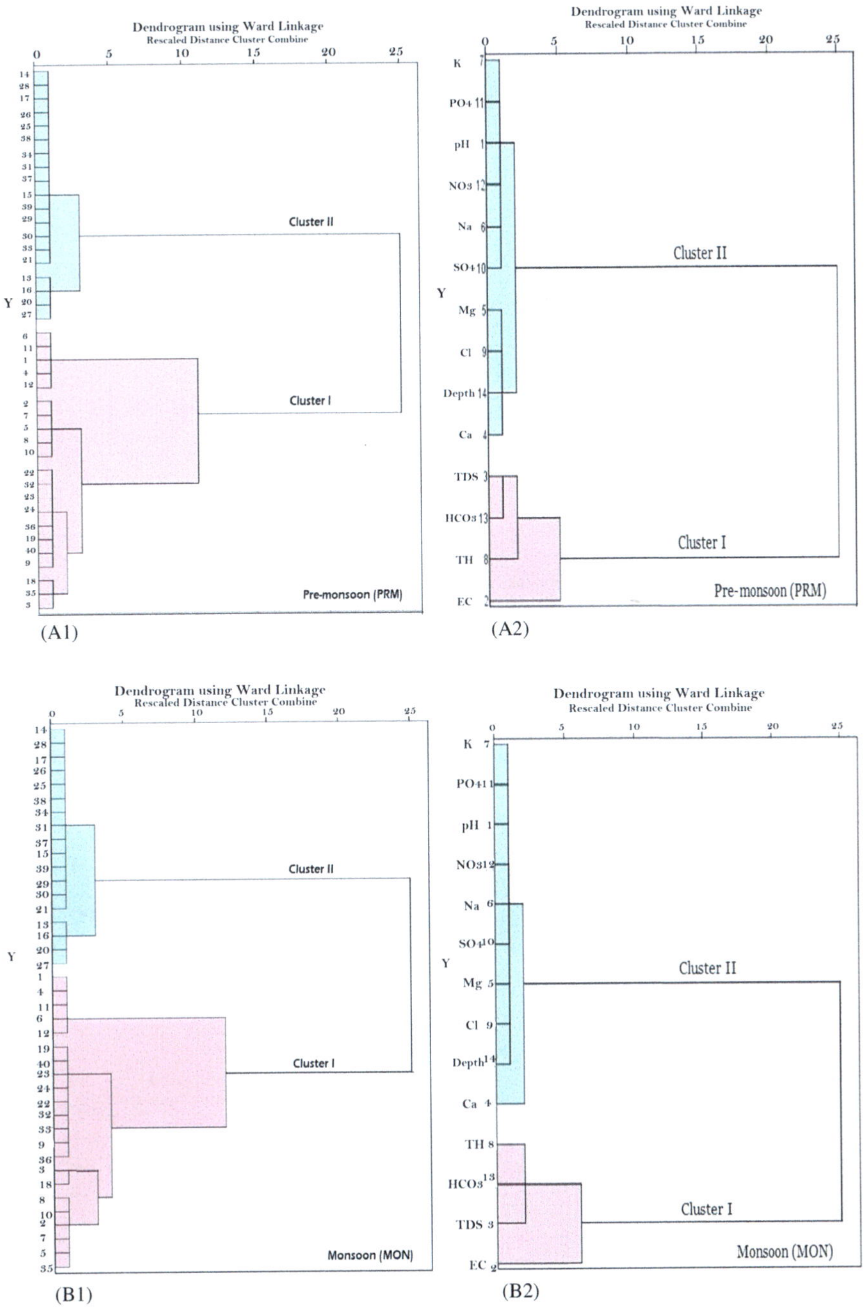

Fig. 4.8 Hierarchical cluster analysis of 40 sampling sites for the PRM, MON, and POM sampling rounds. The dendrogram of Q-mode hierarchical cluster analysis (HCA) displays connotations between samples from different regions of the hydrologic system. Figures A1, B1, and C1 denote the sample ID-based clusters, and Figures A2, B2, and C2 represent the water parameter-based clusters

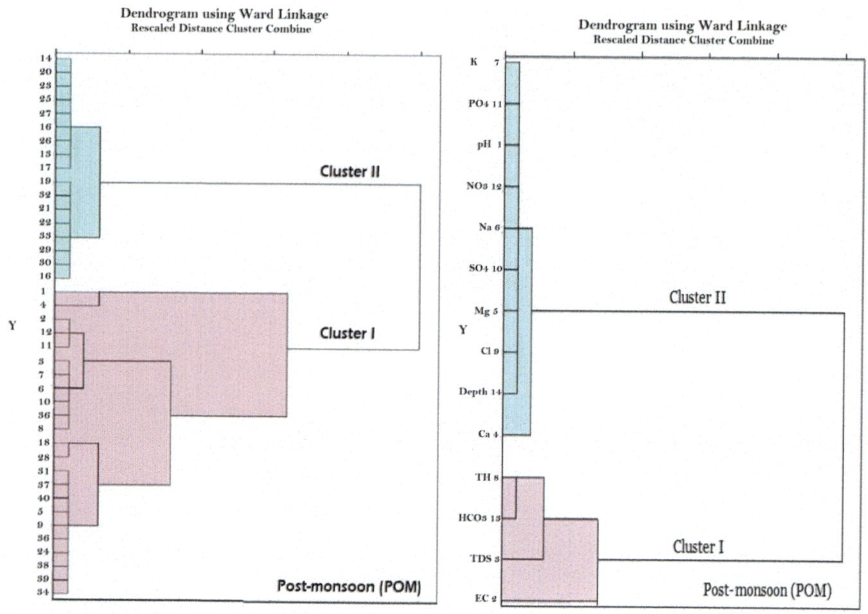

(C1) (C2)

Fig. 4.8 (continued)

Table 4.6 The sample ID included water parameters in two different clusters of three sampling periods

Sampling period	Cluster I		Cluster II	
	Sample ID	Parameter	Sample ID	Parameter
Pre-monsoon (PRM)	S1, S2, S3, S4, S5, S6, S7, S8, S9, S10, S11, S12, S18, S19, S22, S23, S24, S32, S33, S36, S40. (See Fig. 4.8 A1)	EC, TDS, HCO_3^-, and TH. (See Fig. 4.8 A1)	S13, S14, S15, S16, S17, S20, S21, S25, S26, S27, S28. S29, S30, S31, S35, S34, S37, S38, S39. (See Fig. 4.8 A1)	pH, Ca, Mg, Na, K, Cl^-, SO_4^{2-}, PO_4^{3-}, NO_3^-, depth. (See Fig. 4.8 A2)
Monsoon (MON)	S1, S2, S3, S4, S5, S6, S7, S8, S9, S10, S11, S12, S18, S19, S22, S23, S24, S32, S33, S35, S36, S40. (See Fig. 4.8 B1)	EC, TDS, HCO_3^-, and TH. (See Fig. 4.8 B1)	S13, S14, S15, S16, S17, S20, S21, S25, S26, S27, S28. S29, S30, S31, S34, S37, S38, S39. (See Fig. 4.8 B1)	pH, Ca, Mg, Na, K, Cl^-, SO_4^{2-}, PO_4^{3-}, NO_3^-, depth. (See Fig. 4.8 B2)
Post-monsoon (POM)	S1, S2, S3, S4, S5, S6, S7, S8, S9, S10, S11, S12, S18, S22, S23, S24, S28. S32, S35, S36, S40. (See Fig. 4.8 C1)	EC, TDS, HCO_3^-, and TH. (See Fig. 4.8 C1)	S13, S14, S15, S16, S17, S19, S20, S21, S25, S26, S27, S29, S30, S31, S33, S34, S37, S38, S39. (See Fig. 4.8 C1)	pH, Ca, Mg, Na, K, Cl^-, SO_4^{2-}, PO_4^{3-}, NO_3^-, depth. (See Fig. 4.8 C2)

Fig. 4.9 Morphological changes in the Padma River (Bangladesh) in (**a**) 1988 and (**b**) 2018: decrease the river size and increase the density of green plants. (Due to the Green Revolution in the country from 1985)

moving in location for at least the earlier 35 years (source NASA: https://earthobservatory.nasa.gov/world-of-change/PadmaRiver). Millions of inhabitants depend on this river for transportation and irrigation in farmlands. Once a year, hundreds and sometimes thousands of hectares of land erode and drop into the Padma River. Since 1967, over 60,000 ha of cultivable lands has been lost, which is roughly equal to the area of Chicago City. These natural-colour satellite images display the changes to the shape and width of the Padma since 1988, and each twist and zigzag tells a different geologic story about the river (Fig. 4.9). The images were acquired by the Landsat satellites: the Thematic Mapper on Landsat 5, the Enhanced Thematic Mapper Plus on Landsat 7, and the Operational Land Imager on Landsat 8. They include an amalgamation of shortwave infrared, near-infrared, and visible light to highlight differences between land and water. Two representative images were developed in January and February during the dry period.

Figure 4.10 shows the spatial distribution of two different groundwater clusters in the study zone for both sampling seasons. The first cluster originated near the two banks of the Ganges and *Gorai* rivers, and most of the cluster I sampling stations were on both banks. In contrast, the cluster II sampling points were located far from the banks of both rivers. The main river, the Ganges (Padma), and two small rivers, *Kaliganga* and *Gorai*, flow through the sampling point. Groundwater flows within the study area from east to north and west to south. The upper basin is mostly the recharge zone, and the lower basin is the runoff zone. The largest sampling points for cluster I and cluster II were in the reload and unload regions, respectively. The return of irrigation water will be distributed similarly to the agricultural areas of both

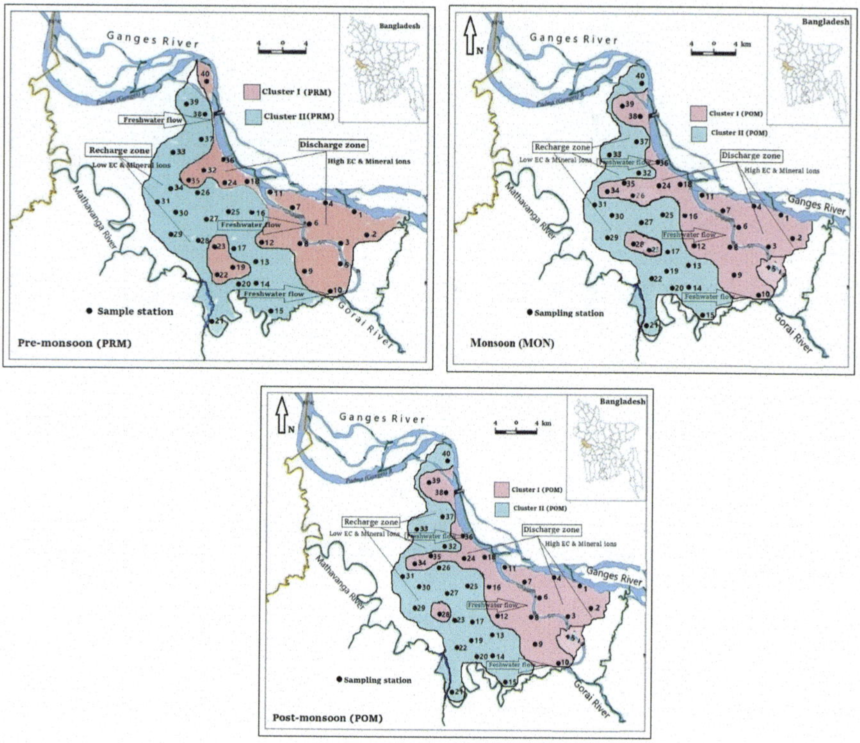

Fig. 4.10 Spatial variation in clusters and recharge/discharge zones in the study area for different sampling periods

clusters. Geogenic and/or anthropogenic causes separated samples 19, 22, 23, and 28 from cluster I in the PRM or POM period and sample 5 from cluster II in the POM (Fig. 4.10). Therefore, apart from spatial causes, other factors may also be relevant to the process of cluster divergence.

4.2.1.4 Bicarbonate Index (BCI)

BCI indicates the ratio of the concentration of carbonate mineral components and silicate mineral components in the groundwater:

$$\text{Bicarbonate index (BCI)} = \frac{[\mathrm{Ca}^{2+}] + [\mathrm{Mg}^{2+}] + [\mathrm{HCO}_3^-]}{[\mathrm{Na}^+] + [\mathrm{K}^+] + [\mathrm{Cl}^-]} \tag{4.3}$$

The bicarbonate index (*BCI*) measured (Eq. 4.3) is the comparable ratio between the earth metal bicarbonate and base metal chloride. The elevated *BCI* value designated the domination of Ca–Mg–HCO_3 over sodic ions that led to the importance of carbonate rock–water interactions rather than silicate, halite, or other rock weathering. It also measured inland seawater intrusion. The *BCI* values of the sampling sites included in cluster I were higher than those in cluster II in the three sampling rounds (Table 4.6). The EC values also showed the same trends (Fig. 4.11). In the dry PRM season, the *BCI* values were found to be very dispersed in both clusters compared to the POM. During this season, river flow did not influence the mineralogy of the water samples. However, after heavy rain, i.e. the MON and POM periods, the *BCI* values were somewhat steady in both clusters (Fig. 4.12). So, the river flow significantly influenced the geochemical features of the groundwater in the study zones. The calculated *BCI* average values were 10.82 and 10.11 for clusters I and II in the PRM, 12.72 and 12.06 for clusters I and II in the MON, and 13.65 and 12.83 for clusters I and II in the POM season, respectively (Table 4.7). These high BCI values directed that the carbonate mineral of earth metals (Mg and Ca) is dissolutely related to the chloride of base metals (Na and K). Table 4.7 and Fig. 4.11 show that the EC/TDS and other ions, particularly Ca^{2+}, Mg^{2+}, HCO_3^-, and SO_4^{2-}, were suggestively higher in cluster I in both seasons than in cluster II. The maximum sampling sites of cluster I are positioned on the river bank or near the riverside of the main two rivers. Therefore, the impact of river flows on the mineral/rock weathering processes as well as the mineralization of water in the aquifer basement is expected. These expectations are illustrated in the subsequent section.

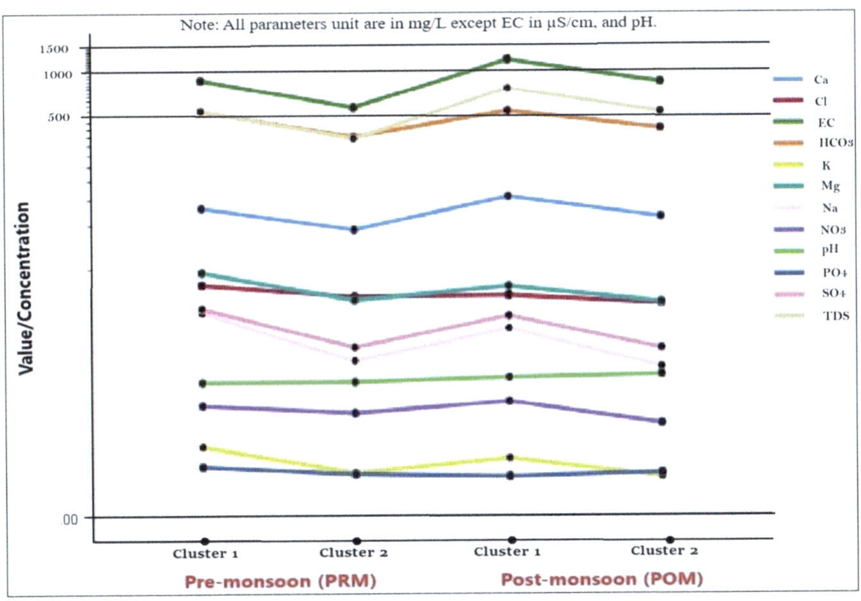

Fig. 4.11 The concentration/value variation of water variables in a separate cluster of the PRM and POM sampling seasons. The maximum values of the variable are higher in cluster I than in cluster II in both seasons

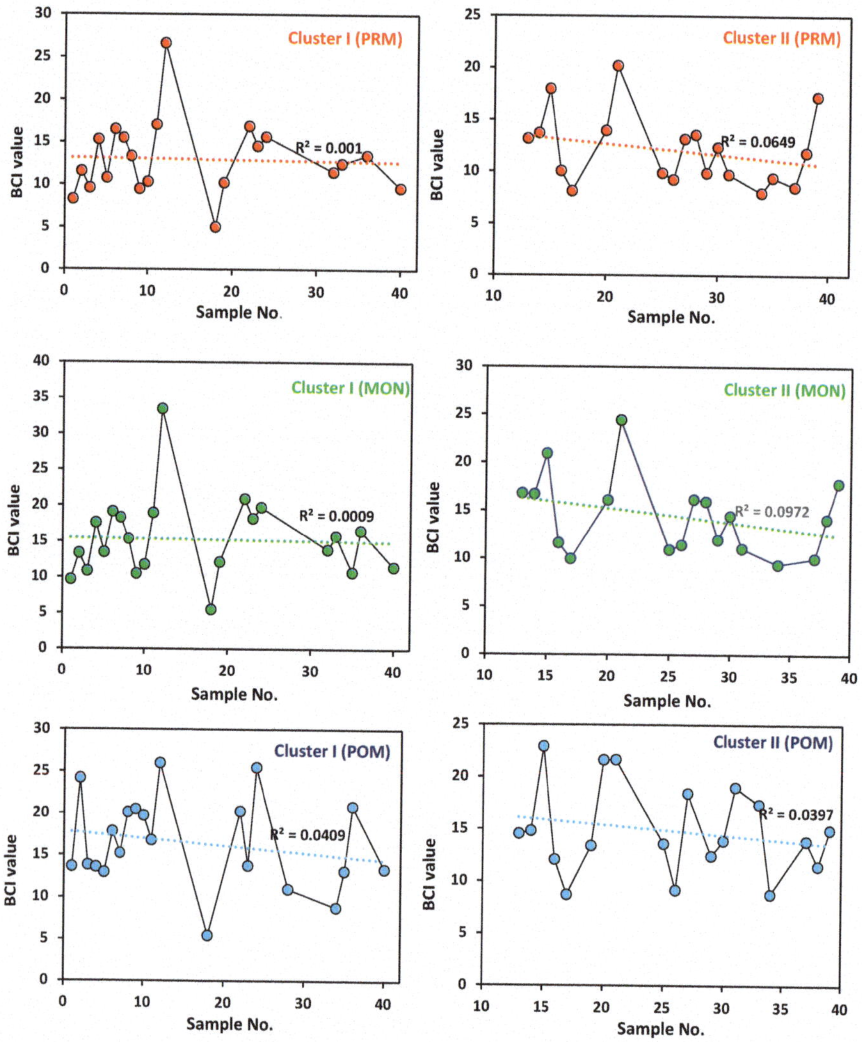

Fig. 4.12 Sample-by-sample values of BCI in the three different sampling seasons

4.2.1.5 Chloro Index (CI)

The chloro index (CI) designates the ratio of the concentration of silicate mineral or sodic components and carbonate mineral components in the groundwater. It is the inverse of the bicarbonate index and essentially CI applied for sodic water in coastal or some arid groundwater where water is mainly of the Na–Cl type:

Table 4.7 Average water variable values of two clusters according to Q-mode hierarchical analysis for three sampling periods

	BCI	pH	EC	TDS	Ca^{2+}	Mg^{2+}	Na^{+}	K^{+}	Cl^{-}	HCO_3^{-}	SO_4^{2-}	NO_3^{-}	PO_4^{3-}
Pre-monsoon (PRM), *n* = 40													
Cluster I	10.82	6.98	876.7	542.6	118.7	43.01	22.48	1.94	35.38	447.4	24.14	4.57	1.15
Cluster II	10.11	7.03	570.4	350.9	84.51	27.61	10.10	0.95	29.17	361.4	12.79	3.93	0.91
Monsoon (MON), *n* = 40													
Cluster I	12.72	7.41	996.8	640.1	120.9	35.9	16.96	1.33	29.67	487.2	26.46	5.14	0.98
Cluster II	12.06	7.46	709.4	458.2	97.49	25.20	8.77	0.82	27.11	382.1	13.95	3.02	0.81
Post-monsoon (POM), *n* = 40													
Cluster I	13.65	7.58	1204	764.7	141.3	34.35	17.38	1.44	29.72	537.5	25.28	4.89	0.84
Cluster II	12.83	7.94	850.7	530.9	102.8	26.62	9.07	0.83	25.97	410.7	14.50	3.18	0.94

Note: Excluding BCI, all parameter units are in mg/L except EC in μS/cm and pH

$$\text{Chloro index (CI)} = \frac{[Na^+] + [K^+] + [Cl^-]}{[Ca^{2+}] + [Mg^{2+}] + [HCO_3^-]} \tag{4.4}$$

For the bicarbonate-rich water, in the present samples, the CI obtained a very low value (~0.1), but the BCI value was found to be over 10. In coastal and arid groundwater, the CI value was found to be near or over 1. Here, consider the five different water data points, of which three are coastal and two are arid groundwater, for the calculation of CI. The calculated CI values for the coastal groundwater of Khulna (Bangladesh) [20], Tamil Nadu (India) [3], and Djibouti [5] are 3.54, 1.34, and 1.04, respectively. On the other hand, these values are 1.05 and 1.44 for arid areas such as Egypt [1] and N. China [2], respectively.

4.2.2 *Groundwater Mineralization*

The water–rock interface and mobility of solutes are the vital factors controlling the groundwater mineralization procedures. Bivariate plots of the main ions against TDS were used to express the geochemical classifications that took part in the mineralization of groundwater. The bivariate plots of some selected water parameters with TDS are shown in Fig. 4.13. The correlation factor $R > 0.5$ or $R^2 > 0.25$ between two variables indicated a strong association with each other. Figure 4.13 shows that the correlation factors of TDS with Ca^{2+} (PRM: $R^2 = 0.5998$, MON: 0.569, and POM: $R^2 = 0.3864$), K^+ (PRM: $R^2 = 0.2926$, MON: $R^2 = 0.3294$, POM: $R^2 = 0.1125$) and HCO_3^- (PRM: $R^2 = 0.570$, MON: $R^2 = 0.5357$, POM: $R^2 = 0.1616$) are significantly high. This correlation represents the dominant components that donate to groundwater mineralization in aquifer systems. Regarding concentration, Ca^{2+}, Mg^{2+}, and HCO_3^- are the most dynamic ions in groundwater samples from the study areas (see Table 4.1), and these results endorse the unremitting addition of these ions in the direction of the groundwater flow path, which also contributes to groundwater mineralization. They may originate from similar sources. Consequently, weathering or dissolution is the geogenic geochemical process measured by the salt concentration along the groundwater movement route. Carbonate dissolution results from rainwater saturated with CO_2 and grows rich in carbonic acid (H_2CO_3). This weak acid affects the dissolution of carbonate minerals (calcite/dolomite) in groundwater systems. The shallow aquifer in the study area, when interacting with groundwater, undertakes calcite/dolomite mineral dissolution. Analogous findings were made by several researchers in groundwater chemistry investigations of northern and north-western Bangladesh. During dissolution and water movement in rocks, chemical elements in ionic forms leach out and dissolve in groundwater. The concentrations of SO_4^{2-} in the samples are relatively higher than the groundwater of other region in the country, and it is significantly associated with TDS (see Table 4.3). The SO_4^{2-} loadings may be caused by the

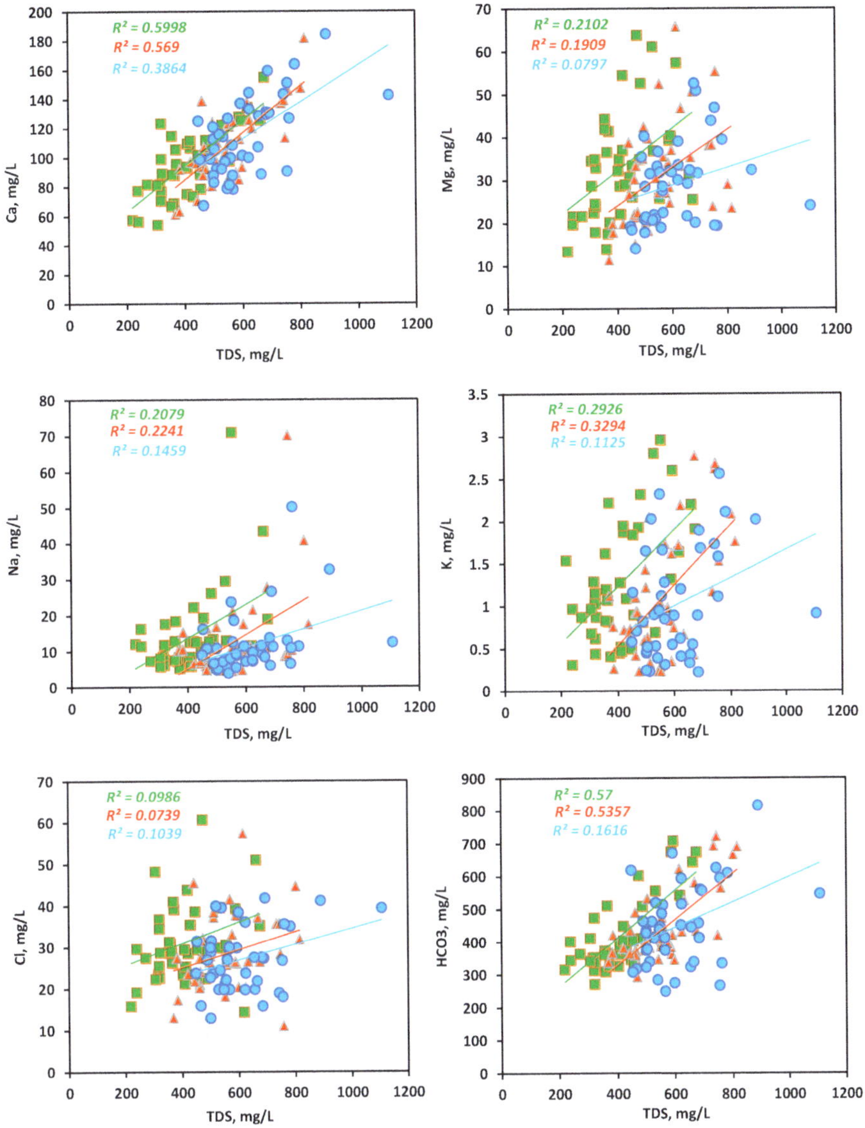

Fig. 4.13 Bivariate plots of most important ions against TDS values (square, triangle, and round shapes denote the pre-monsoon. monsoon and post-monsoon sampling seasons, respectively)

heavy oxidation of pyrite rock under anaerobic conditions or by rock weathering. Due to the higher concentration of acid sulphate, the pH values of the collected samples are below 7, particularly in the PRM season. The molar ratio Ca^{2+}/SO_4^{2-} is very high related to the concentration of SO_4^{2-} in the present study, which designates that the role of the pyrite oxidation process is unimportant. It is expected that SO_4^{2-} originates from the dissolution of sulphate-bearing Ca or Mg minerals. On the other hand, the concentrations of Na^+, Mg^{2+}, and Cl^- are positively associated with TDS in most of the samples, but the relations are less strong and scattered (Fig. 4.13). Thus, these ions are not the controlling components of the entire geochemical process. Again, the plots illustrate that along with TDS, the concentrations of Ca^{2+}, Mg^{2+}, HCO_3^-, Na^+, and Cl^- increase and become more scattered from the PRM to POM seasons. Figure 4.11 displays that the R^2 values in the POM season are less than those in the other two sampling periods. This is because heavy precipitation and vast agricultural activities occurred just before the POM season.

In Salima District, Malawi [24], the linear relationship for cations with TDS was very strong, especially for Ca^{2+} ($R^2 = 0.865$) and Na^+ ($R^2 = 0.823$); however, Ca^{2+} had a higher rate of increase in the form of the slope than Na^+. In the Nkhotakota District (Malawi), the scenario was similar, with the association between the levels of Ca^{2+} and TDS ($R^2 = 0.869$) being very high and a higher rate of increase than the other cations. In the Nkhotakota District, the correlation of Mg^{2+} and TDS ($R^2 = 0.855$) was equally as high as that for Ca^{2+} and TDS but had a lower slope value, whereas Na^+ and TDS ($R^2 = 0.607$) had a medium value correlation. For anions in the Salima District, SO_4^{2-} and TDS showed the highest correlation ($R^2 = 0.689$), followed by the correlation between HCO_3^- and TDS ($R^2 = 0.579$). In the Nkhotakota District, the anion correlation showed a high relationship between HCO_3^- and TDS ($R^2 = 0.952$), followed by the correlation between Cl^- and TDS ($R^2 = 0.488$). On the other hand, Islam et al. [20] conducted a geochemical investigation of coastal groundwater in the Khulna District, Bangladesh. The results of that study illustrate the elevated level of salts, and their resulting Cl^- in groundwater originates from geogenic minerals and the mixing of seawater with upland freshwater. Though there are some minor amounts of other ions (K^+, Mg^{2+}, SO_4^{2-}, NO_3^-), Na^+' and Cl^- are present in approximately 90% of all seawater ions. Na^+ and TDS are other significant parameters that can be used to detect the impact of key components and groundwater salinity. In this study, the levels of Na^+ and Cl^- in groundwater were plotted compared to TDS (Fig. 4.13). The diagrams displayed that most Na^+ and Cl^- ions of the groundwater were strongly correlated ($R^2 = +0.75$ and +0.76, respectively) with TDS. As stated by the WHO's [25] cataloguing of groundwater based on TDS, 60% of the sample falls in the undesirable category, 35% is poor, and only 5% falls in the excellent class of the spatial distribution of TDS, which is presented in several figures. [20] All other components, i.e. Na^+, K^+, Ca^{2+}, and Mg^{2+}, were also well connected with Cl^- with R^2 values of 0.82, 0.79, 0.78, and 0.58, respectively, denoting that they originated from the same sources. Therefore, the degree of correlation between common ions and TDS in water is different for different water types. Thus, the mineralization process in groundwater depends on sampling positions and geographical variations.

4.2.3 Geochemical Evaluation

The present study exposed that the order of the concentration of cations is Ca^{2+-} > Mg^{2+} > Na^{+} > K^{+} and that of anions is HCO_3^{-} >> Cl^{-} > SO_4^{2-} > NO_3^{-} > PO_4^{3-} in the groundwater samples of the study area. Multivariate analysis provides general information about the sources, distribution, and occurrence of solutes in groundwater. In this section, hydrogeochemical studies that assess the rock–water interactions, influencing factors, and pollution pathways are conducted with the aid of multivariate analysis, numerous bivariate plots, and computer programmes.

4.2.3.1 Water Facies

Statistical distribution diagrams (Piper trilinear plot) were utilized not only to achieve better understanding of the hydrochemical processes operating in the groundwater systems but also to characterize the water types present in the study area. It is a convenient technique to categorize different hydrochemical facies or origins of groundwater by plotting the content of major cations and anions in groundwater samples, representing the origin, source of dissolved solutes, and processes that influence the features of these natural freshwaters. The water class or facies depends on the solute concentration in the water. This class provides information on rock sources and mineral dissolution processes. The present study plotted Piper's and Chadha's diagrams to describe the origin of geochemistry and water classes of the local groundwater. The Piper diagram of cations Ca^{2+}, Mg^{2+}, Na^{+}, and K^{+} and anions HCO_3^{-}, Cl^{-}, and SO_4^{2-} was used to determine the water classes for the three sampling seasons (Fig. 4.14). On the basis of symbolic area in the Piper plot, most water samples are considered as absolutely Ca and HCO_3 types, and the water category is symbolized as Ca–HCO_3. However, for the PRM season, a small number of samples lay in the nondominant-type area. These samples may be unevenly classified as the Ca–Mg–HCO_3 type. However, for the coastal groundwater of Bangladesh or any other coast of the world, the Piper diagram did not give a simple type of water facies. For example, in the Khulna District [20], south coast of Bangladesh, the groundwater samples plotting on the Piper diagram reveal that four classes such as Na–Cl (35%), Na–Cl–HCO_3 (55%), Na–Mg–Cl (5%), and Ca–Mg–Cl (5%) was found and Na–Cl–HCO_3 are the major facies types. This points to the dominance of Na^{+} in the cations, the interplay of HCO_3^{-} and Cl^{-} in anions, and the effect of sea water in the above study area.

Some Case Studies: Application of Piper Diagrams to Identify the Water Class

(a) *Bougaa area, north-eastern Algeria* [14] To characterize water types, the concentrations of cations and anions (in mEq/L) are drawn in two base triangle Piper diagrams, which are expected further into the central diamond field. The water class is measured along with its placement near the four angles of the diagram.

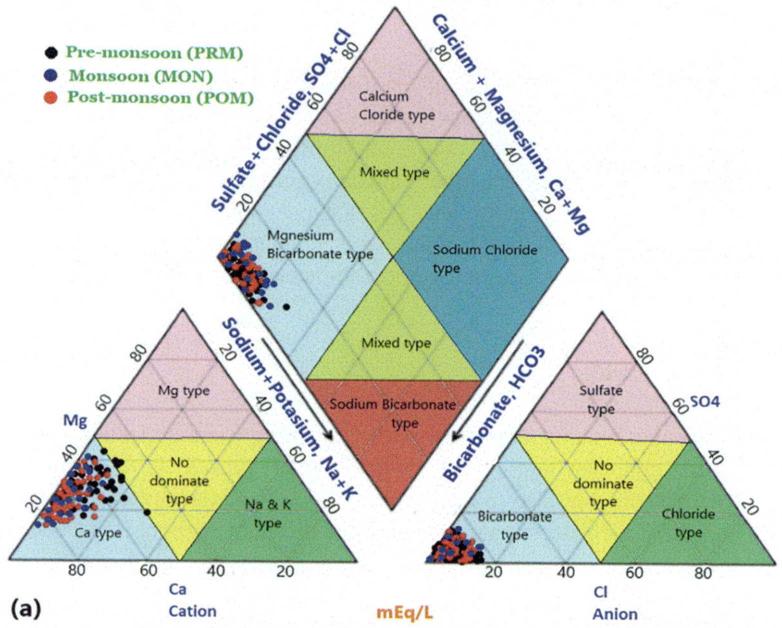

Fig. 4.14 Piper diagram for groundwater classification

The chemical data values from the groundwater samples in that study area are plotted on the diagram, which demonstrates that maximum water samples are in the class of Ca–Mg–Cl and Na–Cl.

(b) *Kashmir Valley, India* [26] In this study area, the groundwater facies were very complex. The study revealed that Ca–HCO_3, Mg–HCO_3, Na–HCO_3, and Ca–Mg–HCO_3 are the common hydrochemical classes, suggesting that man-made activities and geology have played a substantial role in controlling the water chemistry in this valley. The Ca–Mg–HCO_3 water class is considered recharged water from sources correlated to rainfall, snowmelt, and the dissolution of carbonate minerals. This kind of classes signifies 95% of the central zone, 80% of the southern part, and 90% of the northern zones of the study area. This result postulates that the chemical properties as well as the factors affecting the geochemistry of these areas are similar. The Na–HCO_3 types are present at 20% in the southern area, 10% in the northern area, and 5% in the central area of the valley. These water facies are a strong indicator of the cation exchange procedure in groundwater. The higher concentration of Na^+ providing by clay minerals substituting the accessibility of Ca^{2+} ions from the recharged Ca–Mg–HCO_3 over the cation exchange process, in which Ca^{2+} ions are involved in the aquifer material and Na^+ is released to water, leads to the creation of Na–HCO_3 water facies. In this study, this was established by the chloro-alkaline index (CAI) calculation.

(c) *Mokopane Area, Limpopo, South Africa* [16] Using the trilinear Piper diagram, the study recognized that all groundwater samples fell under three water classes, viz. 50% of Na–HCO_3, 42% of Mg–HCO_3, and rest of the Na–Cl. These water classes designate chemical reactions throughout rock–water interactions within the geological context. The results demonstrate the domination of very weak acids (i.e. HCO_3^-) over hard bases (i.e. SO_4^{2-} and Cl^-). Besides, this investigation reported that dissolution of magnesite/gypsum, alumina–silicate/carbonate weathering, cation exchange, arid climate, and alkaline conditions of the water are key factors accountable for high levels of Na^+, Mg^{2+}, and HCO_3^- in groundwater, though, with poor management of contamination sources such as untreated sewerage and agrochemical runoff, there is a movement propensity from the HCO_3^- pole to the Cl^- pole.
(d) *Halabja Saidsadiq Basin, Iraq* [18] The graphical plot of chemical analysis on a Piper diagram for the dry and wet period, this study confirms that a large portion of the groundwater tests in the study area characterized the groundwater as alkaline with existing HCO_3^- and with SO_4^{2-} and Cl^-. Nevertheless, water–rock interaction processes and groundwater movement direction have been accountable for the leading water class of Ca–Mg–HCO_3. In this study, the effect of carbonate rocks on the properties and facies of groundwater within these classes is quite clear. So, the high concentration of substances such as soluble earth alkaline metals could be attributed to groundwater recharge from carbonate rocks, which signifies the surroundings of the Avroman, Balambo, and Jurassic aquifers that were measured within the studied catchment areas.
(e) *The urban area of Zamora, Mexico* [6] A hydrochemical Piper diagram was constructed in this study using the AquaChem software. A calcite water class (Ca–HCO_3) was detected in maximum samples, representing that all sampling spots belonged to the same aquifer. These facies of water are categorized by low alkalinity, low mineralization, and low residence time in the aquifer. The groundwater with the short residence time is usually the HCO_3 type, then it is sulphate, and the water with the maximum durability is chlorinated. In the case of a cationic arrangement, a similar order would be Ca^{2+} with the less residence time, then Mg^{2+}, and, last, Na^+ with the maximum residence time in the aquifer basements.

To categorize the groundwater and find the hydrochemical courses, a new hydrochemical illustration, the Chadha diagram, can be used. This diagram is a modified version of the Piper plot and the expanded Durov diagram. The difference is that the two equilateral triangles are absent, and the form of the main study field is different. The main study subfields of the diagram describe the overall characteristics of water.

From Chadha's classifying diagram (Fig. 4.15), the linear plots of [HCO_3–(Cl +SO_4+NO_3+PO_4)] vs. [(Ca+Mg)–(Na+K)] show strong positive correlations ($R^2 = 0.2850$ in PRM, $R^2 = 0.3119$ in MON, and $R^2 = 0.5151$ in POM). This designates the occurrence of Ca–HCO_3 class and exposes that alkali earth metals (Ca^{2+} and Mg^{2+}) suggestively crossed base metals (Na^+ and K^+) and a strong

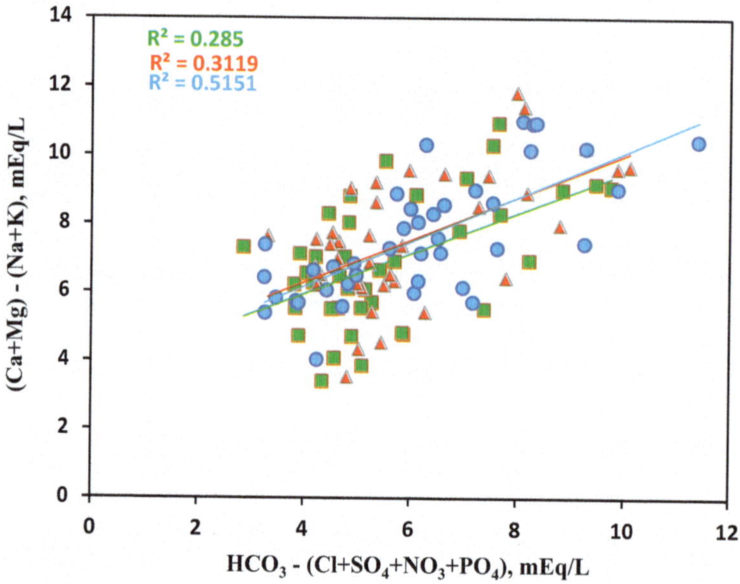

Fig. 4.15 Chadha's plot for groundwater classification

conjugate base (HCO_3^-) ruled over a weak conjugate base (Cl^-, SO_4^{2-}, NO_3^-, and PO_4^{3-}). Both Piper's and Chadha's plots signify the Ca–HCO_3 type of water class in the groundwater samples.

Similar to the present study, the groundwater of different geological settings did not show the simple facies of water. The groundwater of coastal and arid regions exhibited a complex type of water, which was confirmed by Chadha's diagram. In the central part of Telangana (semiarid area), India [27], Chadha's diagram demonstrates that majority of the groundwater samples showed that alkali earth metals exceeded base metals and weak acids exceeded strong acids. Moreover, groundwater facies belong to the Ca–Mg–HCO_3 and Na–HCO_3 classes, and a limited sample shows the Ca–Mg–Cl and Na–Cl types. On the other hand, in the coastal region of the Cuddalore district of Tamil Nadu, India [3], samples falling in the Ca–Mg–Cl region in the same diagram are prominent between the western side of the region and the coastal region, perhaps representing cation exchange reactions or a hydrochemical evolutionary route from Ca–Mg–Cl facies water to Na–Cl mixed seawater; however, Ca–Mg–Cl characterizes the reverse ion exchange procedures. Samples representing the Na–Cl facies in another part of the diagram indicate seawater mixing, which is typically constrained in coastal regions. This diagram demonstrates that the groundwater samples of an arid area in North China [2] exhibited alkaline earth cations (Ca^{2+} + Mg^{2+}) and weak acidic anions (CO_3^{2-} + HCO_3^-) exceeding both base metals (Na^+ + K^+) and the conjugated base of strong acidity (Cl^- + SO_4^{2-}). This group is chemical of Na–HCO_3-type water (77%) or Ca–HCO_3-type water (23%) and is descriptive of all groundwater throughout the study area and typically observed in the young alluvium plain. This group is categorized by freshwater with low salinity (TDS = 291–773 mg/L) in the same study area. Another group shows that the alkali metals crossed alkaline earth metals and anions of strong acidity exceeded anions of weak acidity. This group contains relatively high-salinity groundwater (TDS = 1299–2579 mg/L) of Na–Cl-type water (43%) or Na–SO_4-type water (57%). In the northern zone of the study area, samples from the old alluvium plain show high concentrations of SO_4^{2-} and Cl^-.

Islam et al. [20] conducted a study in the coastal groundwater of Khulna District, Bangladesh, in which samples were placed in four quadrants of Chadha's graph and obtained a very complex type of water class as follows:

- Type 1: Ca–HCO_3 facies (recharge water)
- Type 2: Ca–Mg–Cl facies (reverse ion exchange water)
- Type 3: Na–HCO_3 facies (base ion exchange water)
- Type 4: Na–Cl facies (end class of waters, i.e. seawater)

Type 1, when water goes into the sub-layers from the surface layer, transports dissolved carbonate salts in the form of HCO_3^- and geochemically moveable Ca^{2+} ions. Type 2 may present groundwater where Ca^{2+} and Mg^{2+} are more abundant than Na^+ and K^+ whichever is due to the special release of Ca^{2+} and Mg^{2+} from the mineral dissolution of exposed bedrock or perhaps reverse cation exchange processes of Ca^{2+} and Mg^{2+} into groundwater solution and following adsorption of Na^+ on the mineral solid surface. Maximum samples falling in type 3 waters are characteristic of seawater mixing. Type 4 waters signify base metal ion exchange reactions, but unexpectedly no sample falls in this field. From this, it is clear that the water suitability of seaside areas containing high amount of Na^+ and Cl^- with characteristic seawater mixing in type 3 and with no illustration in type 2 and type 4 indicates the nonappearance of cation exchange.

When seawater encroaches into the freshwater baring coastal aquifer of this study area, (Ca–Mg–Cl) facies water may be originated. In this regard, Na^+ in seawater is substituted with more Ca^{2+} or Mg^{2+} in the clay minerals, where Na^+ is adsorbed on the surface of clay mineral as stated by the following reactions:

$$(Ca^{2+} + Mg^{2+}) + HCO_3^- + NaX \rightarrow CaX_2 + Mg(HCO_3)_2 + NaHCO_3$$

$$NaCl + MgX_2 \rightarrow NaX + MgCl_2$$

where X points to the exchanger. Therefore, seawater intrusion is not a fact of salinization in the study area. When seawater is added to freshwater, notable geochemical features emerge. Changes in the geochemistry of this brine are caused by water–rock interactions. This may involve the following three mechanisms: (a) base exchange reaction with clay minerals, (b) adsorption to clay minerals, and (c) carbonate dissolution precipitation.

4.2.3.2 Source Rock Weathering

The weathering of rocks is of key importance in controlling groundwater chemistry. Numerous processes are involved in controlling groundwater chemistry, including rock–water interactions, evaporation, salinization, and precipitation. The Gibbs diagram is generally used to determine the connection among aquifer lithology and water chemistry. The diagram is separated into zones based on the involvement of recharge precipitation, rock weathering, and evaporation in geochemistry. Gibbs ratios for both cations and anions were projected to categorically determine the main controls on the hydrogeochemistry using Eqs. 4.5 and 4.6, respectively:

$$\text{Gibbs ratio 1, GR1} = \frac{\text{Cl}}{(\text{HCO}_3 + \text{Cl})} \tag{4.5}$$

$$\text{Gibbs ratio 2, GR2} = \frac{\text{Na}}{(\text{Na} + \text{Ca})} \tag{4.6}$$

where the concentration of those ions is in mEq/L.

Two plots illustrate the connection between TDS and two Gibbs ratios Na/(Na+Ca) and Cl/(Cl+HCO_3) (Fig. 4.16). The figure shows that all the water samples of the three seasons fell under the rock weathering dominance section. The cationic (a) and anionic (b) diagrams designate the existence of rock weathering reactions in the study area.

North-eastern Algeria (nonarid) [14] Gibbs diagram displays that the major samples of this area fall in the evaporation dominance and rock–water interaction fields, which suggests that groundwater quality is influenced by mineral dissolution and rock weathering, although the groundwater samples in the evaporation dominance area illustrate a rise in salinity by the collective ions of Na^+ and Cl^- with cumulative values of TDS.

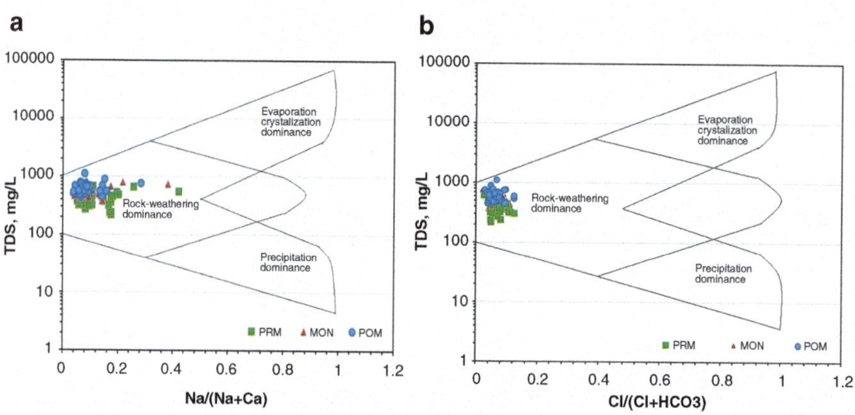

Fig. 4.16 Gibbs diagrams for groundwater samples of the study area

Ejina Basin, north-western China (semiarid) [2] From the Gibbs diagram, one more dominant process of this study area determining the water composition is precipitation and evaporation. Moisture in the unsaturated areas in the subsurface and evaporation of surface water are major processes in the evolution of the chemical structure of groundwater. Evaporation concentrates the remaining water and the percolated water, resulting in precipitation and deposition of evaporates that are ultimately percolated into the saturation layer. It is anticipated because evaporation significantly increases the level of ions formed by rock weathering, resulting in higher salinity in the TDS. Groundwater mineral balance helps predict the presence of active minerals and estimate mineral reactivity in groundwater systems. Given that certain carbonate minerals are generally found in groundwater, it is rational to undertake that these minerals are active in typical groundwater systems and that solution concentrations can be controlled.

Coastal aquifers of Ghiss-Nekkor in Al Hoceima State, Morocco [28] The maximum groundwater samples are nearby the seashore and are plotted graphically in the upper right angle of the Gibbs plot, representing that the groundwater of the study area is mostly impacted by evaporation processes and adding with seawater through the ground layer. To define and explain the processes concerned in groundwater salinization, numerous chemical connections were recognized to demonstrate an association between Cl^-/Br^- and Cl^-. These two anions are expected to be conservative groundwater constituents, while they do not interrelate in redox reactions or ion exchange and do not produce insoluble precipitated salts. High levels of Cl^- in the wells are normally near the coastal periphery. The salinity of the samples is perhaps attributed to seawater intrusion. Aquifer base saltwater intrusion may be caused by deep faults cutting through impervious layers of clays and silts or by lateral facies change in shallow deposits, which might explain the existence of saltwater. A higher Cl^-/Br^- ratio is at all times measured as a good interpreter of domestic water impact. The presence of polluted discharges without previous treatment can also explain the high Cl^- level in region far from the threat of seawater intrusion. In addition, maximum samples are spread between agricultural contamination and septic tank waste, with a propensity to leaching evaporated rocks.

Bougaa area, north-eastern Algeria [14] For this study, the Gibbs diagram demonstrates that maximum samples fall in the evaporation dominance and rock–water interaction regions, which recommends that groundwater chemistry is impacted by rock weathering, although the samples in the evaporation dominance region illustrate an increase in salinity by the higher concentration of Na^+ and Cl^- concerning rising TDS. The concentrations(mg/L) of Cl^- ion in groundwater vary relative to those of Na^+, which means that the presence of other possible sources of Cl^-, which can be associated with man-made contamination from miscellaneous sources such as industrial, agricultural, and domestic origins.

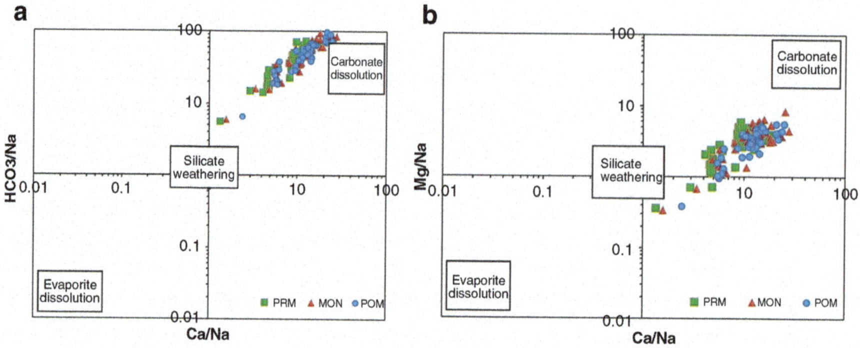

Fig. 4.17 Bivariate plot of (**a**) Ca/Na vs. HCO_3/Na and (**b**) Ca/Na vs. Mg/Na to categorize the mineral weathering of groundwater in the study zone. Previous observations have shown that the major cations and anions result from rock weathering rather than other processes

Now, it is imperative to determine the features of rock weathering and the course of rock–water interaction. The dissolution processes of rock–water interactions are partly affected by evaporation, evaporate dissolution, silicate weathering, carbonate dissolution, etc. The feature of dissolution processes demonstrated in the bivariate plots of Ca^{2+}/Na^+ vs. HCO_3^-/Na^+ indicates fully carbonate mineral dissolution throughout the PRM, MON, and POM sampling seasons (Fig. 4.17a). In contrast, another bivariate diagram of Mg^{2+}/Na^+ vs. Ca^{2+}/Na^+ (Fig. 4.17b) reveals higher degrees of Ca^{2+}/Na^+ and Mg^{2+}/Na^+ ratios for groundwater (average, PRM: 7.75, 4.40; MON: 9.98, 4.68; and POM: 11.42, 4.83, respectively). These values are much greater than those of the coastal groundwater in Bangladesh. These ratios of the groundwater of the sea belt area are much less than 1. The observed higher values of Ca^{2+}/Na^+ and Mg^{2+}/Na^+ in the groundwater samples are attributed to the impact of carbonate dissolution rather than silicate weathering. So, carbonate dissolution was the main process controlling the solute loads in groundwater for the three sample rounds.

In addition, Fig. 4.17a, b demonstrates that this rock weathering was characterized as carbonate-based mineral weathering. A major quantity of these ions could result from the weathering of crystalline calcite/dolomitic limestones and Ca–Mg silicates but mostly from calcite, dolomite, and gypsum. Now, it is important to know which mineral(s) are recognized as the dominant ions in the aquifer water body.

To better understand the origin of solutes in groundwater, biplots of major ions that willingly dissolve or react with other ions in groundwater were plotted (as Fig. 4.18). The most abundant cations of all groundwater types are typically dominated by Na^+, Ca^{2+}, and Mg^{2+}. These are thought to be related to the weathering and/or dissolution of carbonate (calcite, dolomite), silicate (albite), sulphate minerals (gypsum), and many other minerals, as described in the following reaction. Maximum trace metals are generally weathered from mixed rocks:

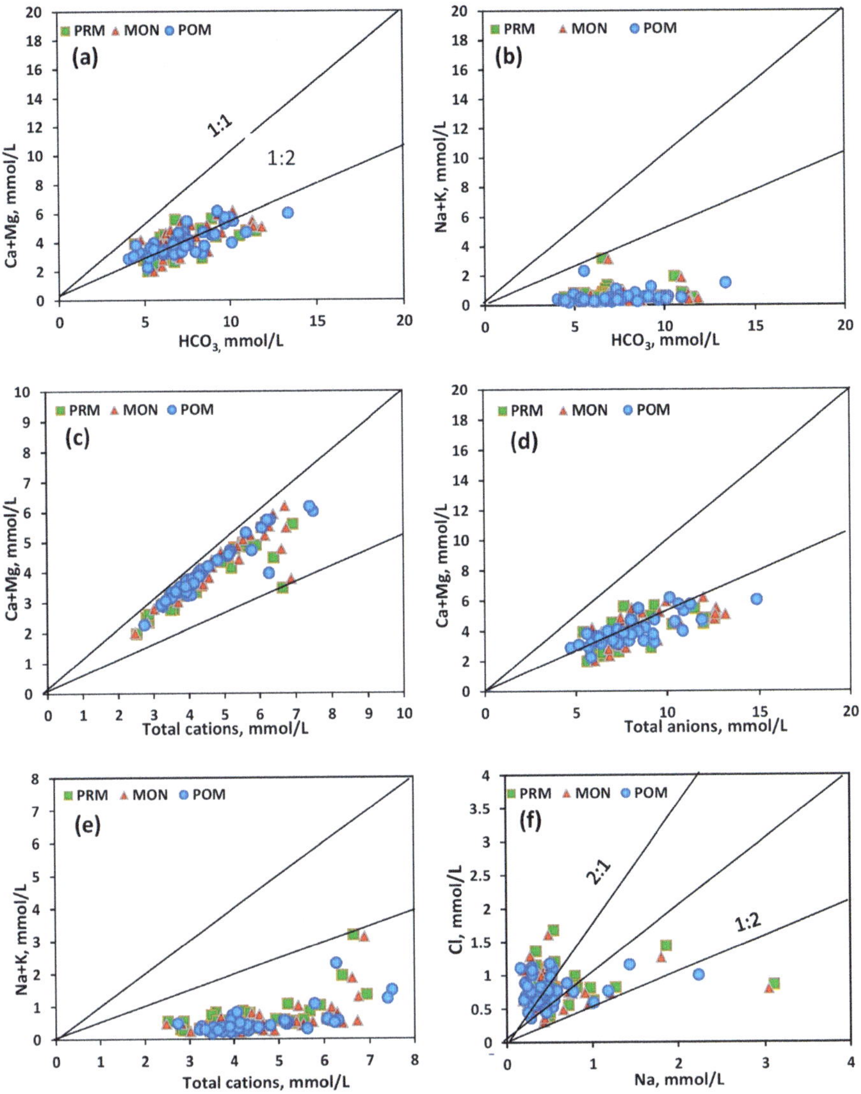

Fig. 4.18 Bivariate plots of (**a**) Ca^{2+} + Mg^{2+} vs. HCO_3^-, (**b**) Na^+ + K^+ vs. HCO_3^-, (**c**) Ca^{2+} + Mg^{2+} vs. total cations, (**d**) Ca^{2+} + Mg^{2+} vs. total anions, (**e**) Na^+ + K^+ vs. total cations, and (**f**) Na^+ vs. Cl^-

$$CO_2 + H_2O \rightleftharpoons H_2CO_3(\text{carbonic acid})$$

$$CaCO_3(\text{calcite}) + H_2CO_3 \rightleftharpoons Ca^{2+} + 2HCO_3{}^- \,(\text{calcite dissolution})$$

$$CaMg(CO_3)_2(dolomite) + 2H_2CO_3 \rightleftharpoons Ca^{2+} + Mg^{2+} + 4HCO_3^{-}(dolomite\ dissolution)$$

$$CaSO_4.2H_2O(gypsum) + H_2O \rightleftharpoons Ca^{2+} + SO_4^{2-} + 3H_2O(gypsum\ dissolution)$$

$$2NaAl_2Si_3O_8(albite) + 2H_2CO_3 + 9H_2O \rightleftharpoons Al_2Si_2O_5(OH)_4(kaolinite) + 2Na^{+} + 4H_4SiO_4 + 2HCO_3^{-}(silicate\ weathering)$$

The mineralization processes were analysed utilizing bivariate plots, which demonstrate relations among dissolved components in groundwater. These diagrams deliver important information about the probable processes that account for groundwater quality. If Ca^{2+} and Mg^{2+} could be derived from the weathering of carbonate rocks in the aquifer, the HCO_3^- over divalent cation (Ca^{2+} + Mg^{2+}) ratio would be below 0.5. Figure 4.18a shows the ratio for all sampling seasons less than 0.5, i.e. under the 1:1 equi-line. Therefore, the ratio plot designated that these two divalent cations mostly come from Ca- and Mg-carbonate rock sources. Instead, the HCO_3^- vs. (Na + K) plot demonstrates that all samples of the three sampling seasons fall below the 1:2 line (Fig. 4.18b). This indicated the very low concentration of Na and K compared to HCO_3^-, and the very amount of these metals does not come from carbonate salts. The plot of gross anions vs. (Ca^{2+} + Mg^{2+}) shows that all the data fall below the 1:1 line, which reflects the requirement of cations from the dissolution of carbonate minerals (Fig. 4.18d). Besides, the excess of alkaline earth ions showed an extra source of Ca^{2+} and Mg^{2+} and was balanced by HCO_3^- and SO_4^{2-}. This statement is buoyed by Fig. 4.18c, where the total cation vs. (Ca^{2+-} + Mg^{2+}) plot illustrates that the data are somewhat below the 1:1 line, directed by a slight involvement of other cations such as Na^+ and K^+ in water samples.

The bivariate design of Na^+ vs. Cl^- is typically used to control the mechanism of rock–water interface, total salinity, and sodic water interruptions from peripheral sources. The average values of the molar concentration ratio of [Na^+]/[Cl^-] in water samples were less than 1 (~0.7 in three sampling periods). In addition, the maximum samples of the three seasons are positioned over the 1:1 line of the Cl^- vs. Na^+ plot, i.e. the samples contained a lower amount of Na^+ corresponding to Cl^- (Fig. 4.18f). These findings indicated that no silicate rock (feldspar) weathering occurred in the aquifer systems in the study area. Nevertheless, this finding does not support the ion exchange process. Besides, the concentration ratio of (Na^+ + K^+) vs. total cations was 0.14 in the PRM and 0.11 in the POM periods and positioned much below the 1:2 equi-line, which obviously shows the very poor domination of alkali ions (Na^+ and K^+) over alkaline earth metal ions (Ca^{2+} and Mg^{2+}) (Fig. 4.18e). Therefore, the study results confirmed that the earth metal carbonate rocks were the major source of bicarbonate and earth metal ions in the samples.

For the groundwater of north-eastern Algeria [14], the diagrams of Ca^{2+-} vs. HCO_3^- and (Ca + Mg) vs. HCO_3^- recommend other noncarbonate sources of Ca, which is the dissolution of evaporites (mainly gypsum), as revealed in the plot of Ca^{2+} vs. SO_4^{2-}. The plot of (Ca + Mg) vs. (HCO_3^- + SO_4^{2-}) demonstrates that most of the sample clusters around the 1:1 uniline, indicating that the dissolution of calcite, dolomite, and gypsum is the leading reaction in groundwater, whereas the sampled waters shift to the left relative to an excess of (HCO_3^- + SO_4^{2-}), displaying the effect of silicate weathering and ion exchange. The coastal and arid groundwater showed dissimilar and complex plots when the dominant ions were plotted.

In the case of coastal groundwater, when halite dissolution is accountable for Na^+ and Cl^-, the Na^+/Cl^- concentration ratio is almost equal to 1. Once this ratio is greater than 1, it characteristically directs Na^+ release from the reactions of silicate rock weathering. In a study of the Goda coastal system, Republic of Djibouti [5], the Na^+ vs. Cl^- diagram for groundwater illustrates a linear trend among these elements but exposes an excess of Na^+ over Cl^- concentrations. The maximum number of samples falls above the seawater line ($Na^+ = 0.86 \times Cl^-$) and overhead the halite dissolution mark (1:1). Coastal rainfall and atmospheric deposition appear to be key sources of both Na^+ and Cl^-, whose levels increase under evaporation action. In the seaside alluvial aquifer of the Goda coastal areas, saline water intrusion owing to overwater mining from the aquifer can also be a source of Na^+ and Cl^-. In this study, the function of evaporation is further established by analysing the plot of Cl^- against Br^-. Both are conventional ions that do not contribute to ion exchange reactions in aquifer systems, are not influenced by redox reactions, and do not form any solid substances. In seawater, Cl^- is generally the most abundant chemical component, and the level of Br^- is much lower. In this investigation, the molar ratio Cl^-/Br^- varies from 125 to 755, without any halite dissolution. The diagram of Cl^- against Br^- demonstrates that maximum samples (90%) are near to the seawater line ($Cl^- = 649 \times Br^-$), again representing that the source of these ions in groundwater comes from seaside rainfall and evaporation effects. On the other hand, the Na^+/Cl^- ratio is unstable in favour of Na^+, and the excess of this ion is due to the weathering of the several volcanic rocks. Na^+ is a significant ion of alkali feldspars in volcanic formations, such as albite (Na-feldspar), anorthite, and phlogopite. The hydrolysis reactions of silicate are slow and occur in the presence of CO_2 and produce amorphous kaolinite, Na^+ and Ca^{2+} cations, and bicarbonate ions (HCO_3^-):

$$NaAlSi_3O_8\ (Na-feldspar) + CO_2 + H_2O \rightarrow Al_2Si_2O_5(OH)_4\ (kaolinite) + Na^+ + HCO_3^- + H_4SiO_4$$

$$Ca(Al_2Si_2)O_8\ (anorthite) + CO_2 + HO_2 \rightarrow Al_2Si_2O_5(OH)_4 + Ca^{2+} + HCO_3^-$$

$$KMg_3AlSi_3O_{10}(OH)_2\ (phlogopite) + H_2CO_3 + H_2O \rightarrow Al_2Si_2O_5(OH)_4 + Mg^{2+} + K^+ + HCO_3^- + H_4SiO_4$$

When silicate weathering is a main process, the bicarbonate (HCO_3^-) ion is leading in groundwater (above reactions). In the study area (Goda coastal system), HCO_3^- is indeed the dominant anion, which highlights the key role of silicate hydrolysis with carbonation in groundwater mineralization processes.

In the China arid area [2], groundwater showed a very scattered diagram for the most significant parameters. In these areas, a major quantity of HCO_3^-, Ca^{2+}, and Mg^{2+} was probably the result of the weathering of crystal dolomitic limestones and Ca–Mg-silicate rocks, typically from calcite, gypsum, plagioclase, and feldspar. The diagram of (Ca^{2+} + Mg^{2+}) vs. (HCO_3^- + SO_4^{2-}) displays that most of the values fall below the 1:1 uniline, reflecting the obligation of cations from the silicate rock weathering. The diagram of (Ca^{2+} + Mg^{2+}) vs. HCO_3^- illustrates the upper limits of HCO_3^- input from the dissolution of carbonate minerals. The sources of Ca^{2+} and Mg^{2+} in aquifer water can be expected from the (Ca^{2+} + Mg^{2+})/HCO_3^- ratio. As this ratio rises with salinity, Ca^{2+} and Mg^{2+} are adding to the water at a higher rate than HCO_3^-. If Ca^{2+} and Mg^{2+} were added individual from the carbonate dissolution in the aquifer basements and from the weathering of additional amphibole minerals, this ratio would be around 0.5. The low (Ca^{2+}+Mg^{2+})/HCO_3^- ratios (<0.5) could be the result of whichever Ca^{2+} + Mg^{2+} reduction by cation exchange or HCO_3^- enhancement. At the alkaline conditions (pH > 7), though high ratios cannot be attributed to HCO_3^- depletion and HCO_3^- does not form H_2CO_3. High ratios propose that the additional alkalinity of these water samples is balanced by alkalis (Na^+ + K^+). Additionally, the plot of (Ca^{2+} + Mg^{2+}) vs. total cations of the samples displays that the data are far below the 1:1 trend, indicating an increasing supply of Na^+ and K^+ in groundwater as TDS increase.

4.2.3.3 Probability of Ion Exchange Processes

Cation exchange is one of the significant geochemical courses that plays a vigorous role in controlling variations in groundwater chemistry and quality. It was found that not enough cation exchange happened in the groundwater of the study zone, which can be clarified by the graph of [(Na + K) − Cl] vs. [(Ca + Mg) − (HCO_3 + SO_4)] (Fig. 4.19a). If effective cation exchange within Na^+ and (Ca^{2+} + Mg^{2+}) were active in an aquifer, the slope would be a negative value (i.e. $m = -1$ in $y = -mx + c$). In the very scattered plot of Fig. 4.19a, a very weak negative slope ($m = -0.006$ in PRM and −0.0272 in MON) of the plot was found in the PRM and MON sampling periods. Hence, this fact showed that very slight cation exchange may be occurring in both seasons. But in the POM period, the slope presented a little positive value ($m = +0.0121$), representing that no cation exchange occurred in that period. So, alkaline and alkali earth metal concentrations typically depend on mineral dissolution or weathering courses. On the other hand, the bivariate figure of (Ca + Mg) vs. (Na + K) gives a very scattered plot, and the value of correlation factors (*R*) among them is very low (Fig. 4.19b). The concentration of earth metals (Ca and Mg) does not depend on basic metals (Na and K) in the samples. Therefore, these groups of metals are added to groundwater independently.

The majority of the groundwater samples from southern Punjab, Pakistan [21], show reverse ion exchange processes, and a small number show direct ion exchange. In this study, the dominance of (HCO_3^- + SO_4^{2-}) vs. (Ca^{2+} + Mg^{2+}) could be connected to cation exchange and silicate rock weathering. Only limited samples

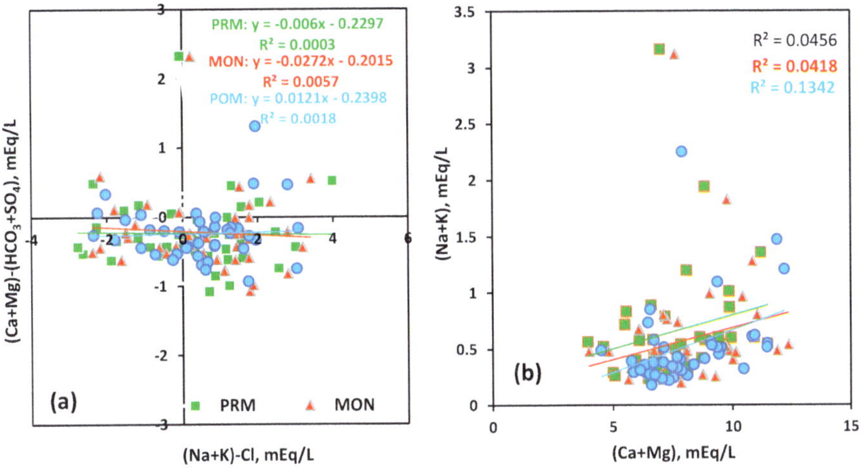

Fig. 4.19 (**a**) Bivariate plot of Cl^- correlated (Na^+ + K^+) and (Ca^{2+} + Mg^{2+}) correlated (HCO_3^- + SO_4^{2-}) to assess the cation exchange of water in the study area; (**b**) concentration (mEq/L) of earth metals over basic metals: bivariate plot of (Ca + Mg) vs. (Na + K) to determine the inter-dominance of major cations

were discovered above the 1:1 line of the plot, indicating that reverse cation exchange is significant in groundwater. Likewise, if cation exchange is the primary action in groundwater, the linear connection of (Na^+ + K^+ + Cl^-) vs. (Ca^{2+} + Mg^{2+}) – (HCO_3^- + SO_4^{2-}) has a slope of −1. In this plot, the linear line had a slope of −1.002 (near to −1), representing cation exchange in the groundwater systems. Furthermore, in the groundwater, reverse cation exchange processes should occur. Another finding of this study is the lithogenic Na^+ in the groundwater was supposed and accessible from the meteoric origin, which was well adjusted by equivalent Cl^-, and this Cl^- was removed from equivalent Na^+. The R^2 value of Na^+ against the (Na^+ + Cl^-) meq/L plot was 0.99, preferring Ca^{2+} and Mg^{2+} precipitation during ion exchange processes.

In contrast, the plotted points of the scatter plots of (Na–Cl) vs. (Ca + Mg–HCO_3–SO_4) for the groundwater of Precambrian hard rock aquifers of Kerala in India [22] have slopes of —0.849 for the weathered area and —1.011 for the deep fractured area. This indicates that reverse ion exchange has an important role in determining the sources of Ca^{2+}, Mg^{2+}, and Na^+ in groundwater in both aquifers and more so in fractured aquifers. Both chloro-alkaline indices (Sect. 4.2.3.5) have negative values in this study, signifying the prevalence of reverse ion exchange in this aquifer. In the weathered aquifer samples, nearly 50% of the samples each have positive and negative indices, indicating that both ion exchange (CAI 1 and CAI 2) and reverse ion exchange are active in this groundwater. The same results were found in the groundwater of Talensi District, northern Ghana [11], but only cation excess occurred in the Bougaa area, north-eastern Algeria's groundwater [14].

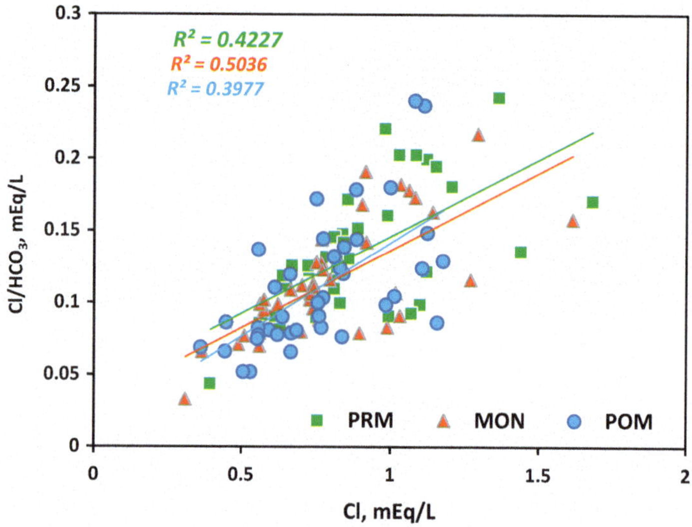

Fig. 4.20 Bivariate plot of Cl^- vs. Cl^-/HCO_3^- to determine saltwater intrusion

4.2.3.4 Seawater Intrusion

The Cl^-/HCO_3^- ratio indicates that the impact of salinization may be owing to seawater contact with groundwater. This ratio ranges from 0.066 to 0.223. This value is very low compared to another study of the coastal area where chloro-salt is a big issue in groundwater quality. Figure 4.20 displays that the linear line was strongly connected between Cl^-/HCO_3^- and Cl^- ($R^2 \geq 0.4$) in the three sampling periods. All groundwater samples showed a Cl^-/HCO_3^- ratio lower than 0.5, which means that the seawater did not affect the groundwater. Particularly in the dry period, the seawater of the Bay of Bengal blowup to the southern region is just far from the 100 km distance. Thus, the shallow aquifer of the study area is safe from seawater intrusion.

4.2.3.5 Chloro-alkaline Indices (CAI)

It is crucial to recognize the several changes in chemical arrangement undergone by groundwater throughout its journey on the subsurface runway. The chemical reactions in which ion exchange occurs among groundwater and the aquifer environment can be understood through the study of alkaline chlorine indices (CAI). The interface among the (Na + K-Cl) and (Ca+Mg) is broadly used to recognize cation exchange procedures. Additionally, the chloro-alkaline index (CAI) can be utilized to measure whether the cation exchange is direct or reverse. If the CAI value is greater than 0, it indicates that groundwater Ca^{2+} is exchanged for Na^+ in the aquifer, while the value of CAI is greater than 0 designates converse cation exchange. Positive CAI 1 and

CAI 2 values recommend that Na^+ and K^+ ions are exchanged for Mg^{2+} and Ca^{2+} in water. Besides, when the Mg^{2+} and Ca^{2+} ion index is negative, they are exchanged with Na^+ and K^+ from solid rocks. This is a suggestion of an alkaline-chloride disproportion.

The chloro-alkaline indices CAI 1 and CAI 2 (Eqs. 4.7 and 4.8) designate ion exchange processes between the groundwater and its host environment throughout residence as it permits through the aquifer. When there is an exchange among Ca^{2+} and Mg^{2+} with Na^+ and K^+ in the groundwater, both the above indices would be negative, and if there is a reverse ion exchange, then both these indices would be positive. These indices are computed for the groundwater samples in the study area, and it was observed that CAI 1 values vary between 0.403 and 1.710 with average values of 0.878, whereas CAI 2 values remain between 0.042 and 0.237 with average values of 0.125 in the PRM sampling season (Table 4.8 and Appendix IV). Similar to PRM, both index values were found in the other two sampling periods. It is also detected that 100% of the water samples of the three sampling periods show positive ratios. In addition, Table 4.8 demonstrates that the CAI 1 value is very high compared to the coastal saline groundwater. For the elevated concentration of HCO_3^-, the CAI 2 values of all sampling seasons are much lower than those in the above two studies:

$$\text{Chloro}-\text{alkaline index, CA–1} = \text{Cl}^- - \frac{\text{Na} + \text{K}}{\text{Cl}} \tag{4.7}$$

$$\text{Chloro}-\text{alkaline index, CA–2} = \text{Cl}^- - \frac{\text{Na} + \text{K}}{\text{HCO}_3 + \text{CO}_3 + \text{SO}_4 + \text{NO}_3} \tag{4.8}$$

where all concentrations of ions are stated in mEq/L.

In the study area, CAI 1 and CAI 2 values differ concerning sampling periods, but they are positive for all of the periods, and no wells have negative values. It was found that the correlation between these two indices is significantly high ($R > 0.65$) (Fig. 4.21). So, both index values give the same result for the samples. Additionally, the observation of Table 4.8 designates that reverse ion exchange is the leading process in the groundwater, while no normal ion exchange is also noticed in any sampling wells throughout the study periods. In addition, when reverse ion exchange is a significant hydrogeochemical process for controlling the composition of groundwater, the association between $[(Na^+ + K^+) - Cl^-]$ and $[(Ca^{2+} + Mg^{2+}) -$

Table 4.8 Chloro-alkaline index (CAI 1 and CAI 2) value of groundwater for the three sampling seasons

	Pre-monsoon (PRM)		Monsoon (MON)		Post-monsoon (POM)	
Statistics	CAI 1	CAI 2	CAI 1	CAI 2	CAI 1	CAI 2
Minimum	0.403	0.042	0.310	0.031	0.363	0.048
Maximum	1.710	0.237	1.615	0.211	1.177	0.234
Average	**0.878**	**0.125**	**0.803**	**0.109**	**0.763**	**0.105**
St. deviation (±)	0.264	0.040	0.257	0.036	0.214	0.042

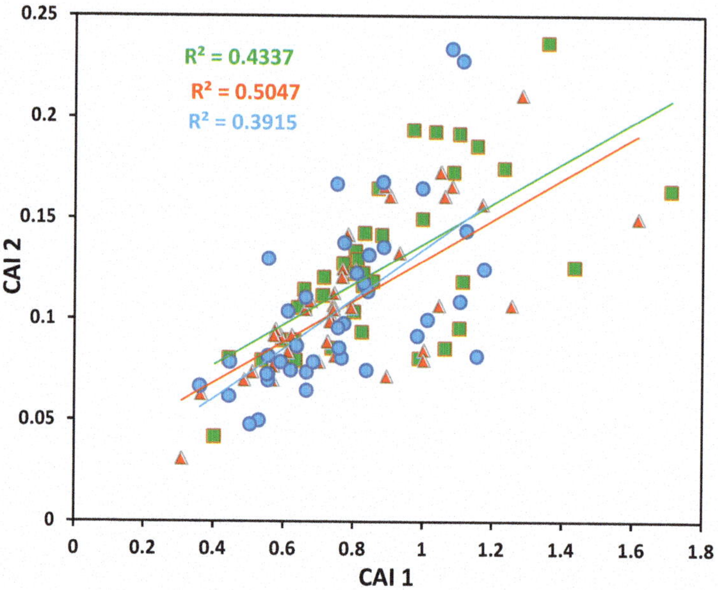

Fig. 4.21 Test of cation exchange: CAI1 vs. CAI2 in the three sampling rounds

$(HCO_3^- + SO_4^{2-})$] should be linear with a negative slope. Figure 4.19a demonstrates that all the groundwater samples of the study zone describe a straight line with a negative slope, which designates the incidence of reverse ion exchange.

Chloro-alkaline indices are a well-established method to differentiate the ion exchange between groundwater and its host rock–water interface environment throughout residence or travel in the aquifer. Several studies used these techniques for that determination; some are included in Table 4.9.

4.2.4 *Mass Balance of Ca^{2+} vs. Mg^{2+}*

The correlation coefficient (r) between Ca^{2+} and Mg^{2+} was found to be 0.45, 0.43, and 0.54 in the PRM, MON, and POM periods, respectively (see again Table 4.3 a, b, c). Besides, Fig. 4.22 shows that the correlation factors (R^2) between Ca^{2+} and Mg^{2+} are not more than the value of strong correlation in the PRM and MON periods. In POM, these metal ions are strongly correlated ($R^2 = 0.2957$). On the other hand, we have seen from the component analysis that these ions are not strongly accompanied by any component number (see again Table 4.4 a, b, c). These results indicated that both cations are not significantly correlated or associated. Thus, the cations Ca^{2+} and Mg^{2+} are not driven into groundwater from the similar rock source. The possibility of calcite ($CaCO_3$) and dolomite ($CaMg(CO_3)_2$) weathering can be revealed by the molar ratio of Ca^{2+} and Mg^{2+} ions in groundwater.

Table 4.9 CAI−1 and CAI−2 values for groundwater from different places

Sampling place	CAI−1			CAI−2			Comments	Reference
	Min.	Max.	Mean	Min.	Max.	Mean		
Telangana, S. India	−3.64	0.94	**0.12**	−0.78	3.36	**0.34**	Na^+ and K^+ are exchanged for Mg^{2+} and Ca^{2+}	[17]
Zamora, Mexico	−1.49	−0.58	**−1.11**	−0.26	−0.12	**−0.19**	Mg^{2+} and Ca^{2+} exchanged with Na^+ and K^+	[6]
Kerala, India	−2.00	0.60	**0.01**	−0.40	2.60	**1.72**	Na^+ and K^+ are exchanged for Mg^{2+} and Ca^{2+}	[22]
Goda, Rep. of Djibouti	−1.06	0.53	**0.31**	−0.05	0.44	**0.10**	Na^+ and K^+ are exchanged for Mg^{2+} and Ca^{2+}	[5]

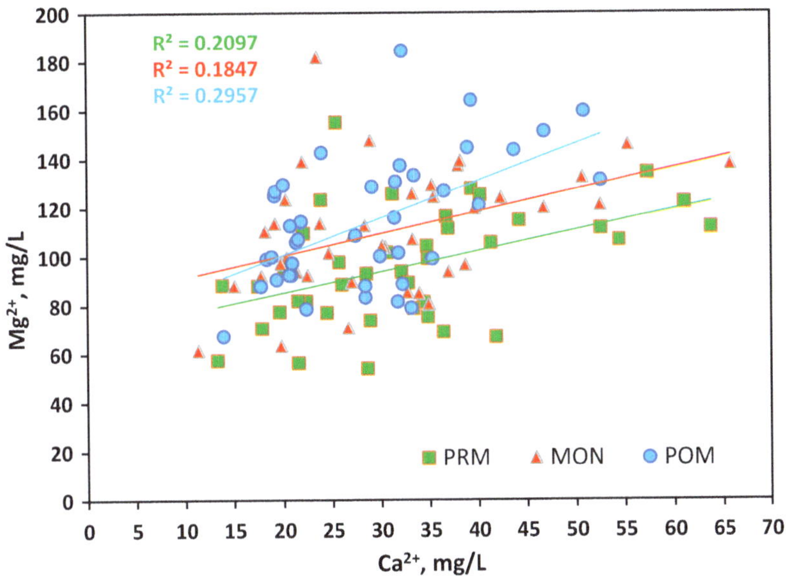

Fig. 4.22 Bivariate correlation between Ca^{2+} and Mg^{2+} concentrations in the samples

If the molar ratio of Ca^{2+}/Mg^{2+} in the groundwater sample is equal to 1 (one), both cations originate from the identical source (e.g. dolomite rocks), whereas a ratio greater than 1 (one) may signify a more dominant calcite involvement from the rocks. This concentration ratio of Ca^{2+} and Mg^{2+} in samples was found to be from 0.96 to 3.82 with an average of 1.94 in the PRM period, and the values varied from 1.26 to 4.64 with a mean value of 2.35 and 1.43 to 3.94 with an average of 2.52 for the MON and POM periods, respectively. The detailed datasets exhibited that both cations invented from a distinct source, i.e. Ca^{2+} from calcite and Mg^{2+} from dolomite in 97.5% of samples and the same source (dolomite) in 2.5% of samples in the PRM sampling period. But these actions were driven into groundwater from a separate source of 100% samples in the MON and POM seasons. The spatial design in the ratio of Ca^{2+} and Mg^{2+} presented variation from the southwest to northeast portions of the study area. The ratio was the highest between the lower floodplain and the upper floodplain area (cluster I) owing to the increase in Ca^{2+} level through the dissolution of carbonate minerals in the discharge area (see Fig. 4.10). The ratio decreased with the distance from the bank of the river to the recharge zone.

4.2.5 Saturation Index (SI) and Mineral Solubility

The value of the saturation index (SI) of a definite mineral/rock is used to identify the exact rock source and its dissolution mechanisms that control the total geochemical processes of the study area. The interaction between groundwater and minerals controls the geochemistry of groundwater. The thermodynamic calculation of the mineral–water equilibrium reaction can expect the thermodynamic control of the chemical composition of groundwater. The saturation index (SI) was used to predict the sensitive mineralogy of the ground aquifers. It was performed without collecting samples of the solid minerals or rocks and studying the mineralogy. An index value indicates whether the water would tend to dissolve or precipitate in a specific mineral. The value of SI is zero when the water and mineral are at chemical equilibrium, negative when the mineral may be dissolved, and positive when it may be precipitated in aqueous phase. This index is calculated by comparing the chemical activities of the dissolved mineral ions, i.e. ion activity product (*IAP*), with their solubility product (K_{sp}) at a specific temperature (Eqs. 4.9 and 4.10). It was computed using the geochemical computer software PHREEQC−3v for groundwater, which can be defined as:

$$A_aB_b(\text{mineral}) \rightleftharpoons [A]^a \times [B]^b \tag{4.9}$$

$$SI = \text{Log}\,\frac{[A]^a_{\text{actual}} \times [B]^b_{\text{actual}}}{[A]^a_{eq} \times [B]^b_{eq}} = \text{Log}\left(\frac{IAP}{K_{sp}}\right) \tag{4.10}$$

For calcite mineral dissolution, [*A*] and [*B*] are denoted as $[Ca^{2+}]$ and $[CO_3{}^{2-}]$ in Eqs. 4.9 and 4.10. The positive value of *SI* indicates supersaturation in water with minerals and a tendency for the minerals to settle down from the water phase. A negative *SI* pointed to undersaturation, tending the minerals to dissolve into groundwater. The *SI* values in a range of −0.5 to +0.5 for a specific mineral can be taken as demonstrative of equilibrium mixing in groundwater, which indicated the tendency of minerals to neither dissolve nor precipitate in water. The saturation index (*SI*) values of the major mineral solid phases, including calcite, aragonite, dolomite, anhydrite, gypsum, and halite, in the analysed water samples are shown in Table 4.10.

Table 4.10 Statistical data of saturation indices (*SI*s) of minerals using the PHREEQC−3v programme in groundwater samples

	Anhydride	Aragonite	Calcite	Dolomite	Gypsum	Halite	log$p$CO$_2$
	Pre-monsoon (PRM)						
Min.	−5.134	−1.190	−0.165	−1.541	−4.067	−8.870	−2.461
Max.	−2.043	0.832	2.098	0.781	−1.901	−5.453	−0.880
Mean	−2.941	−0.363	+0.976	−0.264	−2.437	−7.054	−1.453
SD(±)	0.549	0.198	0.171	0.412	0.467	0.423	0.261
	Monsoon (MON)						
Min.	−5.819	−1.211	0.011	−1.587	−4.321	−8.990	−2.915
Max.	−2.087	0.902	2.671	0.634	−1.750	−6.118	−1.553
Mean	−2.710	−0.353	+1.201	−0.302	−2.886	−7.587	−1.875
ST(±)	0.509	0.219	0.356	0.423	0.514	0.455	0.441
	Post-monsoon (POM)						
Min.	−5.876	−0.980	−0.041	−1.654	−4.754	−9.033	−3.512
Max.	−2.451	0.823	2.760	0.671	−1.860	−6.098	−1.259
Mean	−2.800	−0.286	+1.121	−0.301	−3.003	−7.561	−2.272
SD(±)	0.632	0.209	0.217	0.530	0.504	0.510	0.446

The detailed dataset revealed that about 90%, 20%, and 10% of the groundwater samples in the PRM period exceeded the saturation index value (>0) for calcite, dolomite, and aragonite, respectively. The results showed that the groundwater was oversaturated with these minerals. The study showed that majority of the samples was supersaturated with calcite mineral ($CaCO_3$), which led to a higher concentration of Ca^{2+} and HCO_3^- in groundwater samples in three sampling rounds in the study area. It was detected that the *SI* value of entire minerals was to some extent higher in the POM compared to that in the PRM period. For this reason, the concentrations of Ca^{2+}, Mg^{2+}, and HCO_3^- in the POM were higher than those in the PRM period. The higher *SI* value for calcite directed the prospect for a further increase in Ca^{2+}, Mg^{2+}, and HCO_3^- concentrations in the groundwaters of the study zone due to the extra dissolution of this type of mineral. Besides, the value of the partial pressure of CO_2 gas (pCO_2) in rock dissolution areas has a vital role in the rock dissolution process. Equilibration with atmospheric pCO_2 creates oversaturation regarding calcite and dolomite. However, the precipitation kinetics of dolomite is slower than calcite. An increase in pCO_2 throughout water–rock interactions affects the process of mineral dissolution. The following reactions for the dissolution of carbonate minerals occur in natural aquifer systems:

$$CO_2\ \text{(leached)} + H_2O\ \text{(saturation zone)} \rightarrow H_2CO_3$$

$$CaCO_3\ \text{(calcite)} + H_2CO_3 \rightarrow Ca^{2+}(aq) + HCO_3^-\ (aq)$$

$$CaMg(CO_3)_2\ \text{(dolomite)} + H_2CO_3 \rightarrow Ca^{2+}(aq) + Mg^{2+}(aq) + HCO_3^-\ (aq)$$

$$H_2CO_3 \rightarrow H_2O + CO_2\ \text{(degas)}$$

These reactions consumed CO_2 and increased the concentrations of Ca^{2+}, Mg^{2+}, and HCO_3^- through H_2CO_3 formation. As a result, the pCO_2 would decrease and produce basic materials that increase the pH. Although, the lower saturation of the minerals has impacted the chemical composition of the groundwater, the supersaturation of groundwater recommended that these carbonate minerals were the major component in the host mineral. Evaporated minerals such as halite, gypsum, and anhydrite were undersaturated in all groups, indicating that the concentrations of the soluble ions Na^+, Cl^-, and SO_4^{2-} were not inadequate by mineral equilibrium.

The value of partial pressure of CO_2 gas acting an important role in geochemistry. The $\log_{10}(pCO_2)$ reflects the geochemical procedure and relates the saturation index (*SI*) of the carbonate minerals. The typical air temperature of the study area was 29 ° C (Table 4.1), which was adequately high to reduce the pCO_2. The overall temperature dependence of the estimated pCO_2 is expected to be small.

The partial pressure of CO_2 of 40 groundwater samples in the study area was determined by the PHREEQC−3v computer programme, and the mean values of both sampling seasons are presented in Table 4.10. The $\log_{10}(pCO_2)$ values in groundwater vary from −2.461 to −0.880 with a mean value of −1.953 (±0.261) in the pre-monsoon (PRM) season, −2.915 to −1.553 with a mean value of −1.875 (±0.441) in the monsoon (MON) season, and −3.512 to −1.259 with an average of −2.272 (±0.446) in the post-monsoon (POM) season, which are higher than the air partial pressure of CO_2 ($pCO_2 = 0.0004$ atm. or $\log_{10}pCO_2 = -3.397$ atm., equivalent to 0.04% CO_2 in the atmosphere) under equilibrium conditions. The origins of CO_2 in groundwater are (a) percolated atmospheric CO_2, (b) dissolution of carbonate minerals, and (c) microbial decomposition of organic carbon in oxic conditions and CO_2 from plant respiration in the root zone. Relatively higher $\log_{10}(pCO_2)$ values for groundwaters perhaps designate the roles of sources (b) and/or (c) in addition to air CO_2. This result showed that the above (b) and (c) sources of CO_2 (soil CO_2) in groundwater are major in the study zone, which controls the entire geochemical process. Meanwhile, the dissolved CO_2 pressure of the water is higher than that of the atmosphere, and the water is supersaturated with carbonate minerals. Additionally, the soils of the study zone are sandy, gravel, and clay–sandy types that make the topsoil very loose, and additional atmospheric CO_2 penetrates the soil with rainwater, increasing the extra dissolution of carbonate rocks through reaction with H_2CO_3. Along with CO_2, atmospheric O_2 may penetrate the ground level and be used as an oxidizing agent of organic matter.

It was found that the $\log_{10}(pCO_2)$ value of samples is inversely proportional to groundwater pH and proportional to HCO_3^- concentration. However, the molar concentration of HCO_3^- is very unimportant to pCO_2 for groundwater related to the pH value. For the calculation of $\log_{10}(pCO_2)$, a modelled equation, such as PHREEQC-3v, was used, in which the term $\log_{10}[HCO_3^-]$ is small related to the pH term because concentration is on the order of 10^{-3} molar (M) and pH values typically vary from 6.5 to 8.5, signifying a negligible effect on the estimated CO_2 values. In the PRM period, most of the groundwater samples were slightly acidic (pH < 7) with lower pH values, and HCO_3^- was found comparatively at a lower

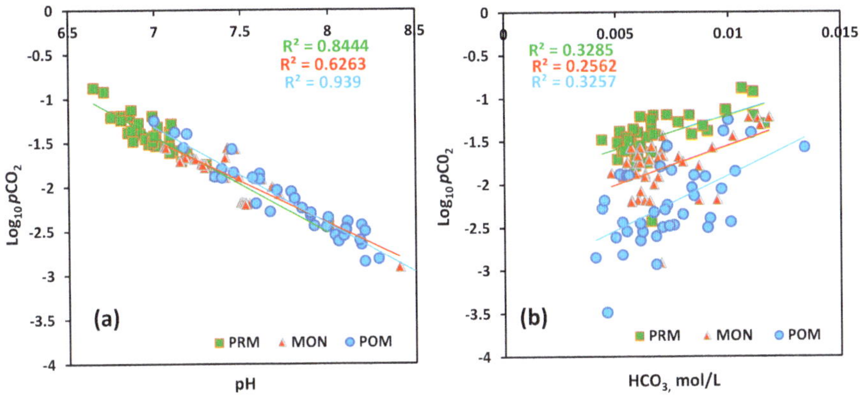

Fig. 4.23 Geochemical relationship of (**a**) pH vs. $\log_{10}(pCO_2)$ and (**b**) $[HCO_3^-]$ vs. $\log_{10}(pCO_2)$ in groundwater

concentration than in the MON and POM seasons. As predicted, the average values of $\log_{10}(pCO_2)$ were inversely proportional to pH and HCO_3^-, respectively (Fig. 4.23a, b).

Numerous studies have assumed very complicated roles of pCO_2, water pH, and concentration of HCO_3^- on the potentiality of rock saturation in groundwater, especially in the carbonate rock weathering process. In addition, the value of $\log_{10}(pCO_2)$ mutually depends on water pH, the amount of organic matter in water, the equilibrium constant for the association and dissociation reaction of H_2CO_3, and the concentration of HCO_3^- in water.

Figure 4.21(a) illustrates that the $\log_{10}(pCO_2)$ value regularly decreases with increasing pH for both periods. The high partial pressure of CO_2 in the saturation zone leads to the production of more H_2CO_3 through the addition of H_2O, and the association constant of CO_2 is adequately high ($K_{CO2} = 10^{-1.5}$ at 25 °C). This weak acid has a lower dissociation constant ($K_{H2CO3} = 10^{-6.40}$, at 25 °C), indicating that a small amount of H_2CO_3 dissociation or degassing of CO_2 occurred. However, at higher pH or lower H^+ concentrations, the rate of the forwarding reaction of H_2CO_3 dissociation ($H_2CO_3 \rightleftharpoons H^+ + HCO_3^-$) becomes high. For this reason, the saturation rate of CO_2 gas in H_2CO_3 ($CO_2 + H_2O = H_2CO_3$) and, obviously, the pCO_2 value would be decreased. The excess H_2CO_3 would react with Ca^{2+} and Mg^{2+} and may be precipitated in a calcite supersaturation solution. Therefore, $\log_{10}(pCO_2)$ governs the recrystallization rate of carbonate minerals. This recrystallization rate depends on the amount of organic carbon (OC) in water to block the active sites on the calcite crystals with increasing $\log_{10}(pCO_2)$. This process is encouraged by the increase in negative charge on the calcite surface and the decrease in bonding capacity of the humus acid of soil with decreasing pH of the water solution [29]. So, the specific amount of dissolved organic carbon decreases owing to an increase in the ionic strength of the solution. For the above-mentioned reasons, at high groundwater pH in the post-monsoon (POM) season, the $\log_{10}(pCO_2)$ value of water samples is higher than during the monsoon (MON) and pre-monsoon (PRM) seasons. Hence, the overall HCO_3^- level of the samples in the POM season was higher than that in the PRM.

According to hierarchical cluster analysis (HCA), sampling stations of cluster I are located on the riverbanks, and these samples transport a higher level of HCO_3^- and lower pH value. River flow can affect this situation. The pCO_2 in river water is typically out of equilibrium with the air. This water contains a substantial portion of high CO_2 gas, and the rate of re-equilibrium with the atmosphere is comparatively slow. Particularly in POM, huge amounts of river water infiltrate the groundwater and cause H_2CO_3 supersaturation, and the HCO_3^- concentration becomes high with lower pH.

Several researchers have used the concept of the saturation index (SI) to evaluate the local geochemical history of groundwater. Other chemical and statistical scenarios of groundwater combined with SI values for numerous minerals obviously deliver the conclusion about the source of water solutes with involved reactions. For the groundwater of the Bougaa area, north-eastern Algeria [14], saturation indices were computed from the chemical data of samples using the geochemical modelling programme PHREEQC. The results showed that all samples were undersaturated in anhydrite, gypsum, and halite, whereas maximum groundwater samples were supersaturated in calcite, aragonite, and dolomite. These findings recommend that the dissolution of evaporite minerals and precipitation/dissolution of carbonate minerals (calcite and dolomite) are the leading processes controlling the hydrogeochemistry of groundwater in the area. On the other hand, the computed values of SI for calcite and dolomite in the groundwater samples from the Ejina Basin, north-western China [2], vary from −0.19 to 2.66 and −0.31 to 1.32, with averages of 1.08 and 0.64, respectively. Around 90% of the SI values for calcite and dolomite are greater than 0 in the groundwater samples, which are oversaturated with regard to these minerals owing to evaporation; thus, they are deposited. Moreover, the hydrogeochemical study results for the groundwater of the Mokopane Area, Limpopo, South Africa [16], illustrate that maximum groundwater samples are oversaturated regarding calcite, aragonite, and dolomite minerals, clearly depiction the source of HCO_3^- through carbonate dissolution. Detecting the SI for anhydrite, it was found that total water samples were unsaturated regarding this mineral, indicating that both Ca^{2+} and SO_4^{2-} do not come from anhydrite. Additionally, SO_4^{2-} mostly comes from man-made sources rather than geogenic sources in the aquifer.

4.3 Isotope Investigations

Environmental tracers can deliver consistent datasets that can be used to identify groundwater recharge sources and their flow directions over an extensive range of spatial and temporal scales. Groundwater table fluctuations and lysimeters provide estimates of local recharge over a few times to a few periods; environmental tritium (^3H) and chlorofluorocarbons (CFCs) in groundwater naturally constrain recharge rates over years to decades, whereas ^{14}C and ^{36}Cl constrain average recharge on extended timescales and over larger zones. With a half-life of 12.32 years, 3H is a potential candidate for dating groundwater recharge over the past 50–100 years. The ^3H isotope is part of the water molecule, and its abundance in groundwater sequestered from the atmosphere is affected only by radioactive decay, not by reactions between the water and the aquifer matrix. Due to the production of ^3H during atmospheric nuclear testing, the ^3H input function in precipitation exhibits a distinct peak in the 1950s to the 1960s. This 'bomb ^3H pulse' was used to track the recharging of the water currents during this period. Since the ^3H input function is not constant, this single ^3H concentration measurement did not provide accurate groundwater dating. However time-series measurements of ^3H allow for quantitative dating. Most commonly, the presence of ^3H in groundwater is interpreted as modern recharge. Radioactive tracers containing tritium are also used in many industrial applications.

Stable isotopes ($\delta^{18}O$ and δ^2H), on the other hand, are widely used to identify relationships between surface water and groundwater systems and are widely recognized as useful tracers for providing insight into water dynamics. Isotopes are powerful integrated records of important processes such as evaporation, transpiration, recycling, and mixing. Recharge from direct precipitation, runoff, lakes, snow, and glaciers can be distinguished by their characteristic stable isotope signatures. Various other hydrological processes that can alter the isotopic composition of groundwater include mixing with various springs, enrichment of heavy isotopes by evapotranspiration, isotopic fractionation during rainfall, and enrichment of oxygen isotopes during water–rock interactions.

Isotopic studies as a supporting tool are the latest techniques for assessing groundwater hydrogeochemical processes. The isotope compositions $\delta^{18}O$ and δD are sensitive tracers and are extensively used to study natural water cycles and groundwater dynamics. Oxygen ($\delta^{18}O$) and deuterium (δD) isotopes in groundwater samples were determined using an isotope ratio mass spectrometer or a MAT−252 mass spectrometer. Standard mean ocean water (SMOW) has been adopted as a reference for measuring H and O isotopic compositions, with analytical accuracies for δO and δD within ±0.5‰ and ±1‰, respectively. The isotope ratios of CO_2 and H_2 were determined on a Finnigan Thermo Quest Delta plus XL mass spectrometer, and the results are expressed in deviations (‰) from the Vienna Standard Mean Ocean Water (VSMOW) standard. Radio-isotopic ^3H compositions were determined with a Quantulus−1220 liquid scintillation counter. The results are reported in δ units (per-mile deviation of isotopic ratios from the international standard V-SMOW), where δ is defined as:

$$\delta = \left[\frac{(R_{\text{sample}} - R_{SMOW})}{R_{SMOW}}\right] \times 1000 \tag{4.11}$$

where $R = D/H$ or $^{18}O/^{16o}$.

Wen et al. (2005) conducted an isotope investigation of groundwater in the Ejina Basin in north-western China [2]. They found that the δD and $\delta^{18}O$ compositions of two river water samples are −8.0‰, 56‰ and −7.9‰, 57‰, respectively. The δD and $\delta^{18}O$ compositions of the confined groundwater range from −86‰ to −82‰ and −9.6‰ to −8.8‰, respectively. The δD and $\delta^{18}O$ compositions of the shallow groundwater vary from −74‰ to −44‰ and −8.6‰, −5.5‰, respectively. From the diagram of δD vs. $\delta^{18}O$ it was found that isotopic compositions of groundwater and river water are situated between the world meteoric waterlines. The world meteorite waterline uses the following equation (Eq. 4.12) to represent the regression line for global continental precipitation samples:

$$\delta D(‰) = \left[8.0 \times \delta^{18}O + 10\right]‰ \tag{4.12}$$

A meteorological waterline is calculated for a specific area called a local meteorological waterline (LMWL) and used as a baseline within that area. Local waterlines may differ in slope and cross section from global waterlines. Such tilts and misaligned intersections are primarily due to moisture. In this study, surface water and groundwater (LMWL) follow the world meteoric waterline. This shows that precipitation is the main source of groundwater recharge resources. Under dry backgrounds, the regeneration of precipitation into groundwater occurs only if some surface water leads the penetration process. The pattern of the water cycle gives water sources in arid regions a unique and identifiable isotopic composition. Regarding the distribution of $\delta^{18}O$ and δD values in groundwater and river water, it is clear that the average $\delta^{18}O$ and δD values tend to increase from river water to shallow groundwater and deep groundwater. Both stream water and groundwater exhibit a sample shift to higher values of δO, a typical pre-replenishment phenomenon in semiarid regions. Water can be lost through evaporation from unsaturated zones or the water table. Majority of the samples shows evaporative loss.

Several isotopic compositions of groundwater samples are plotted near the global meteorological waterline and to the lower left of the trend. This indicates that groundwater is largely unaffected by evaporation. Groundwater is located within 1 km of the riparian zone. Some isotopic compositions of groundwater samples deviate from waterlines in the weather world. Groundwater comes from precipitation that occurs under climatic conditions in different regions or regions subject to conditions or biochemical changes. The groundwater isotopic compositions $\delta^{18}O$ and δD are at the water evaporation line. Groundwater isotopic compositions $\delta^{18}O$ and δD produce much larger isotope fractions during the process of replenishing groundwater with river water, leading to deviations from the global meteoric waterline over a long history. These samples are plotted along a line with a slope of 5. Its slope is less than 8, which is the slope of the global meteoric waterline. Such low gradients can be created by evaporation. The isotopic $\delta^{18}O$ and δD compositions of the confined groundwater samples are close to the global meteoric waterline and are less affected by evaporation. This study uses tritium ^{3}H measurements to analyse the age of groundwater. Tritium 3H, with a half-life of 12.43 years, is often used to estimate residence time in groundwater. It is directly incorporated into water molecules, making it the only radioactive isotope that actually comes from groundwater. In this study, the $^{3}H_{1}$ levels in groundwater can be separated into three groups. The first group belongs to shallow groundwater samples within a 1 km width of the river water impact zone. Groundwater tritium levels are over 10 TU, close to river levels of 43.32 TU and 45.74 TU. They contain some residual bomb tritium. This shows that the groundwater has been recharged by recent Heihe water, and the hydrological cycle is active. The second group belongs to near-surface groundwater samples from 1 km to 10 km of the river water impact area. Tritium levels in groundwater range from 1 Tu to 10 TU. This designates that the groundwater belongs to a mixture of groundwater between sub-modern (the 1950s) and recent recharge. The third group belongs to strained groundwater samples, with tritium values less than 1 TU. This designates that the groundwater source originated from a relatively long recharge period, such as groundwater before the 1950s.

Another study [3] was conducted on coastal groundwater in southeastern India. The local meteorological waterline (LMWL) in Tamil Nadu has the equation in [4.11] and is close to the derived global meteorological waterline (GMWL). Small variations are due to various climatic effects, such as temperature, secondary evaporation, and variations in precipitation and humidity. The resulting LMWL for δD and δO analysis indicates that the isotopic composition of the groundwater ranges from −7.7 to −2.1‰ for δO and −55.6 to −18.5‰ for δD. Comparing rainwater stable isotopes with groundwater samples reveals that most samples comply with the LMWL. LMWL shows recharging due to local precipitation. Very few samples show evaporative/seawater intrusion enrichment. A heavier isotope was also observed in five samples. This may be due to near-shore precipitation or saltwater intrusion. Since the fractionation of stable isotopes is primarily temperature dependent, the origin of groundwater salinity is commonly determined using the δD–δO relationship. Group B samples also show terminal elements with isotopic compositions indicative of seawater intrusion. It is clear that considerable mixing was detected between the meteorite samples and the seawater final elements.

This study attempted to evaluate the geochemical characteristics, interpret the water quality data, and assess the water quality for drinking and irrigation prepossess of 40 monitoring sites in the Kushtia District (the upper Ganges River basin) of Bangladesh. Samples were collected for three periods, i.e. the pre-monsoon (PRM), monsoon (MON), and post-monsoon (POM) periods, from each year of 2019–2020 and 2020–2021. To analyse the water chemistry and evaluate the water quality and water facies, numerous sophisticated techniques, computer programmes, and statistical methods have been utilized. The major aims of this study are to assess the hydrogeochemical process and assess the water type of groundwater to determine its appropriateness for drinking and irrigation purposes. For this purpose, we characterize the groundwater samples, explore the numerous factors controlling the water chemistry and facies, recognize the hydrogeochemical processes, and assess the water quality for drinking and irrigation purposes in the study zone.

4.4 Summary

The study conducted geochemical analyses to recognize the origin of solutes and assess the geochemistry of groundwater in the investigated area. Hydrogeochemical water parameters were determined through widely accepted laboratory methods and reported methodically. The most neutral pH (PRM, 7.02; MON, 7.43; and POM: 7.83) and higher EC (PRM, 669.6; MON, 867.5; and POM, 956.8 μS/cm), TDS (PRM, 413.2; MON, 558.2; and POM, 601.5 mg/L), and total hardness (PRM, 362.2; MON, 396.1; and POM, 4.4.7 mg/L) of groundwater samples were the key geochemical features in the study area. The results presented that the values of Ca, HCO_3^-, and TH crossed the national and international tolerable values for drinking and irrigation water quality standards. The study explored that the order of abundance of cations in samples is $Ca^{2+} > Mg^{2+} > Na^+ > K^+$ and that of anions is $HCO_3^- >> Cl^- > SO_4^{2-} > NO_3 > PO_4^{3-}$. The correlation coefficient values of water parameters designated the relations among numerous parameters that found the different natural geochemical processes. Principal component analysis (PCA) exposed that the chemical characteristics of the groundwater were the cause of rock–water interactions, mixing, or external impacts. This study confirmed that a few humans caused contamination (fertilizer deposition) sources of groundwater with NO_3^- and PO_4^{3-}. Hierarchical cluster analysis revealed two separate groundwater zones where the groundwater may be more affected by neighbouring rivers' flow rate, irrigation return flow, and freshwater intrusion from the river stream. It was observed that the samples of cluster I transport more HCO_3^- of earth metals with high electrical conductivity than the cluster II samples. Diagram methods showed that the analysed samples were mainly Ca–HCO_3 type, and rock weathering was found to be the dominant natural process in the investigated samples. The mineralization and rock–water interaction were controlled by several geochemical processes in the aquifer, which were explained by numerous bivariate plots. The saturation index (*SI*) and mass balance values indicated that about 80% and 20% of groundwater were supersaturated with calcite and dolomite, respectively, in the PRM sampling season. On the other hand, nearly 100% of samples were supersaturated with calcite in the POM period. The results of the study showed that the CO_2 in water controlled the pH, and later the pH of the water and excess aquifer CO_2 controlled the water–rock equilibrium reactions. Therefore, the chemical characteristics of groundwater in the study area are dependent on the water–rock interaction, carbonate-based mineral dissolution, local lithological conditions, and most neighbouring river morphology. The study findings may be utilized for sustainable water resource management.

The chemical features of groundwater in the study area were dependent on the water–rock interaction, carbonate-based mineral dissolution, and most adjacent river morphology. The study findings may be used for sustainable water resource management on a regional scale and may be put in another area with similar topography. Accordingly, it is expected that the results of this investigation will deliver a useful understanding for future groundwater monitoring and management of the study area. Also, the findings of this investigation deliver guidelines for agricultural activists, policymakers, public health departments, and water managers of shallow aquifers. In addition, the study suggests that groundwater should be drunk after proper treatment to remove trace heavy metal contaminants from potable water. Besides, deep wells should be drilled near river valleys and avoid areas with a thick clay layer. This study measured higher levels of some trace metals in samples that make the water very harmful for every purpose. Elevated concentrations of iron, manganese, and lead seriously impact drinking water quality. Further study should be considered using modern and sophisticated investigative approaches with extensive parameters over a wide area. The study found that the lack of public awareness of water quality, inadequate modern water treatment practices, absence of water quality monitoring, and reluctance to apply the law are the main challenges to ensuring a safe water supply in the study area. Furthermore, public consciousness building and publicity programmes for groundwater pollution are imperative to ensure safe water for all. Also, advanced research and survey-based works in these areas need to be enhanced to explore the safe water status and water-related problems.

References and Further Study

1. Alfy, M. E., Abdalla, F., Moubark, K., & Alharbi, T. (2019). Hydrochemical equilibrium and statistical approaches as effective tools for identifying groundwater evolution and pollution sources in arid areas. *Geosciences Journal, 23*(2), 299–314. https://doi.org/10.1007/s12303−018−0039−7
2. Wen, X., Wu, Y., Su, J., Zhang, Y., & Liu, F. (2005). Hydrochemical characteristics and salinity of groundwater in the Ejina Basin, Northwestern China. *Environmental Geology, 48*, 665–675. https://doi.org/10.1007/s00254−005−0001−7
3. Chidambaram, S., Sarathidasan, J., Srinivasamoorthy, K., et al. (2018). Assessment of hydrogeochemical status of groundwater in a coastal region of Southeast coast of India. *Applied Water Science, 8*(27), 1–14. https://doi.org/10.1007/s13201−018−0649−2
4. Inim, I. J., Affiah, U. E., Asuaiko, E. R., et al. (2017). Hydrogeochemical evaluation of groundwater in coastal alluvial aquifer of Akwa Ibom, Southeastern Nigeria. *Journal of Coastal Sciences, 4*(2), 1–8.
5. Ahmed, I. M., Jalludin, M., & Razack, M. (2020). Hydrochemical and isotopic assessment of groundwater in the Goda Mountains Range System. Republic of Djibouti (Horn of Africa). *Water, 12*(2004), 1–22. https://doi.org/10.3390/w12072004
6. Reyes-Toscano, C. A., Alfaro-Cuevas-Villanueva, R., Cortés-Martínez, R., et al. (2020). Hydrogeochemical characteristics and assessment of drinking water quality in the urban area of Zamora, Mexico. *Water, 12*(556), 1–26. https://doi.org/10.3390/w12020556

7. Gibrilla, A., Osae, S., & Akiti, T. T. (2010). Hydrogeochemical and groundwater quality studies in the Northern part of the Densu River Basin of Ghana. *Journal of Water Resource and Protection, 2*, 1071–1081. https://doi.org/10.4236/jwarp.2010.212126
8. Okoroh, D. O., & Ibuot, J. C. (2022). Hydrogeochemical assessment of groundwater quality: a case study of Federal College of Education (Technical), Omoku, Rivers State. *Water Practice & Technology, 17*(7), 1458–1469. https://doi.org/10.2166/wpt.2022.077
9. Islam, M. S., & Mostafa, M. G. (2021). Chapter 4: Groundwater Status and Challenges in Bangladesh. In E. Lichtfouse (Ed.), *Sustainable agriculture reviews* (Vol. 52, pp. 79–146). Springer. https://doi.org/10.1007/978−3−030−73245−5_4
10. Sethy, S. N., Syed, T. H., Kumar, A., & Sinha, D. (2016). Hydrogeochemical characterization and quality assessment of groundwater in parts of Southern Gangetic Plain. *Environment and Earth Science, 75*(232), 1–15. https://doi.org/10.1007/s12665−015−5049−4
11. Chegbeleh, L. P., Akurugu, B. A., & Yidana, S. M. (2020). Assessment of groundwater quality in the Talensi District, Northern Ghana. *Scientific World Journal, 2020*(8450860), 1–24. https://doi.org/10.1155/2020/8450860
12. Vespasiano, G., Muto, F., & Apollaro, C. (2021). Geochemical, geological and groundwater quality characterization of a complex geological framework: The case study of the Coreca Area (Calabria, South Italy). *Geosciences, 11*(121), 1–22. https://doi.org/10.3390/geosciences11030121
13. Kumar, M., Ramanathan, A. L., Rao, M. S., & Kumar, B. (2006). Identification and evaluation of hydrogeochemical processes in the groundwater environment of Delhi, India. *Environmental Geology, 50*, 1025–1039. https://doi.org/10.1007/s00254−006−0275−4
14. Kouadra, R., & Demdoum, A. (2020). Hydrogeochemical characteristics of groundwater and quality assessment for the purposes of drinking and irrigation in Bougaa area, Northeastern Algeria. *Acta Geochimica, 39*(5), 642–654. https://doi.org/10.1007/s11631−019−00393−3
15. Peter, S. K. (2020). *Hydrogeochemical analysis and modelling of groundwater in Mbeere south subcounty, Embu county, Kenya*. PhD Thesis, The School of Pure and Applied Science, Kenyatta University.
16. Molekoa, M. D., Avtar, R., Kumar, P., Minh, H. V. T., & Kurniawan, T. A. (2019). Hydrogeochemical assessment of groundwater quality of Mokopane Area, Limpopo, South Africa Using Statistical Approach. *Water, 11*(1891), 1–18. https://doi.org/10.3390/w11091891
17. Gugulothu, S., Subbarao, N., Das, R., & Dhakate, R. (2022). Geochemical evaluation of groundwater and suitability of groundwater quality for irrigation purpose in an agricultural region of South India. *Applied Water Science, 12*(142), 1–13. https://doi.org/10.1007/s13201−022−01583-w
18. Abdullah, T. O., Ali, S. S., Al-Ansari, N. A., & Sven Knutsson, S. (2019). Hydrogeochemical evaluation of groundwater and its suitability for domestic uses in Halabja Saidsadiq Basin, Iraq. *Water, 8*(312), 1–13. https://doi.org/10.3390/cells8040312
19. Mohammed-Aslam, M. A., & Rizvi, S. S. (2020). Hydrogeochemical characterisation and appraisal of groundwater suitability for domestic and irrigational purposes in a semi-arid region, Karnataka state, India. *Applied Water Science, 10*(237), 1–37. https://doi.org/10.1007/s13201−020−01320−1
20. Islam, S. M. D., Bhuiyan, M. A. H., Rume, T., & Azam, G. (2017). Hydrogeochemical investigation of groundwater in shallow coastal aquifer of Khulna District, Bangladesh. *Applied Water Science, 7*, 4219–4236. https://doi.org/10.1007/s13201−017−0533−5
21. Iqbal, J., Su, C., Abdur Rashid, A., et al. (2021). Hydrogeochemical assessment of groundwater and suitability analysis for domestic and agricultural utility in Southern Punjab, Pakistan. *Water, 13*(3589), 1–23. https://doi.org/10.3390/w13243589
22. Nandakumara, N. P., & Balakrishnan, K. (2020). Groundwater quality variations in Precambrian hard rock aquifers: A case study from Kerala, India. *Applied Water Science, 10*(2), 1–13. https://doi.org/10.1007/s13201−019−1084−8

23. Ghani, J., Ullah, Z., Javed Nawab, J., et al. (2022). Hydrogeochemical characterization, and suitability assessment of drinking groundwater: Application of geostatistical approach and geographic information system. *Frontiers in Environmental Science, 10*(874464), 1–16. https://doi.org/10.3389/fenvs.2022.874464
24. Nyirenda, T. M., Mapoma, H. W. T., Msiska, A., Dzonzi-Undi, J., & Sella Jumbo, S. (2015). Hydrogeochemical assessment of groundwater quality in Salima and Nkhotakota Districts, Malawi. *International Journal of Science and Research, 4*(9), 2319–7064.
25. WHO. (2022). *Guidelines for drinking-water quality*. 4th edition incorporating the first and second addenda. World Health Organization.
26. Jeelani, G. H., Shah, R. A., & Hussain, A. (2014). Hydrogeochemical assessment of groundwater in Kashmir Valley, India. *Journal of Earth System Science, 123*(5), 1031–1043.
27. Adimalla, N., Li, P., & Venkatayogi, S. (2018). Hydrogeochemical evaluation of groundwater quality for drinking and irrigation purposes and integrated interpretation with Water Quality Index studies. *Environmental Processes, 5*, 363–383. https://doi.org/10.1007/s40710–018–0297–4
28. Yousfi, Y. E., Himi, M., & Ouarghi, H. E. (2022). Hydrogeochemical and statistical approach to characterize groundwater salinity in the Ghiss-Nekkor coastal aquifers in the Al Hoceima Province, Morocco. *Groundwater for Sustainable Development, 19*, 100818. https://doi.org/10.1016/j.gsd.2022.100818
29. Islam, M. S., & Mostafa, M. G. (2022). Evaluation of hydrogeochemical processes in groundwater using geochemical approaches and geostatistical models in the upper Bengal basin. *Geofluids, 2022*(9591717), 1–21. https://doi.org/10.1155/2022/9591717

Chapter 5
Trace Metals in Groundwater: *Sources and Mobilization*

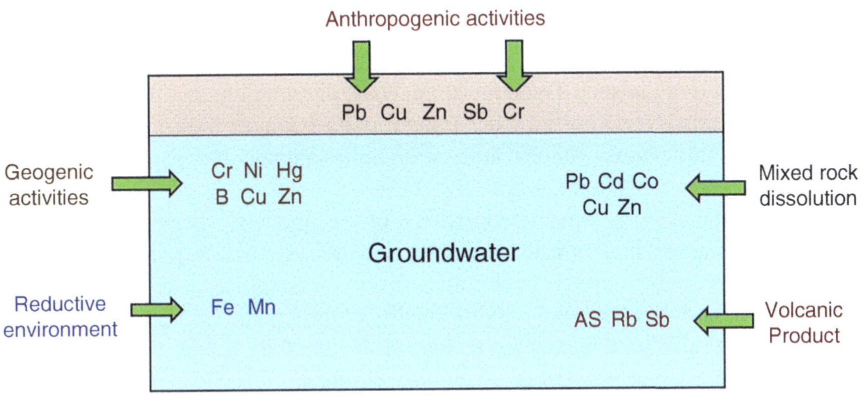

Abbreviations

BADC	Bangladesh Agricultural Development Corporation
BDWS	Bangladesh Drinking Water Standard
DO	Dissolved oxygen
DOC	Dissolved organic carbon
DPHE	Department of Public Health and Engineering (Bangladesh)
FAO	Food and Agriculture Organization
LD	Lethal dose
MTL	Maximum threshold limit
US-EPA	United States Environmental Protection Agency
WHO	World Health Organization

M. S. Islam, *Hydrogeochemical Evaluation and Groundwater Quality*,
https://doi.org/10.1007/978-3-031-44304-6_5

5.1 Occurrence of Trace Metals in Groundwater

At present, groundwater quality and quantity have radically deteriorated worldwide due to overexploitation, overexploitation of water, excess use of agrochemicals, vast industrialization, and urbanization. Besides, frequent changes in geogenic courses and some anthropogenetic causes have greatly deteriorated the groundwater quality regarding trace and heavy metal contamination in some areas of the country throughout the last three to four decades. Thus, the proper evaluation of sources and mobilization mechanisms of harmful trace metals such as Fe, Mn, Co, Cd, Pd, Cr, Hg, As, Zn, Cu, etc. (at a certain concentration), in aquifer basements is very important to ensure safe water for all-purpose uses. Groundwater contamination through trace elements is a global problem and a serious threat to public health, the natural ecosystem, food resources, and facilities. Due to the consumption of groundwater, the presence of toxic elements in human physiology can cause numerous problems. The trace elements in small amounts in the body may preserve cells, though the presence of more than the threshold limit of these elements may cause damage to biotic organisms. Trace metal analysis, distribution, and source identification in groundwater samples of the Bengal delta (Ganges basin) are the major descriptions in this chapter. Some comparable studies among other parts of the world and the present investigation are also discussed.

Trace metals have the properties of ecosystem perseverance and bioaccumulation, and these metals pass in the aquatic system through several routes. These elements can be found on the Earth's crust in their regular form. Some of them are very toxic, and they harm not only the quality of the aquatic environment but also human well-being. These metals are so hazardous that they cannot be degraded or decomposed in ecosystems, and they can bioaccumulate. Trace elements once enter the environment through the air, drinking water, or numerous chemicals and products that are man-made. The toxicity of these elements in the human body system decreases energy levels; disturbs brain functioning; disrupts the operation of numerous other organs, such as the kidney, lungs, brain, and liver; and hampers blood composition.

Trace metals have two major sources in groundwater: (a) natural or geogenic and (b) anthropogenic or man-made sources. Natural sources include the dissolution of parent rocks at the aquifer level. At first, these metals or their minerals accumulate in the soil layer, which is leached with rain or flood water and finally reaches the subsurface water layer. The sources of trace metals in groundwater through human activities are well defined, but natural or geogenic activities are not easy to identify. These metals are found in high levels in the environment; they are formed by natural processes such as volcanic eruptions, dissolution of rocks/minerals in the water phase, and discharge into rivers, lakes, and oceans owing to the action of water. Loads of trace metals in groundwater depend on the groundwater's local lithology or geology, hydrogeology, and geochemistry. One of the chief geogenic sources of metal contamination is the dissolution of metal-laden rocks in groundwater. The dissolution of sedimentary rocks such as calcite, dolomite, shale, or mixed rock pollutes the water through high levels of toxic metals. When the interaction of water with rock elements occurs, it leads to the addition of these metals into the water phase; thus, pollution arises. Examples of such rocks are basalt, gabbro, granite, siderite, calcite, cuprite, calamine, azurite, smithsonite, malachite, chromite, arsenic trioxide, orpiment, kaolinite, montmorillonite, arsenopyrite, pyrolusite, nepheline, andesite, etc. [1] Sulphide ore deposition also increases because it is connected with the mineralization of gold and hydrous iron oxide ($Fe_xO_y.nH_2O$) rocks. In this investigation, to evaluate factors affecting the level of trace metals in groundwater and identify their possible sources and mineralization processes, multivariate statistical techniques, including Pearson's correlation coefficient, principal and robust component analysis, and dendrogram cluster analysis, are used.

5.2 Trace Metal Distributions in Groundwater Samples

Trace metals are a widespread limit of contamination (mg/L) and continue to be a human health concern due to their toxicity dimensions even in low concentrations and can show an opposite effect on living existence and a tendency to bioaccumulate and bio-magnify in lipids/fat and tissues of biotics over time. Metals such as Cr, Co, Cd, Pb, Hg, and As have no essential function in the human body. Furthermore, long-term exposure may cause acute disruptions in the normal operations of the human organ systems where those metals are deposited. Several trace metals, such as Fe, Mn, Cu, and Zn, act as micronutrients and are required in the human body in limited quantities for metabolic actions, but at a higher level, they can cause opposite health effects. The main human-caused sources of trace metals in groundwater are natural substances percolated into the soil system or rocks, the residue of agrochemicals, precise release from the sewage treatment plant and industrial runoff, and uncontrolled releases or escape from landfill spots and chemical accidents or disasters. Groundwater is contaminated with excess trace metals, particularly arsenic, and has become an alarming situation in Bangladesh. A total of 11 trace metals were investigated during three sampling seasons, and the obtained results are stated in Table 5.1 and Appendices V–VII. The average metal concentrations of the three sampling seasons are shown in Fig. 5.1. The results presented that the metal concentrations were usually obtained in the order of the PRM<MON<POM sampling periods. The standard deviation (±SD) of all data is found to be very poor, and a low standard deviation means data are clustered around the mean value. On the other hand, the skewness values showed that the data were normally distributed. The measured trace metal ion concentrations are uses for the identification of possible sources of metal ions in aquifer basement and the calculation of health risk assessment of drinking water.

Chromium (Cr)

Chromium (a form of Cr^{3+} and Cr^{6+}) is a naturally occurring trace metal that is found in very trace levels in groundwater and is not prejudiced by point-source pollution. The major sources of chromium are the discharge from tannery industries and landfills or other solid waste in Bangladesh. Chromium (VI) effortlessly arrives at

Table 5.1 Statistical analysis of trace metals in the groundwater samples of the pre-monsoon (PRM), monsoon (MON), and post-monsoon (POM) sampling seasons in 2019–2020 and 2020–2021 (also presented in Appendices V–VII)

	B	Fe	Mn	Cr	Pb	Co	Ni	Cd	As	Cu	Zn
Pre-monsoon (PRM)											
Min.	0	1.14	0.66	0	0	0	0	0	0	0.03	0.79
Max.	1.25	12.62	4.84	0.088	0.096	0.088	1.095	0.051	0.019	3.11	6.63
Mean	**0.632**	**7.998**	**2.726**	**0.044**	**0.036**	**0.044**	**0.305**	**0.008**	**0.008**	**1.593**	**3.945**
SD (±)	0.394	3.156	1.227	0.024	0.034	0.023	0.351	0.011	0.006	0.831	1.618
Skewness	−0.416	−0.438	0.115	−0.397	0.482	−0.476	1.132	2.131	0.077	−0.108	−0.273
Monsoon (MON)											
Min.	0	1.87	0.46	0	0	0	0	0	0	0.14	0.79
Max.	1.235	13.73	5.66	0.088	0.092	0.097	1.185	0.042	0.019	3.32	6.72
Mean	**0.662**	**8.522**	**2.873**	**0.049**	**0.039**	**0.046**	**0.326**	**0.008**	**0.009**	**1.733**	**4.048**
SD (±)	0.395	3.273	1.373	0.023	0.031	0.025	0.358	0.011	0.007	0.865	1.584
Skewness	−0.374	−0.308	0.143	−0.665	0.328	−0.362	1.168	1.979	−0.089	−0.091	−0.365
Post-monsoon (POM)											
Min.	0	1.985	0.540	0	0	0	0	0	0	0.090	0.670
Max.	1.450	14.440	5.605	0.095	0.099	0.093	1.190	0.045	0.022	3.370	6.770
Mean	**0.693**	**8.784**	**2.945**	**0.052**	**0.042**	**0.048**	**0.330**	**0.008**	**0.009**	**1.752**	**4.086**
SD (±)	0.419	3.3503	1.360	0.023	0.031	0.024	0.355	0.012	0.007	0.880	1.607
Skewness	−0.170	−0.443	0.09	−0.550	0.269	−0.570	1.145	2.008	0	−0.110	−0.410

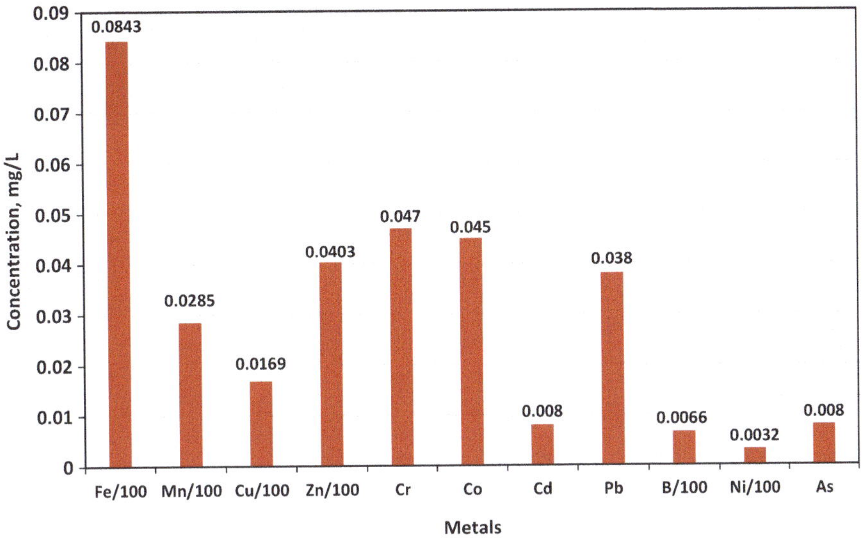

Fig. 5.1 Average metal concentrations (mg/L) of three sampling seasons. The concentrations of Fe, Mn, Cu, Zn, B, and Ni are denoted by 100 × mg/L

cell membranes, and Cr(III) does not. In the human body, the maximum amounts of chromium accumulate in the lymph nodes, liver, lungs, kidneys, and spleen, and enduring exposure can damage the kidneys and liver. BDWS, WHO, and US-EPA recommended the tolerable highest value of chromium in drinking water of 0.05 mg/L (Table 5.1). The average concentration of Cr in the samples of three sampling periods is stated in Table 9. The analysed results displayed that approximately 50% of the samples contained more overloaded chromium than the threshold limits (Table 5.2).

Iron (Fe)

Iron is a serious issue in rural drinking water in the maximum regions in the world, including Bangladesh. It is the third greatest abundant constituent in the Earth's crust, and it remains as ferrous (Fe^{2+}) and ferric (Fe^{3+}) oxides, sulphides, hydroxides, and carbonates with some mixed rocks. Though a low amount of iron is essential in the human body and plant metabolism and cannot cause much impairment, it inspires offensive bacterial growth (iron bacteria) inside a waterworks and supply system, resulting in the deposition of a slushy coating on the piping system. Moreover, high iron content (over 0.3 mg/L) leads to a surplus that can cause stomach problems, nausea, vomiting tendency, diabetes, and hemochromatosis in the human body. Numerous studies have shown that the iron concentration in the groundwater of Bangladesh is very high. The groundwater of the maximum areas of the country exceeded the permissible concentration limit of about six (6) mg/L of Fe [4]. The Bangladesh National Drinking Water Quality Standard (BNDWQS)

Table 5.2 Drinking and irrigation water quality standards

Parameters	Standard values for drinking water				Standard values for irrigational water		
	BDWS[a]	*WHO*[b]	*US-EPA*[c]	*INDIA*[d]	*BIWS*[e]	*US-EPA*[f]	*FAO*[g]
B (mg/L)	1.00	–	–	0.01	<1.00	0.75	0–2.00
Mn (mg/L)	0.10	0.50	–	0.10	–	0.20	0.20
Fe (mg/L)	0.3–1.0	0.30	0.30	0.30	–	5.00	5.00
Ni (mg/L)	0.10	0.02	–	–	0.50	0.20	0.20
Co (mg/L)	0.05	–	–	–	–	0.05	0.05
Cr(6) (mg/L)	0.05	0.05	0.10	0.05	0.01	0.10	0.10
Cd (mg/L)	0.005	0.03	0.005	0.01	0.01	0.01	0.01
Pb (mg/L)	0.05	0.01	0.015	0.05	0.10	5.00	5.00
As (mg/L)	0.05	0.01	0.01	0.05	1.00	0.10	0.10
Cu (mg/L)	1.00	2.00	1.30	0.05	0.20	0.20	0.20
Zn (mg/L)	5.00	3.00	–	–	–	2.00	2.00

[a]Department of Public Health and Engineering, Bangladesh (2017)
[b]HWO-Drinking water standard, 4th ed. (2011)
[c]US-EPA-Drinking water standard (2018)
[d]Drinking water standard for India (IS10500, 2012)
[e]Bangladesh irrigation water standard (2009)
[f]US-EPA-Guidelines for water reuse (2004)
[g]FAO-Water quality for agriculture (1985)

illustrates that approximately 50% of underground water carries excess quantities of iron, and the study also presented the high average level of Fe present in the deep tube well (1.37 mg/L) and shallow (2.65 mg/L) water all over the country. The study results revealed that an average of 100% of water samples carried an excess level of Fe in all sampling periods (Table 5.3). The mean iron concentration of the three sampling periods was found to be 8.435 mg/L, which is 28 times greater than the permissible limit. This mean value is much higher than that of other alluvium aquifers in the world (Table 5.4).

Manganese (Mn)

Manganese is an element essential to human, animal, and plant metabolism. It is compulsory for the function of numerous cellular enzymes. It can activate hydrolases, transferases, kinases, and decarboxylases. However, excess ingestion (>1.8 mg/L) of Mn-loaded water could show neural symptoms that are similar to Parkinson's disease. There is no primary recommendation for Mn in drinking water supplies set by the US-EPA and WHO because they have no serious health threats. While, the secondary thrilling contaminant level of 0.5 mg/L was observed for Mn because higher levels yield offensive taste, colour, odour, staining, and corrosion. Table 5.1 displays that the concentrations of Mn are 2.726, 2.873, and 2.945 mg/L in the PRM, MON, and POM, respectively. The study results exposed that 100% of the water samples carried an excess level of Mn in all sampling periods (Table 5.3). So, Mn-contaminated water could pose a serious threat to public health and irrigation in the study area.

Table 5.3 Statistics of trace metal concentrations in groundwater during the PRM and POM sampling rounds

Metal	Pre-monsoon, PRM ($n = 40$)		Monsoon, (MON) ($n = 40$)		Post-monsoon, POM ($n = 40$)		Mean value of three seasons	Average % samples exceed the WHO std.	The standard value for drinking water		
	Mean	*% Samples exceed the WHO std.*	*Mean*	*% Samples exceed the WHO std.*	*Mean*	*% Samples exceed the WHO std.*			WHO (2011)	USEPA (2011)	DPHE[a] (2017)
B	0.632	0	0.662	0	0.693	0	0.662	0.0	2.4	None	1.0
Fe	7.998	100	8.522	100	8.784	100	8.435	**100.0**	0.3	0.3	0.3–1.0
Mn	2.726	100	2.873	100	2.945	100	2.848	**100.0**	0.4	0.05	0.1
Cr	0.044	52.5	0.049	47.5	0.052	37.5	0.048	45.8	0.05	0.1	0.05
Pb	0.036	60	0.039	70	0.042	80	0.039	**70.0**	0.01	0.015	0.05
Co	0.044	42.5	0.046	47.5	0.048	55	0.046	48.3	0.05	None	None
Ni	0.305	50	0.325	55	0.330	55	0.320	53.3	0.1	0.1	0.1
Cd	0.008	47.5	0.008	50	0.008	47.5	0.008	48.3	0.003	0.005	0.003
Cu	1.593	32.5	1.733	35	1.752	35	1.693	34.2	2.0	1.3	1.0
Zn	3.945	72.5	4.048	72.5	4.086	67.5	4.026	**70.8**	3.0	5.0	5.0
As	0.008	47.5	0.009	50	0.009	52.5	0.009	50.0	0.01	0.01	0.05

*All parameter units: mg/L

[a]Bangladesh guideline

Table 5.4 Trace metal concentration in groundwater of different regions

Region	Fe	Mn	Co	Cd	Cr	Pb	As	Cu	Zn
Andhra Pradesh, India [5]	1.153	0.1065	–	0.0832	0.0543	0.0610	<0.001	0.501	0.307
Bafoussam, Cameroon [6]	0.026	0.032	0.0023	0.00017	0.0028	–	0.0001	0.0547	0.1618
Rangpur, Bangladesh [7]	7.725	0.684	–	–	–	–	0.0088	–	0.0333
Tripura, India [8]	1.36	0.10	–	0.010	0.010	0.020	0.011	0.010	0.020
Uganda [9]	0.57	0.34	–	0.065	0.007	1.865	–	2.141	16.08
S. China [10]	0.512	0.056	–	0.0001	–	0.005	0.001	0.012	0.0015
Zamora, Mexico [11]	0.084	0.021	0.0005	–	–	–	0.002	0.016	–

Nickel (Ni)
Nickel typically occurs in the hydrated form [Ni $(H_2O)_6^{2+}$] in natural freshwater at pH 5–9. The primary cause of Ni in potable water is the dissolution of metals in contact with water, such as fittings and piping. A small quantity of Ni is harmless and required for human health. Ni contamination in water can cause serious health effects, including lung cancer, nasal sinus, chronic bronchitis, and lung function disorder. According to the US-EPA and WHO guidelines, the level of nickel should not exceed 0.1 mg/L in drinking water. The study presented that the usual concentration of Ni in water samples for three periods was 0.32 mg/L, and around half of the samples exceeded the permissible limits (Table 5.3).

Cadmium (Cd)
Cadmium is a highly toxic heavy metal with a very extended half-life and short LD_{50}, and it exists naturally mainly with zinc-bearing rock. This metal can be released into groundwater from supressed wastes containing metal refinery by-products and electronic goods and by the coal-burning station. It can consume by eating vegetables grown in contaminated soil and fish or other seafood from contaminated or drinking water holds cadmium. Severe exposure can cause anaemia, bone marrow disorders, nausea, cancer, diarrhoea, kidney failure, muscle cramps, and liver injury in the human body. In Bangladesh, the mean concentrations of Cd were 0.015–0.034 mg/L in groundwater, and the values were five to ten times higher than the acceptable value (0.003 mg/L). While the samples contained a mean value of 0.008 mg/L, about 50% of the sample exceeded the standard tolerable range of Cd concentration (Table 5.3).

Lead (Pb)
Lead is another universal toxic heavy metal and an extensive public health concern worldwide. It can cause diverse biochemical belongings when exposed to it for a comparatively short time. These effects may include increases in the blood pressure of adult persons, interference with red blood cell chemistry, interruptions in typical physical and mental development in a child, hearing and learning capacities of an infant, shortage in attention span, cancer, kidney disease, and stroke. The highest permissible concentrations of Pb in drinking water recognized by the BDWS and WHO are 0.05 and 0.01 mg/L, respectively. The LD_{50} value of Pb is 4665 mg/kg of body weight for males and 5610 mg/kg of body weight for females [2]. The average level of Pb in the water samples in the three sampling periods was 0.039 mg/L (Table 5.1). Besides, the results presented that approximately 70% of the samples crossed the maximum permissible range (Table 5.3).

Copper (Cu)
Copper is an essential component for creatures and plants and plays a substantial role in metabolism of human body. It is extensively used in plumbing pipes and fittings and may dissolve from water supply pipes in an acidic environment. Copper sulphate $(CuSO_4)$ salt is infrequently mixed in water tanks to subdue undesirable algal bloom,

and copper compounds have been utilized in large volumes in agrarian insect killer sprays. Provisional exposure to Cu in drinking water can lead to gastrointestinal problem, and long-term exposure can lead to Cu toxicosis, which results in kidney and liver function damage, hepatic cirrhosis, anaemia, and worsening of the basal ganglia. An excessive amount of copper in water bodies causes serious harm to fish and other aquatic life. The guideline value of Cu in drinking water is 2 mg/L [3]. Countrywide investigation results exposed that the Cu level in groundwater varies from 0.0002 to 0.39 mg/L and is almost safe (<2 mg/L). In this study, on average, 34.2% of groundwater samples crossed the tolerable ranges of Cu concentration (Table 5.3).

Zinc (Zn)

Zn is a naturally abundant trace metal and a vital nutrient for metabolism and development of human body, mostly for newborns and children. Higher Zn concentrations in drinking water can lead to stomach pains, vomiting, and neurological problems. Chronic exposure to Zn is accountable for iron shortages, low copper ingesting, and lower values of high-density lipoprotein (HDL) cholesterol. Water with a Zn level of higher than 3 mg/L tends to be opalescent, forms an oily film when boiled in the water, and has an objectionable loathful taste. Maximum collected groundwater samples from numerous areas of Bangladesh exhibited that the concentrations of Zn (0.01 to 0.941 mg/L) were much lower than the highest satisfactory ranges. Nonetheless, the average level of Zn in groundwater in the study zone was over 4 mg/L, and 70% of samples crossed the tolerance level of this metal (Table 5.3).

Arsenic (As)

Arsenic, a first-category carcinogenic material, occurs naturally in groundwater stores throughout maximum portions of Southeast Asia. In the Ganges delta of Bangladesh portion and northern part of India, severe contamination of groundwater by geogenic occurring As. It affects approximately 25% of shallow hand tube wells in the two local aquifers. It was first documented in the 1990s in Bangladesh. The concentration level generally crossed the maximum threshold level of 0.01 mg/l recommended by the WHO. Over 75 million people from 59 out of a total of 64 districts were heavily at risk of drinking As-contaminated water in Bangladesh [4]. Each year, an estimated 43,000 people are occupied by As poisoning in the country. The study exposed that the groundwater samples in the study area, Kushtia District, an upper Ganges flood plain, are free from arsenic toxicity. Same as arsenic, the boron concentration (mean: 0.662 mg/L) in the groundwater of the study areas was also below the toxic level (2.4 mg/L) (Tables 5.1 and 5.3).

5.3 Trace Metal Concentration in Groundwater: *World Scenarios*

Trace metal concentrations vary in groundwater with several factors. Groundwater age is one of the issues that can affect the level of trace elements in groundwater. Other factors include geological settings, climate change, and human actions. Climate plays an important role because in areas where rainfall is low and the degree of evaporation is high, there is less water to dilute the products of rock weathering. Geological features play a role because the metals accessible for leakage into groundwater rely on the kinds of minerals present in the sediment and rocks. Last, human activities such as heavy irrigation and pumping can affect levels of trace metals in groundwater, frequently by shifting the geochemical situations, such as pH and redox environments, inside the aquifer. So, the trace elements in groundwater were obtained in different concentrations in different regions of the world. Some results concerning that are stated in Table 5.4.

5.4 Sources and Dissolution of Trace Metals

Groundwater age is just one of the factors that can influence the load of trace metals. Other factors include climate, human actions, and geological formations. Climate plays a role because, in regions where rainfall rates are low and evaporation rates are high, there is less water to dilute the products of rock weathering. Local geology plays an important role because the metals available for discharge into groundwater are subject to the kinds of minerals present in the rocks and sediment. Finally, human actions such as irrigation and pumping can affect levels of heavy metals in groundwater, often by varying the geochemical environments, for example, pH and redox conditions, within the aquifer.

The presence of trace metals in groundwater can be due to geochemical, geogenic, or human activities. In natural or geogenic activities, the dissolution of minerals or rocks that contain these metals, soil–water interactions, explosions from volcanoes, deposition of atmospheric salts, and geological dissolution processes can occur. From these occurrences, trace metals enter dissimilar units of the environment. Trace metals can be found in the forms of sulphates, sulphides, oxides, hydroxides, phosphates, and silicates or mixed types. The rapid increase in industrialization and urbanization is an example of man-made sources through which water is polluted by trace metals.

The local geology, agrochemical leaching, overexploitation of groundwater, etc., were the main causes of toxic elements in drinking water. The metal concentration in groundwater sometimes depends on several physiochemical parameters of water samples, such as pH, electrical conductivity (EC), total dissolved solids (TDS), and total hardness (TH). In addition, water depth is another factor that regulates the mineralization of aquifers and solute concentrations. The values of those four parameters in the three different sample seasons, viz. the pre-monsoon (PRM), monsoon (MON), and post-monsoon (POM) periods are shown in Table 5.5. Additionally, the average concentrations of trace metals and some relevant physicochemical parameters, including water depth, pH, EC, TDS, and TH, of the three seasons are stated in Table 5.5. The dataset designated that the maximum parameter values in the MON and POM seasons were higher than those in the PRM season, which may cause percolation leaching during the rainy season in the study areas (Tables 5.5 and 5.6). In this section, the identification of metal pollution sources is performed through the source rocks in the aquifer and using multivariate statistical techniques.

5.4.1 *Source Rock of Metals in the Aquifer*

The bioavailability of metals in groundwater depends on various physical, lithological, chemical, and bacteriological factors in sedimentary aquifers. The accessibility of metals at the sedimentary level is very closely tied to the quantity of sediment-laden trace elements that are dissolved in adjacent pore water. In oxic sediments, several metal elements may be adsorbed on the solid surface of clay particles, and Fe, Mn, and Ni oxides are layered on clay or organic granules [12]. As the amount of O_2 in sediment decreases, characteristically owing to bacteriological degradation, the metal-oxide coatings initiate to dissolve and deliver adsorbed metals. In anoxic sediments, several metals react with sulphides produced by microbial actions to form unsolvable metal sulphides. Metals may be released from sorbets into sediment pore water in ionic forms during changes in pH and oxidation–reduction potential (Eh). For example, in oxic sediment, Cu and Ni are adsorbed on the surface of organic matter (OM) and Fe/MnO and Fe/MnO, respectively. In anoxic sediment, these two metals (Cu and Ni) are adsorbed on the surface of sulphide rocks of Cu and Fe (Cu_2S, CuS, FeCuS, etc.) and OM/NiS, respectively. Additionally, adsorbed metals may be released into pore water by microbial degradation of organic matter. Certain microbes can methylate some metals, such as mercury, arsenic, and lead, to organic types that are more available than inorganic forms. Except for the geogenic mechanism of metal dissolution in groundwater, metal concentrations can increase due to anthropogenic causes. Based on the average metal concentration of different sampling seasons, some statistical methods are used here to determine the corresponding source of each trace metal.

5.4.2 *Multivariate Analysis*

Various statistical and computer programmes were used for the sources and dissolution of trace elements in local aquifer groundwater. In this connection, multivariate analysis (Pearson's correlation matrix, principal component analysis, cluster analysis, etc.), numerous biplots, diagrams, and programmes are used for this purpose and are discussed below:

Table 5.5 pH, EC (μS/cm), TDS (mg/L), and TH (mg/L) values in three different sample seasons

Period	pH range (**mean**)	EC range (**mean**)	TDS range (**mean**)	TH range (**mean**)
Pre-monsoon (PRM)	6.63–8.04 (**7.03**)	367.5–1043.5 (**670.0**)	220.0–667.5 (**413.4**)	121.5–564.0 (**362.8**)
Monsoon (MON)	7.06–8.41 (**7.43**)	560.5–1237.0 (**867.5**)	369.0–815.9 (**558.2**)	157.1–594.4 (**396.1**)
Post-monsoon (POM)	7.14–8.91 (**7.80**)	663.0–1710.0 (**947.1**)	449.0–1108 (**600.6**)	225.7–613.3 (**404.8**)

Table 5.6 The average concentration of analysed trace metals and some selected parameters in the groundwater samples of the three sampling seasons in 2019–2020 and 2020–2021, with basic statistics

	Depth	pH	EC	TDS	TH	B	Fe	Mn	Cr	Pb	Co	Ni	Cd	As	Cu	Zn
Min.	22	7.03	530.3	351.8	216.6	0	1.71	0.55	0	0	0	0	0	0	0.09	0.75
Max.	125	8.22	1305.3	866.8	576.5	1.3	13.3	5.25	0.089	0.095	0.09	1.137	0.045	0.019	3.27	6.71
Mean	**50**	**7.43**	**831.3**	**528.2**	**387.9**	**0.66**	**8.43**	**2.85**	**0.048**	**0.039**	**0.046**	**0.320**	**0.008**	**0.008**	**1.69**	**4.03**
SD(±)	31.9	0.23	172.8	110.4	84.04	0.40	3.23	1.30	0.023	0.032	0.024	0.353	0.011	0.006	0.85	1.60
Skewness	0.96	0.76	0.82	1.05	0.33	−0.34	−0.41	0.12	−0.51	0.38	−0.45	1.15	2.04	−0.08	−0.12	−0.35

(i) *Pearson's Correlation Matrix*

Pearson's correlation matrix was used to determine possible correlations between these trace metals to determine whether either the presence of specific trace metals in a sample facilitates the occurrence of other metals or if they coexist due to anthropogenic or geogenic activity in the study areas. Table 5.7 illustrates the inter-parameter correlation coefficient in groundwater. The intermetal and metal with other relevant parameter relationships deliver valued information on metal sources and flow paths. The correlation analysis displayed strong and significant positive correlations ($r > 0.7$) in some trace element pairs, such as Fe with Mn, Co, Pb, and Zn; Mn with Fe, Pb, Co, Cu, and Zn; Pb with Fe, Mn, Cd, Cu, and Zn; Co with Fe and Mn; Cu with Fe, Mn, Pb, and Zn; and Zn with Fe, Mn, Pb, and Cu. Other metals are not strongly correlated with others. The maximum metals are strongly correlated with Fe and Mn. In the case of other physical parameters, most all-metal concentrations are negatively correlated with water depth, i.e. the concentration of trace metals decreases with aquifer depth. The total hardness (TH) does not correlate with any trace metal, indicating that Fe or any other trace metal is not responsible for water hardness. The EC and TDS are strongly positively associated with the levels of Fe, Mn, Cd, Cu, and Zn. Thus, these metal concentrations depend on salinity and the mineralization rate of the groundwater. The pH is slightly negatively correlated with all metals, which indicates that at lower pH (acidic medium), the metal concentrations are increased by a higher dissolution rate. The dissolution of metal ions in aquifers mostly is subject to water pH. At a low pH value (acidic medium), trace elements tend to dissolve more due to their greater solubility in acidic water. The pH values were acidic in the maximum water samples in the PRM season, indicating that fewer metal ions were present in higher concentrations compared to the POM season (Table 5.5). But the dataset showed that the average metal concentration of the PRM was lower than that of the MON and POM sampling seasons, indicating other geogenic causes rather than the water pH.

Based on the geological formation of the study area, the coarse sandy alluvium lithology and the deltaic floodplain land are thought to be porous and penetrable enough to allow the passage of trace metal-laden water into the aquifer basement. The correlation matrix exposed that most of the trace elements have substantial associations with each other, indicating the similarity of sources. This illustrated that Fe and Mn have an important correlation with all the trace metals and that the presence of extra Fe and Mn in the samples can be attributed to both geogenic and human-caused processes. The geogenic processes that could release Fe and Mn in groundwater include the redox conditions of the aquifer system and the rusting properties of water supply pipes.

Among the 11 analysed heavy metals, the concentrations of Fe, Mn, Pb, Ni, Cu, and Zn in more than 50% of samples exceeded the WHO standard permissible level for drinking purposes (Table 5.3), and they were strongly correlated with each other. Cd and Pb may be from the same contamination source owing to the strong positive correlation among them. Cd occurs naturally with Pb in the same sulphide ores. B, Cd, Cr, As, and Co have weak correlations with each other, and the dissolution

Table 5.7 Pearson's correlation matrix for trace metals and physicochemical parameters of groundwater samples

	Depth	**pH**	**EC**	**TDS**	**TH**	**B**	**Fe**	**Mn**	**Cr**	**Pb**	**Co**	**Ni**	**Cd**	**As**	**Cu**	**Zn**
Depth	1.00															
pH	−0.22	1.00														
EC	−0.08	−0.40	1.00													
TDS	−0.10	−0.44	***0.98***	1.00												
TH	0.11	−0.35	***0.70***	***0.70***	1.00											
B	0.24	−0.15	0.12	0.14	0.18	1.00										
Fe	−0.05	−0.31	**0.57**	**0.53**	0.29	0.17	1.00									
Mn	−0.03	−0.25	**0.60**	**0.55**	0.26	0.15	***0.89***	1.00								
Cr	0.13	−0.24	0.34	0.35	0.29	0.31	0.40	0.32	1.00							
Pb	−0.11	−0.25	0.46	0.43	0.06	0.10	**0.64**	***0.70***	0.39	1.00						
Co	−0.07	−0.02	0.25	0.19	0.08	0.26	**0.64**	***0.75***	0.20	0.37	1.00					
Ni	−0.11	−0.23	0.41	0.40	0.24	0.25	0.35	0.47	0.05	0.46	0.18	1.00				
Cd	−0.05	−0.35	**0.50**	**0.51**	0.32	0.32	0.41	0.46	0.35	**0.50**	0.25	0.24	1.00			
As	0.17	−0.14	0.42	0.38	0.56	0.31	0.41	0.41	0.29	0.25	0.23	0.34	0.26	1.00		
Cu	0.02	−0.30	**0.58**	**0.56**	0.16	0.04	***0.72***	***0.72***	0.28	***0.78***	0.34	0.44	0.38	0.29	1.00	
Zn	0.17	−0.28	**0.66**	**0.63**	0.37	0.17	***0.77***	***0.75***	0.47	**0.57**	0.41	0.29	0.44	0.33	***0.76***	1

Note: Correlation is strongly significant ($r \geq 0.50$) at the $p = 0.01$ level with 95% *CI* (two-tailed, **bold**, and *italic*)

Table 5.8 Principal component analysis of detected water parameters (sorted by size)

	PC1	PC2	PC3	PC4
Mn	**0.869**	–0.370	0.081	0.093
Fe	**0.853**	–0.288	0.159	–0.054
Zn	**0.838**	–0.101	0.081	–0.252
EC	**0.832**	0.283	–0.365	0.020
TDS	**0.807**	0.319	–0.390	–0.007
Cu	**0.792**	–0.320	–0.112	–0.262
Pb	**0.721**	–0.410	–0.051	–0.146
Cd	0.612	0.132	0.008	0.026
As	0.539	0.337	0.208	0.388
Ni	0.529	–0.053	–0.181	0.342
Co	0.527	–0.457	0.333	0.340
Cr	0.512	0.169	0.338	–0.201
pH	–0.454	–0.372	0.037	0.425
TH	0.540	**0.701**	–0.166	0.186
Depth	–0.041	0.343	0.647	–0.429
B	0.297	0.283	0.610	0.355
% of Variance	***42.174***	***11.501***	***9.123***	***6.969***
Cumulative %	***42.174***	***53.675***	***62.798***	***69.767***
Eigenvalue	***6.748***	***1.840***	***1.460***	***1.115***

Note: Bold values indicate strong loading with component number

processes occurred from different sources. The absence of correlations between the metals suggests that the contents of these metals are not controlled by a single factor but rather by a combination of hydrogeochemical support phases. The loading association and cluster characteristics of the metal concentration are discussed in the PCA and cluster analysis.

(ii) *Principal Component Analysis (PCA)*

Principal component analysis (PCA) was utilized to compare the compositional and spatial patterns between the investigated water samples and the recognized latent factors and to find the possible sources of the trace elements in the groundwater. PCA is widely applied to classify the contribution of anthropogenic and natural pollution sources. Factor analysis (Table 5.8 and Fig. 5.2) gives four PCs of R-mode factors that have eigenvalues greater than one. These four factors have a percentage variance of 42.174%, 11.501%, 9.123%, and 6.969% with eigenvalue >1. FC1 shows that Mn (0.869), Fe (0.853), Zn (0.838), Cu (0.792), and Pb (0.721) have high positive factor values (>0.7), indicating that those metals are strongly associated with each other and dissolve from a similar type of source that is possibly exogenous discharge. Although the origin of Fe and Mn is partially attributed to geogenic processes in the correlation analysis, this class is made up of parameters unusual to anthropogenic sources rather than mineral dissolution. Based on the geological setting of the study area, there have been reports of mineral or soil deposits rich in trace metals that could release these potentially toxic metals into groundwater systems. It was assumed that the abovementioned highly associated metals are

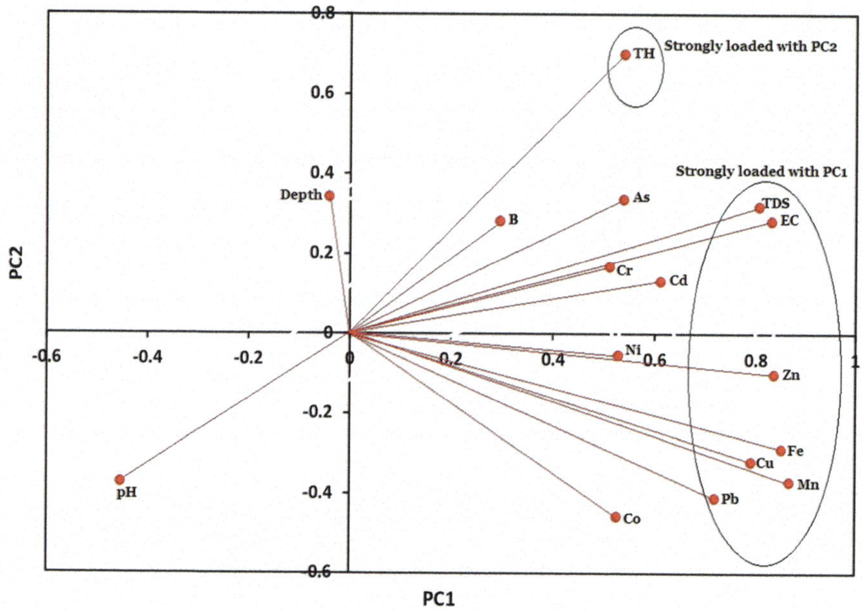

Fig. 5.2 Robust PC of metals and other selected water parameters in the water samples

dissolved with water in aquifer sediment levels from mixed rocks, such as (FeNi)O(OH), (CoFe)AsS, $ZnCrO_4$, $CuFeS_2$, and $(FeZn)6Sb_2S_9$ [1]. Cd, As, Ni, Co, and Cr have moderate factor values in PC1, which indicates that they are moderately associated with each other. In PC2 and PC3, the factor value of the maximum metals is near zero or negative. Only boron (B) has a significant factor value in PC3 (0.610). The negative factor values of water depth and pH indicate a negative association with all metal concentrations, which was supported by Pearson's correlation matrix.

The positive and negative figures in PCA explained that the water samples were affected or unaffected by the presence of extracted loads on a particular constituent. The moderately strong association between EC, TDS, and TH indicates the presence of an enormous ionic form of metal components that is gathered by aquifer rock–water interactions and anthropogenetic sources, such as agrochemicals from agricultural runoff. The findings of the PCA were observed by cluster analysis and TDS vs. metal plot, which is discussed in the next subsections.

(iii) *Cluster Analysis*

The hierarchical cluster analysis method gives three key cluster groups based on a dendrogram using Ward's method for the trace elements with some physicochemical water parameters in the study zone (Fig. 5.3). Parameters that fit in an analogous cluster are likely to have been percolated from the same mineral/rock source. Similar to PCA, cluster analysis places parameters into groups based on the same features and associations with each other. In agglomerative schedule cluster analysis, the

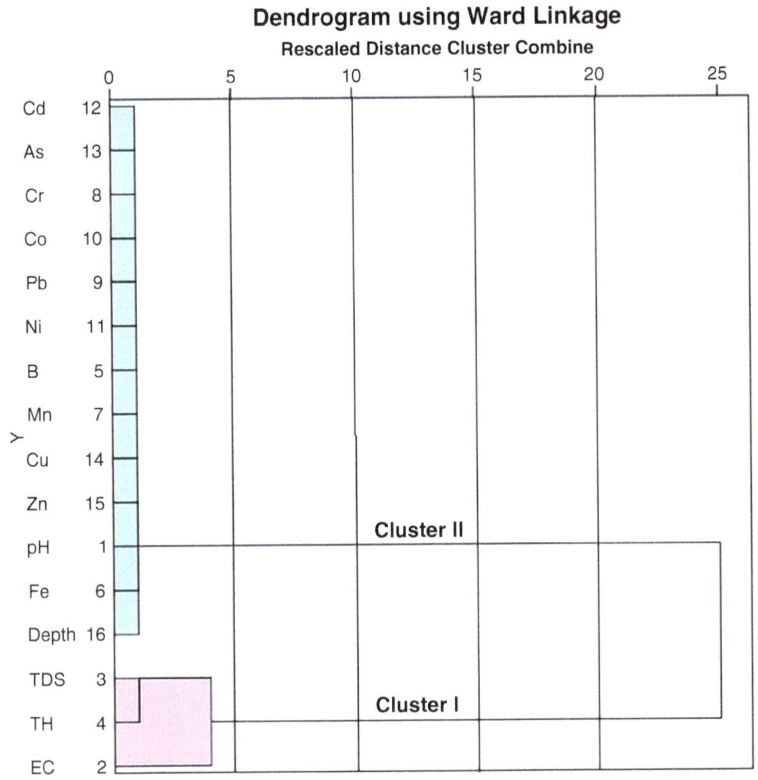

Fig. 5.3 Hierarchical cluster analysis of 40 sampling sites for the average value (mg/L) of trace metals and other selected parameters in three sampling seasons (PRM, MON, and POM). Dendrogram of Q-mode hierarchical cluster analysis (HCA) presenting connotations between samples from different regions of the hydrologic system

maximum comparable variables are mentioned in one cluster and related to a closely accompanying cluster(s) and further from clusters with fewer relatives, all of which are allied to form one large cluster. Cluster 1 is formed by only three variables: TDS, EC, and TH. Cluster II was constructed by all measured metals and well depth. Similar to the findings from PC and correlation analysis, all metal concentrations are dependent on pH and water depth (Cluster II). Low pH and shallow depth are responsible for higher concentrations of metals (mainly Fe, Mn, Pb, Cu, and Zn). The first cluster displays a similar connotation among TH and TDS and proposes the dominion of groundwater by precipitation and the related collaboration with atmospheric CO_2. Figure 5.3 shows that EC, TDS, and TH are not in the same cluster as metals, indicating that the metal loadings are not dependent on other general cations and anions (e.g. Na, K, Ca, Mg, HCO_3, SO_4^{2}, NO_3^{-}, etc.) of groundwater. It may be assumed that those metals are dissolved from rock water independently, and some portion comes from anthropogenic sources such as agrochemical leaching.

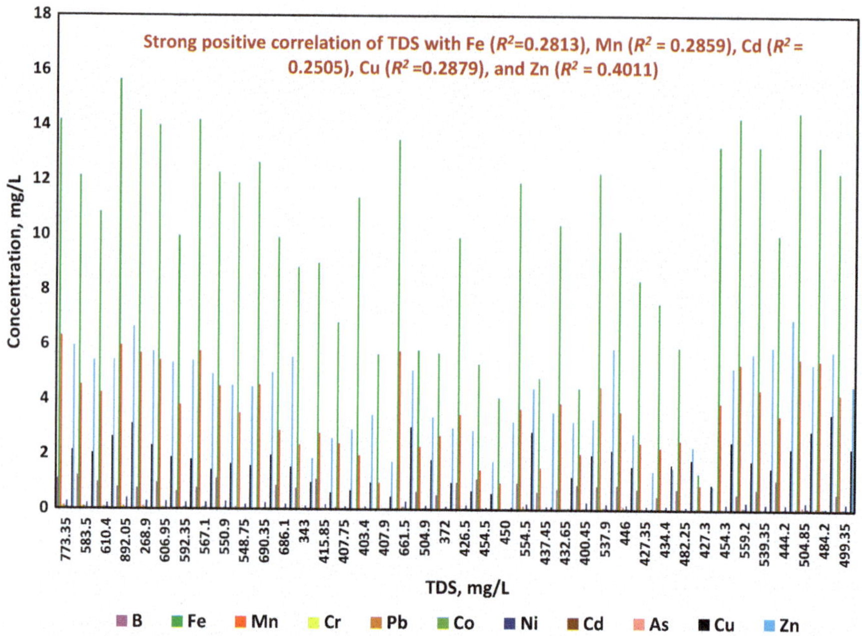

Fig. 5.4 Bivariate diagram of TDS vs. trace metals

5.4.3 Mineralization Process

In aquifer systems, the rock–water interface and flow of solutes are the significant factors controlling groundwater mineralization procedures. A diagram of measured trace metals associated with TDS shows that the geochemical classes took part in the groundwater mineralization (Fig. 5.4). This figure establishes that the analysed trace metal ions, Fe ($r = 0.53$), Mn ($r = 0.55$), Cd ($r = 0.51$), Cu ($r = 0.56$), and Zn ($r = 0.63$), of groundwater are significantly positively correlated with TDS. This result represents the leading trace metals that donate to the groundwater through metal-laden sediment in the aquifer. Water chemistry specified that Pb, Fe, Mn, Cu, and Zn are the most abundant trace metals in groundwater samples of the study area, and these results suggested the continuous addition of these ions along the groundwater movement pathway, which also suggestively donates to groundwater mineralization. The Fe and Mn concentrations crossed the guideline value in the 100% samples (see Table 5.3). These two ions may be invented from similar sources. Thus, dissolution or weathering is the natural geochemical process controlled by the salt contents in the groundwater movement route.

The four statistical methods mentioned above, viz. correlation matrix, PCA, cluster analysis, and bivariate regression correlation, are used for the identification of trace metal pollution sources. Pearson's matrix revealed that all the metals were moderately or weakly correlated in the acidic medium of the water samples. The study area is an agricultural zone and is vastly irrigated using shallow groundwater. No, an industrial plant is located in the study area or adjacent area. Therefore, among anthropogenic sources, only agrochemical leaching is the cause of metal sources in groundwater. Besides, several studies have assumed that trace metal loading in aquifers occurs by local geogenic activities that are positively influenced by heavy water mining. Therefore, except for some anthropogenic sources, the geogenic source is an important factor in increasing the trace metal level in aquifers and was supported by the abovementioned statistical analysis.

5.5 Sources and Dissolution of Fe and Mn

The groundwater of almost all regions of the country contains elevated levels of iron and manganese. The study results show that the average concentrations of Fe and Mn in 100% of samples crossed the standard values of both metals for domestic and irrigation uses. The origin and dissolution of both metals from minerals/rocks into the groundwater are evaluated in this section through the concentration levels of some water parameters, multivariate statistical analysis, biplots, computer programmes, and previous research findings. The findings of the spatial and temporal (PRM and POM seasons) variations are presented here.

5.5.1 Water Variables with Fe and Mn Concentrations

In different sampling periods, the elementary statistics of designated groundwater parameter values in the study zone are shown in Table 5.9 and Appendices VIII and IX. In this present study, Fe concentrations were gated in samples that vary from 1.14 to 12.62 mg/L (mean: 7.998 $\pm$ 3.156) in the pre-monsoon (PRM) and 1.985 to 14.44 mg/L (mean: 8.784 $\pm$ 3.36) in the post-monsoon (POM). In addition, the Mn levels in the PRM and POM were measured to be 0.66 to 4.84 mg/L (average: 2.726 $\pm$ 1.227) and 0.96 to 6.58 mg/L (average: 3.76 $\pm$ 1.54), respectively. The values of Fe and Mn in all samples were higher than the national and international recommendation values for domestic and irrigation water (Table 5.2). Periodical variations in the concentrations of the two metals were found. For the evolution of Fe and Mn dissolution mechanisms and identification of affecting factors on the concentration of these two metals at the aquifer level, some selected water parameters are investigated and included in Table 5.9. All the water parameter concentrations in the POM season were higher than those in the PRM except for the dissolved oxygen (DO) and $\log_{10}pCO_2$.

The pH of the groundwater samples in PRM period was found to be close to the neutral range (pH: 7.02 $\pm$ 0.22), but the water was somewhat alkaline (pH: 7.02 $\pm$ 0.22) in the POM period. Both conductivity (EC) and total hardness (TH) are recorded as higher values during the POM period than the PRM period. In

Table 5.9 Statistical data of the Fe and Mn with some other parameters in groundwater of the pre-monsoon (PRM) and post-monsoon (POM) sampling periods

	Turb.	pH	EC	TH	Fe	Mn	HCO_3^-	SO_4^{2-}	NO_3^-	DO	DOC	$\log_{10} pCO_2$
Pre-monsoon (PRM), n = 40												
Min.	1.93	6.63	367.5	121.5	1.14	0.66	270.8	2.77	0.81	0.68	0.61	−2.458
Max.	19.56	8.04	1043.5	564.0	12.62	4.84	7038.2	45.74	13.86	4.35	13.60	−0.884
Mean	6.26	7.03	670.0	362.8	7.998	2.726	419.6	16.42	4.11	2.13	4.46	−1.952
SD (±)	4.37	0.23	172.3	93.66	3.156	1.227	110.7	9.13	3.69	1.00	3.27	0.260
Skewness	1.57	1.91	0.39	−0.07	−0.438	0.115	1.25	1.22	1.06	0.49	0.39	−0.933
Post-monsoon (POM), n = 40												
Min.	2.29	7.14	663	225.7	1.985	0.540	249.1	2.95	0.80	0.81	0.60	−3.514
Max.	26.09	8.91	1710	613.3	14.44	5.605	815.6	41.71	18.34	4.02	13.71	−1.260
Mean	7.85	7.80	947.1	404.8	8.784	2.945	448.8	15.10	3.70	2.07	5.71	−2.272
SD (±)	5.06	0.40	220.0	93.8	3.36	1.360	124.5	9.24	3.66	0.95	3.22	0.450
Skewness	1.66	0.07	0.93	0.60	−0.443	−0.09	0.60	1.34	2.01	0.31	0.38	−0.009

Note: (a) All parameter units are in mg/L except turbidity in NTU, EC in μS/cm, and pH
(a) *n* number of samples, *Turb.* turbidity, *EC* electrical conductivity, *DO* dissolved oxygen, *DOC* dissolved organic carbon, *TH* total hardness

general, the values of these two parameters crossed national standard guidelines [13]. In addition to other parameters such as HCO_3^-, SO_4^{2-}, and NO_3^-, concentrations of HCO_3^- were in the higher range in both seasons. In this study, low dissolved oxygen (DO) and high dissolved organic carbon (DOC) were substantial groundwater features. The aquifer CO_2 partial pressure ($\log_{10} pCO_2$) was found to be comparatively high compared to other geographic locations. Groundwater $\log_{10} pCO_2$ values were higher in the POM than during the PRM period owing to strong infiltration of rainwater during the monsoon season (Table 5.9). Turbidity is an important indicator of the amount of suspended solids in water that make it cloudy or opaque. Good-quality water should have a turbidity of less than 5 NTU (nephelometric turbidity units), but studies have found higher turbidity values in samples from both sampling periods. This value for the POM sample (7.85 ± 5.06 NTU) was higher than that for the PRM season (6.26 ± 4.37 NTU). Normally distributed data were used for multivariate statistical analysis. The skewness value of the most parameter ranges from 0 to +2 and exhibited a normal distribution.

5.5.2 *Explorative Statistical Approach:* Multivariate Analysis

The analysis of multivariate statistical methods usually uses the distributed data. The data for maximum water parameters has the positive skewness value (Table 5.8). In statistical practices and probability model, skewness is a measure of the asymmetry of the probability distribution of a random variable about its mean. A value between 0.5 and 1 indicated that the data were moderately skewed. If the skewness is less than -1 or larger than 1, the data are very weakly skewed. This indicated that the data were not normally distributed.

Correlation study and factor analysis are comprised in multivariate statistical techniques. Table 5.10 displays the correlation matrix between Fe, Mn, As, and other influencing factors. A small-time variation is observed in the correlation matrix table between PRM and POM seasonal variables. Primarily, Fe and Mn are strongly (positively) correlated in both sampling seasons ($r = \sim 0.8$, $p < 0.01$), indicating the same dissolution mechanism of rock sources in groundwater. Fe and Mn had a weakly significant negative correlation ($p < 0.05$) with pH, suggesting that acidic conditions foster the dissolution of these metals. Furthermore, these two metals had moderate to strong positive correlations with EC over both sampling periods. High EC values indicate high salinity in water, and levels of these metals are strongly related to salinity. A strong and significant positive correlation ($p < 0.01$) between Fe and Mn with TH in the POM designates the influence of the inorganic complex. The significant positive correlations of Fe and Mn with turbidity ($r = 0.7$ and $p < 0.01$) indicate the presence of colloidal iron (III) and manganese (IV) particles in the water. Fe and Mn show weakly significant correlations with HCO_3^- in both sampling seasons, highlighting the consequence of organic decomposition on these two metals in groundwater.

Table 5.10 Pearson's correlation matrix of selected water variables in both the pre-monsoon (PRM) and post-monsoon (POM) periods

Post-monsoon (POM)

	Depth	**pH**	**EC**	**TH**	**Fe**	**Mn**	**HCO_3**	**SO_4**	**NO_3**	**DO**	**DOC**	**log*p*CO_2**	**Turb.**
Depth		0.02	*−0.30**	**−0.51****	*−0.39**	*−0.31**	0.23	−0.09	−0.11	0.21	−0.29	−0.09	**−0.41****
pH	0.02		−0.41	−0.44	−0.38	−0.38	*−0.35**	**−0.47****	−0.04	*0.34**	*−0.38**	**−0.87****	−0.30
EC	*−0.40**	−0.12		**0.59****	**0.68****	**0.70****	**0.44****	**0.60****	**0.50****	**−0.56****	**0.69****	**0.47****	**0.65****
TH	**−0.48****	−0.10	**0.58****		**0.43****	**0.49****	**0.77****	**0.53****	−0.02	**−0.43****	**0.49****	**0.58****	*0.32**
Fe	*−0.38**	−0.09	**0.49****	0.20		**0.79****	0.21	*0.32**	0.22	**−0.84****	**0.93****	0.38*	**0.80****
Mn	*−0.34**	−0.11	**0.60****	0.24	**0.81****		*0.36**	0.25	0.25	**−0.81****	**0.85****	**0.42****	**0.75****
As	−0.21	0.08	0.39*	**0.53****	*0.36**	*0.40**	**0.44****	0.26	0.10	*−0.38**	*0.38**	**0.41****	**0.45****
HCO_3	−0.12	−0.07	**0.72****	**0.58****	0.20	*0.34**		**0.48****	0.03	−0.22	*0.33**	**0.57****	0.21
SO_4	−0.09	−0.27	**0.67****	**0.51****	0.25	0.28	**0.48****		0.27	−0.30	*0.40**	**0.54****	*0.31**
NO_3	−0.10	−0.12	0.24	0.15	0.18	0.30	0.03	0.27		−0.21	0.30	0.04	**0.49****
DO	0.19	0.07	*−0.40**	−0.10	**−0.87****	**−0.83****	−0.06	−0.17	−0.18		**−0.88****	*−0.35**	**−0.76****
DOC	−0.26	−0.09	**0.42****	0.20	**0.79****	**0.80****	0.17	0.24	*0.32**	**−0.89****		**0.42****	**0.77****
log*p*CO_2	−0.11	**−0.82****	*0.39**	*0.32**	0.22	0.29	*0.36**	**0.45****	0.17	−0.16	0.22		*0.32**
Turb.	**−0.43****	0.03	*0.40**	0.19	**0.77****	**0.71****	0.18	0.22	**0.43****	**−0.76****	**0.75****	0.09	

Pre-monsoon (PRM)

*Correlation is significant at the $p < 0.05$ level in 95% *CI* (two-tailed, *italic*)
**Correlation is significant at the $p < 0.01$ level in 95% *CI* (two-tailed, **bold**)

Table 5.11 Principal component loadings of the analysed parameters in the study areas (sorted by size)

Parameters	Pre-monsoon (PRM)			Parameters	Post-monsoon (POM)	
	F1	*F2*	*F3*		*F1*	*F2*
DOC	**0.92**	−0.24	−0.04	DOC	**0.93**	0.25
Mn	**0.91**	−0.27	0.02	Mn	**0.92**	0.18
Fe	**0.87**	−0.36	0.05	Fe	**0.92**	0.27
Turbidity	**0.89**	−0.21	0.15	Turbidity	**0.85**	0.33
EC	0.73	0.32	−0.16	EC	**0.81**	−0.04
TH	0.49	**0.81**	−0.14	TH	0.65	−0.55
pH	−0.13	0.08	**0.97**	pH	−0.53	0.55
Eigenvalue	*4.27*	***1.35***	***1.09***	***Eigenvalue***	*4.98*	***1.04***
% Variance	*53.42*	*16.95*	*13.28*	***% Variance***	*62.16*	*14.11*

Note: Bold values denote strong loading with component number

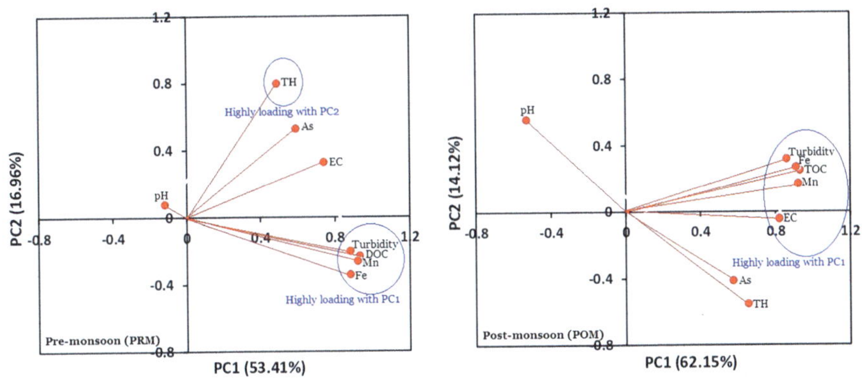

Fig. 5.5 Projection of the factor loading for the different components (PC1 and PC2) for the robust PCA

The oxidizing agents, viz. NO_3^-, SO_4^{2-}, and CO_2, are weakly to moderately and positively accompanying with Fe and Mn. This less substantial association related to DO is perhaps because of the very less favoured electron acceptor after oxygen. But the positive correlations noted among Fe^{2+} and Mn^{2+} with those oxidizing agents imply a reduction process that does not sufficiently encourage Fe and Mn dissolution. Table 5.10 indicated the very high positive and negative relationships ($r > +0.8$ and $r < -0.8$ at $p < 0.01$) of Fe and Mn with DOC and DO, respectively, in both sampling seasons. If the aquifer contains more organic material, the level of DO is consequently low. The roles of those influencing factors on the dissolution procedures of Fe and Mn are discussed in following sections.

The factor analysis exhibited a gross variance of 84.14% with eigenvalue >1 in the PRM period, as determined by three PCs of R-mode (Table 5.11 and Fig. 5.5).

Factor 1 (F1) accounts for 53.41% of the total variation. This factor recommends that with variations in pH and redox environment, the inorganic sources of Fe and Mn dominated and were extremely reliant on organic material loading in groundwater, as shown earlier by the correlation matrix. F2 explains 16.96% of the total variation for the same sampling period. This factor points to the effect of total hardness (TH) in groundwater (Fig. 5.5). This recommends that variations in Ca, Mg, and HCO_3^- levels caused by periodic variation in Fe^{2+} and Mn^{2+} complexation with HCO_3^- release may increase adsorption onto Fe and Mn in groundwater. Factor 3, which accounts for 13.28% of the variation, is the very positive loading of pH. This factor is associated with the pH and redox procedures of Fe and Mn dissolution. The redox potential of these metals is not suggestively correlated with the pH of the aqueous medium.

In the POM sampling period, factor analysis yielded an overall variance of 76.27%, determined by only two PCs in R mode (Table 5.11 and Fig. 5.5). Factor 1 (F1) accounts for 62.15% of the overall variation. This factor recommends that the inorganic iron and manganese sources are very dominant and highly dependent on the organic loading of groundwater as the pH and redox state change significantly after heavy monsoon rains. suggesting. Another factor, F2, explains 14.12% of the total variation. This coefficient indicates a comparatively small effect of pH on the dissolution of Fe and Mn. In addition, this factor presented a negative association between TH and these metals, possibly due to temporal variations.

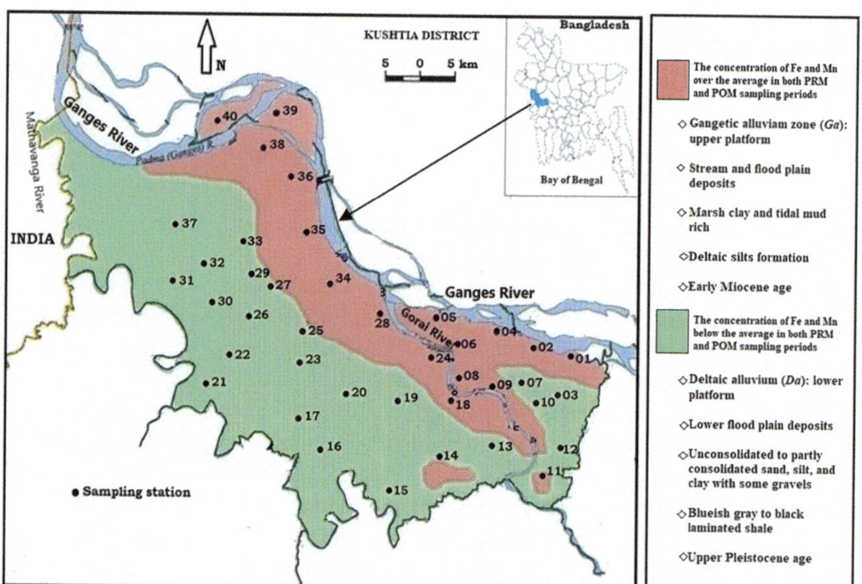

Fig. 5.6 Main lithotypes and soil formations present in the study area and locations of the monitoring points

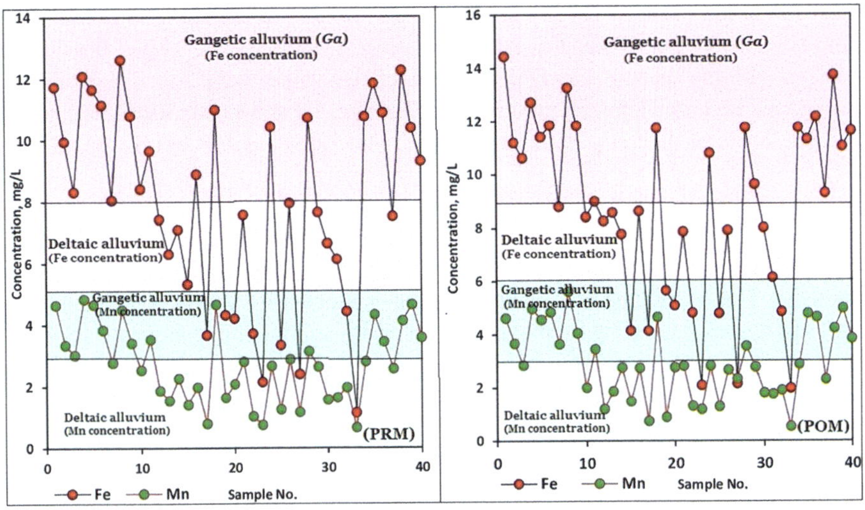

Fig. 5.7 Fe and Mn concentrations in different sampling seasons of the Gangetic alluvial (*Ga*) and Deltaic alluvial (*Da*) platforms

5.5.3 Spatial and Seasonal Distributions of Fe and Mn

Geologically, the study area is divided into two litho-structural platforms, the Gangetic alluviam (*Ga*) platforms and the Deltaic alluviam (*Da*) platforms (Fig. 5.6). The geological and soil foundations of *Da* and *Ga* are significantly different. Both have quaternary strata, but *Ga* is composed of clay with floodplain sediments, fine sandy silts, and silty sediments. It consists mainly of sand, silt, and gravel sediments. The times of *Ga* and *Da* are also different: Holocene (0.0117 ma) and early Pleistocene (>2.6 ma), respectively. The results exposed that both Fe and Mn levels in 95% of the groundwater samples from the *Ga* platform were higher than those from the *Da* platform (Fig. 5.6). The maximum piezometric line (RS2) of 7.6–10.0 m (water table) crossed the *Ga* platform and 5.3–7.5 m crossed the *Da* platform. Therefore, the geomorphology and geology of the study area had a significant impact on the levels of iron and manganese in groundwater. The results of the study also exhibited that the levels of these two metals were slightly dependent on well depth, even if the water samples were taken from the same aquifer.

During the pre-monsoon (PRM) period (March to June), transboundary rivers, the Ganges in Bangladesh and other tributaries, are nearly dry, groundwater recharge from the rivers ceases, and the water table reaches below the surface. In addition, during this dry season, significant water cuts were made due to extensive irrigation of cultivated land in the study area. This situation can affect the water chemistry of the aquifer. However, during the rainy season and post-monsoon period, these scenarios are totally reversed and river flooding occurs in the Ganges basin, with some surface areas flooded with water. During these seasons, large amounts of new water formation occur, accompanied by strong leaching of chemical nutrients from various sources. Therefore, the general water chemistry in the upper floodplain changes with the seasons. Several water parameter values were found to fluctuate during the PRM and POM sampling period, and Fig. 5.7 shows this scenario. The Fe and Mn levels of the Gangetic alluviam (*Ga*) and Deltaic alluviam (*Da*) platforms vary according to the sampling period. Iron and manganese in groundwater samples were recovered in greater amounts on the *Ga* platform than on the *Da* platform during both sampling seasons (Fig. 5.7). Instead, the oxidizing/reducing environment is an important factor that facilitates the dissolution process of these two elements. The alternating redox environment is caused by alternating redox processes in the soil caused by frequent cycles of heavy rainfall inundation and periodic fluctuations in the groundwater table, leading to the formation of Fe–Mn nodules beneath the surface of the soil. Therefore, they are not only the dissimilarity of Fe and Mn dissolution but also another source of seasonal variations in redox potentials.

Dissolved oxygen (DO) is the vital issue in the dissolution of metals at the aquifer basements. However, the recharge of air oxygen in the rainy monsoon is comparatively higher than that in the pre-monsoon dry season. This oxygen-rich water would prevent Fe and Mn from dissolving, and the water pumped from the well would have low concentrations of these elements. After the oxygen was consumed in the recharge water, Fe and Mn dissolved, and the water contained excess Fe and Mn. The study observed higher concentration of Fe and Mn in the post-monsoon period than in the dry pre-monsoon season. This is because the adequately high concentration of DO in the groundwater of the study area does not completely account for the dissolution capability of Fe and Mn. Now, as an alternative of DO, lithological settings are the major cause of elevated concentration of Fe and Mn in the study zone.

5.5.4 Lithological Impacts on Fe and Mn Dissolution

Concentrations of Fe and Mn in groundwater usually vary with geology, groundwater movement outlines, borehole depth, residence time, overburden thickness, and possibly borehole age. Field investigations revealed that colourless groundwater pumped from wells appeared reddish yellow when exposed to air for a short period of time. This makes it possible to confirm that the groundwater is in a reduced state. This reducing environment is accelerated by an amalgamation of fine-grained sediments and organic material. The platform lithofacies in the study area consist of quaternary sediments supporting thick clay layers. In general, Fe and Mn nodules are present in clay layers. The Ganges alluvial (*Ga*) formation consists of river basin loess-like clays (very fine-grained clays), containing abundant Fe and Mn nodules and exhibiting Fe and Mn pollution (Fig. 5.8). The delta alluvium (*Da*), on the other hand, composed mainly of unconsolidated sand, silty sediments, and fine sand with some gravel (Fig. 5.8). The thickness of the clay layer on the platform is obviously different. The clay thickness in this zone differs between about 20 m and 40 m

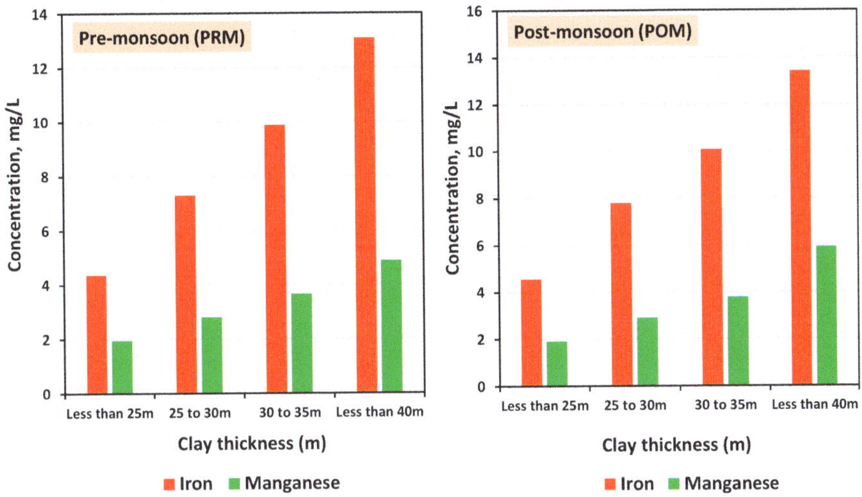

Fig. 5.8 Bar diagram of the clay thickness of sample well and average Fe and Mn concentrations in the PRM and POM sampling periods

[14]. This clay layer is on top of the only sand layer. The presence of dissolved organic carbon in aquifer clays and sub-clays enhances anoxic conditions and facilitates Fe and Mn migration. To demonstrate this point, a diagram showing the relationship between the Fe and Mn content in water samples and the clay thickness of the subsoil of the sampling wells is shown in Fig. 5.8. Clay thickness data estimated from drill hole sampling were collected by the local Bangladesh Agriculture Development Corporation (BADC) office. The clay thickness of the *Ga* platform is thicker than that of the *Da* platform. It was also found that the *Ga* layer has higher Fe and Mn concentrations than the *Da* layer. From this figure, it can be seen that the content of Fe and Mn in the well water increases with increasing clay thickness.

Moreover, the concentrations of Fe and Mn and correlations with each other in PRM and POM periods are quite different. Fe and Mn have different sources on the *Ga* and *Da* platforms. *Ga*-plane deposits are relatively young and contain some organic ingredients. In addition, the lower portion of *Ga* is covered with wetlands for half of the year, adding organic substance to dead vegetation during the dry period. Also, Huang et al. [15] found that most of the organic material originates from dead plants and plant debris in the soil. Thus, the source of Fe and Mn in groundwater of river valleys (*Ga*) is not only clay but also the nature of soils and aquifers' characteristics. The study area has a long history of rice cultivation. Paddy soil is widely used in rice-producing regions. Under conditions of artificial periodic flooding and drainage, paddy soils were subjected to long-term redox fluctuations and experienced a unique series of biochemical changes through several reactions in the aqueous medium [16]. This is the basic formation process of rice soil. Long-term underwater conditions are satisfactory for soil organic substance deposition. With the introduction of organic fertilizers, the organic load on the soil increases steadily. Fe and Mn oxides in the low layer were reduced to low-valence Fe and Mn compounds in water and were attracted to groundwater together with water. In another case, the correlation matrix (see Table 5.10) demonstrates that borehole depth has a negative average correlation with Fe and Mn. A field survey revealed that the average depth of shallow wells in the *Ga* region is shallower than that in the *Da* region. Therefore, water depth is another reason for higher Fe and Mn concentrations.

In the study area, shallow and intermediate alluvium-forming boreholes are rich in Mn and Fe, with low As content. Several studies [10–12] have observed that Fe and Mn loads, dissolution mechanisms, influencing factors, etc., are not the same in different types of aquifers and wells. In the Songnen Plain in north-eastern China, where most wells are gravel wells, the effects of these two elements on groundwater are different from the present study. In particular, in areas with paddy fields and water areas, the concentrations of Fe and Mn in well water are higher than those in other land use areas. The main causes of the high Fe and Mn concentrations in groundwater in this area are Fe and Mn mineral-rich strata and organic matter-rich soils that supply Fe and Mn, as well as in areas with deeper topography and water bodies. These metals are preferably dissolved in groundwater. In addition, the study revealed that pollutant inputs from agriculture increased levels of iron and manganese in groundwater. On the other hand, Fe and Mn loadings in coastal and arid groundwaters are not identical to those in highland groundwaters. In the coastal alluvial plains of Indonesia, Rusydi et al. [17] identified Fe and Mn as natural pollutants of groundwater in the study area. Compounds containing salinity and redox-sensitive parameters of these metals show that brine has a substantial effect on the dissolution of Fe and Mn.

5.5.5 Fe and Mn Relationship and Distribution

For example, locality S08 exhibits significantly higher contents of Fe (14.27 mg/L) joined with higher levels of DOC (13.71 mg/L) and very low DO (0.74 mg/L) in the POM season. On the other hand, in the locality, S01 exhibits higher contents of Mn (4.62 mg/L) coupled with high levels of DOC (11.50 mg/L) and very low DO (0.98 mg/L) in the same period. Therefore, lower redox conditions are very possible. The higher levels are mostly located in the river valley area (Fig. 5.6), where Gangetic deposits and brackish-water agriculture were the dominant geology and land use, respectively. A bivariate plot of Fe and Mn (Fig. 5.9) shows a strong association among their concentrations ($R^2 = 0.8439$ and 0.7941) with a highly significant correlation ($r = 0.81$ and $r = 0.79$ at $p < 0.01$) in different sampling periods. This means that both Fe and Mn behave correspondingly in the studied water samples in terms of solubility and precipitation. This implies homogenous geochemical processes, such as the same content type of minerals in the aquifer

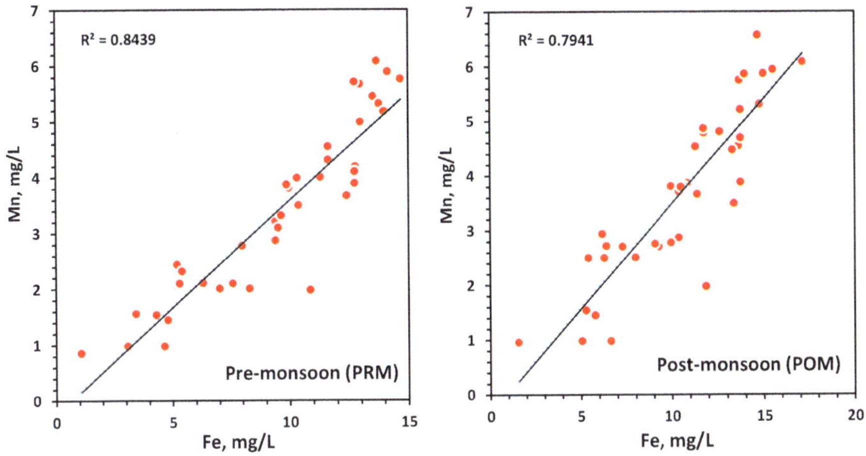

Fig. 5.9 Fe vs. Mn concentration in the pre-monsoon and post-monsoon seasons

systems and the same degrees of mineral dissolution. In addition, Fe is usually present at higher concentrations than Mn in all samples.

5.5.6 *Sources and Dissolution of Fe and Mn*

Multivariate statistical analysis (correlation matrix and factor analysis) showed that the source and mobilization mechanisms of Fe and Mn are just about the same. However, this is not a fact in some different geological settings and places, such as Central Adriatic, Italy [18], and Changchun, north-eastern China [19], in which the dissolution mechanism of these two elements is different. Fe and Mn sources are typically connected to soils, iron–manganese nodules in aquifer clay deposits, and the anthropogenetic sources [20]. Under reducing environments, numerous existing physicochemical parameters of water participated in the dissolution processes of the metal-laden rocks. Reducing conditions at close to neutral pH are known to trigger the weathering of iron oxides and enrichment in groundwater. These redox conditions are elevated by a relatively high content of organic substances and very fine-grained clayey sediments because organic material favours conditions that mobilize Fe and Mn into groundwater, which may be free from the soil or aquifer sediments.

The initial source of Fe in groundwater can be the weathering of iron-rich rocks such as limonite, pyrite, and siderite. Weathering of those rocks in the presence of organic substances, dissolved O_2 and CO_2 gas, and nitrate/sulphate ions at a particular pH results in a rise in the Fe^{2+} concentration in an aqueous medium. In the presence of microbes, this released Fe^{2+} may be converted to Fe^{3+} through the reduction of DO and other anions and create an equilibrium state. Numerous investigations [16, 17, 21] have clarified the dissolution processes of these rocks through quite a few chemical reactions, such as:

$$FeS_2\ (\text{pyrite}) + O_2\ (\text{DO}) + H_2O \rightarrow Fe^{2+} + SO_4^{2-} + H^+$$

$$FeS_2\ (\text{pyrite}) + NO_3^- + CO_2 + H_2O \rightarrow Fe^{2+} + SO_4^{2-} + N_2 + HCO_3^-$$

$$FeCO_3\ (\text{siderite}) + H_2CO_3\ (CO_2 + H_2O) \rightarrow Fe^{2+} + HCO_3^-$$

$$FeOOH.nH_2O\ (\text{limonite}) + H^+ + CO_2 \rightarrow Fe^{2+} + HCO_3^- + e^-$$

Including CO_2, numerous sources of acidity are present in groundwater and affect the Fe and Mn oxidation states. At close to neutral pH (pH = ~7), oxidized Fe^{2+} produces Fe^{3+} oxide/hydroxide solid phases that can function as electron acceptors during the action of Fe^{3+}-reducing bacteria (FRB). Lovley et al. [22] used enormous numbers of *prokaryotic taxa* in the mechanism throughout their metabolism and participated in the oxidation of natural organic substances kept in subsurface conditions. Oxidation of organic matters and the presence of excess H^+ in the aqueous medium comprising Fe^{3+} oxide lead to the liberation of Fe^{2+} into the aqueous phase:

$$Fe(OH)_3 + CH_2O + H^+ \xrightarrow{\text{reducing bacteria (FRB)}} Fe^{2+} + H_2O + HCO_3^-$$

In the presence of the bacterial iron redox cycle is anticipated in numerous redox interfacial environmental situations, such as groundwater iron leakage, the sediment–water interaction in near-neutral pH, and acidic aquatic ecosystems. In the pumping tank, with collective aeration time, both oxide floccules and necrotic iron-reducing microorganisms oxidize the dissolved Fe^{2+} and precipitate as Fe $(OH)_3$. Chemical oxidation of Fe^{2+} to Fe^{3+} happens more gradually at lesser pH values. The complete reaction that defines the formation of insoluble Fe is given as:

$$Fe^{2+}(aq) + O_2(g) + H^+(aq) \rightarrow Fe(OH)_3 \downarrow + H_2O$$

Manganese metal in rocks, soils, and sediments can be found in dissimilar phases (manganese carbonates, chloride, oxides, and silicates), adsorbed on solid iron oxide surfaces, organic composite load, and exchanging forms. Similarly, the reductive weathering of Mn oxides (as MnO_2) by the acetate functional group occurs according to the following overall reaction:

$$CH_3COO^- \text{ (fulvic/humic acid)} + MnO_2(s) + H^+(aq) \rightarrow HCO_3^- (aq) + Mn^{2+}(aq) + H_2O$$

In the case of reverse course, one of the most acceptable methods is that microbes first bend and enzymatically oxidize Mn^{2+} to Mn^{3+}. The formed Mn(III) is converted to Mn(IV), which precipitates as MnO_2 following the equations:

$$Mn^{2+} + O_2 + H^+ \rightarrow Mn^{3+} + H_2O$$

$$Mn^{3+} + O_2 + H_2O \rightarrow MnO_2 \downarrow + H^+$$

The free H^+ ion in the above reaction is reserved to reduce the pH and aids to keep iron in the dissolved state (Fe^{2+}); thus, manganese precipitates out quicker than iron. One more mechanism that can be fitted very well in the present study of the Ganges basin, where the organic matter-rich aquifer is obtained, is that the oxidation of Mn^{2+} may be owing to the microorganisms consuming the carbon from the degradation of organic substances.

5.5.7 *Effect of Water Variables on Fe and Mn Dissolution*

Both iron and manganese participate in oxidation (Fe^{2+} to Fe^{3+} and Mn^{2+} to Mn^{4+}) and reduction (Fe^{3+} to Fe^{2+} and Mn^{4+} to Mn^{2+}) process in groundwater throughout a specific redox environment [17, 19]. Generally, Fe^{3+} and Mn^{4+} keep on the solid phase as oxides and salt accumulates on the sediment in aquifer. Both these elements are weathered from own metal-laden rocks into groundwater through several complicated redox chemical changes in which Fe^{3+} and Mn^{4+} receive electrons from various types of oxidizing agents in the aqueous phase. The order of those redox chemical reactions is described by various measured water parameters, for example, pH, EC, DO, DOC, TH, pCO_2, NO_3^-, SO_4^{2-}, and HCO_3^-. Except for DOC, which has another role in a redox reaction, other chemical water parameters perform as electron acceptors, and physical water parameters control these redox reactions in different ways. The influencing mechanisms of those related parameters on Fe and Mn mobilization in groundwater are discussed below:

(i) *pH and HCO_3^-*

Including CO_2, numerous sources of acidity in water phase are existing in groundwater and influence the oxidation state of Fe and Mn. The influence of pH on Fe and Mn mobilization in aquifer water of the study area is not powerfully significant because of the insignificant connection of both Fe and Mn with water pH (Table 5.10 and Fig. 5.10a). The general inverse correlation designates that Fe and Mn dissolution increases with diminishing pH. This is consistent with other investigations that have found that acidic situations help the dissolution of both metals.

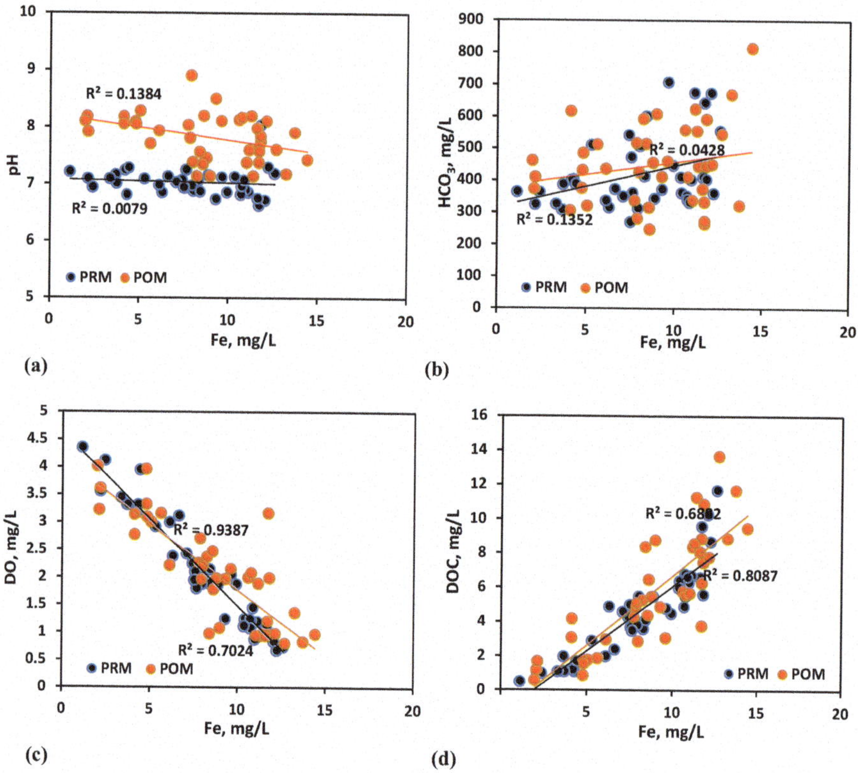

Fig. 5.10 Bivariate diagrams of Fe vs. (**a**) pH, (**b**) alkalinity (HCO_3^-), (**c**) DO, and (**d**) DOC

The rate of Fe dissolution is lesser than that of Mn at a comparatively higher pH value or in an alkaline medium. Meanwhile, HCO_3^- often defines groundwater alkalinity, and a higher concentration of HCO_3^- favours Mn dissolution. Although the average pH value of samples in the POM period was higher than 7 (alkaline nature), the level of Fe was relatively higher than that in the PRM period, in which the pH values were relatively low. This is because water pH is not the only powerful factor that controls the level of Fe in aquifer water.

In the Fe mobilization process, H^+ originated and was used up concurrently, and this process occurred in a somewhat acidic to neutral medium. At low pH value, Fe^{2+} ion produces solid state of Fe^{3+} oxide or hydroxide throughout the action of reducing microorganisms and creates a Fe^{2+}/Fe^{3+} equilibrium. Oxidation of Fe^{2+} to Fe^{3+} happens more gradually at lesser pH values, and the total equilibrium is extremely influenced by pH.

With respect to Fe and Mn dissolution processes, HCO_3^- in groundwater acts in two ways: it decreases the acidity and increases the possibility of metal-complex formation. In this study, the substantial positive associations between HCO_3^- with Fe and Mn may be due to the formation of metal complexes that enhance the metal solubility. Some soluble Fe^{2+} may remain in water as carbonate complexes. In natural waters with carbonate alkalinity >1 mEq/L, Fe^{2+} carbonate complexes such as $FeCO_3$, $Fe(CO_3)_2^{2-}$, and $Fe(CO_3)(OH)^-$ are the predominant forms of Fe^{2+}. In the investigated groundwater samples, the mean concentration of HCO_3^- was about 7 mEq/L, making the formation of metal carbonate complexes likely due to very high alkali loading. Excess bicarbonate does not correlate strongly with dissolved iron (Table 5.10 and Fig. 5.10b). Only a very small quantity of total bicarbonate is used for complex formation.

(ii) *DO and DOC*

Dissolved oxygen (DO) is one of the most delicate substances for the assessment of oxidation–reduction processes. According to McMahon and Chappelle [23] and Khozyem et al. [24], oxic situations may happen when DO values are ≥0.5 mg/L. In judgement, Boyd [25] stated a high opportunity of reductive environments at DO values less than 1 or 2 mg/L. The oxygen concentration of the sample (Table 5.9) is sufficient to provide the oxygen required for oxidation of iron and manganese. Mean DOC levels for the different sampling periods were 4.46 and 5.71 mg/L, above the global groundwater average of 3.8 mg/L. DOC concentrations above 1 mg/L are undesirable in groundwater. High concentrations can lead to dangerous transport of heavy metals. Therefore, DOC played an important role in the large-scale weathering of Fe- and Mn-bearing minerals in the study zone.

Air oxygen is very insoluble in water, only 22 g/L at 25 °C [26]. This oxygen in groundwater comes primarily from leached rainwater. Due to the coarse sandy soil, the hydraulic conductivity of the study area is very high. As a result, oxygen-rich rainwater can easily infiltrate, and the dissolved oxygen (DO) content in groundwater in the study area is satisfactorily high. The low concentration of DO as a preferred electron acceptor suggests that oxygen was used in the decomposition process of organic matter to promote a reducing environment. A reducing environment controls the dissolution of iron and manganese in rocks and minerals. Figure 5.10c shows a very strong negative association between Fe and DO in water samples. A large amount of oxygen was consumed in both the redox reactions $Fe^{2+} \rightleftharpoons Fe^{3+}$ and $Mn^{2+} \rightleftharpoons Mn^{4+}$. Therefore, low oxygen content means high Fe and Mn content in the aqueous medium.

Figure 5.10d shows a very strong association ($R^2 = 0.7988$ and 0.8558 for PRM and POM, respectively) between DOC and groundwater iron concentration. The increased concentration of Fe and Mn in groundwater is not solely due to dissolved free Fe^{2+} and Mn^{2+}. There are various forms of dissolved Fe and Mn in water. In the presence of organic matter (DOC), Fe^{3+} nanoparticles are highly conserved in water, producing suspensions. This unstable solution results from a combination of electrostatic effects (attraction) and steric effects (repulsion) of the negatively charged nanoparticles. This process allows all the iron to bind with flux and humus macromolecules and indeed form soluble complexes. The equilibrium constants for the formation of Fe^{3+} complexes with fulvic and humic groups are much higher than the equivalent constants for Fe^{2+}. When these complexes are formed either by direct contact of Fe^{3+} with organic materials (OM) or by oxidation of pre-existing Fe^{2+}–

OM complexes, Fe^{3+} resists precipitation and contributes to increasing Fe concentration. Like Fe, Mn also forms a kind of complex with an oxidation number of +4. Therefore, higher DOC levels in the sample are used to directly react with Fe and Mn minerals to form colloidal complexes. Formation of colloidal macromolecular complexes is confirmed by sample turbidity values. The average turbidity values of the samples are above normal freshwater standards (approximately 5 NTU) (Table 5.9). The correlation coefficients between turbidity and Fe and Mn were very high ($r = 0.7$–0.8 with $p \sim 0.01$) for both sampling periods (Table 5.10). Sample ID S38 from PRM and S04 from POM showed the highest turbidity values at 19.56 and 26.10 NTU, respectively. These two samples therefore contain higher amounts of iron at 12.26 and 13.72 mg/L. This indicates that some Fe remains as Fe^{3+} colloidal macromolecule complex form.

(iii) NO_3^-, SO_4^{2-}, *and* CO_2

The NO_3^- ion is a stronger oxidant than Fe^{3+} and Mn^{4+} (just behind oxygen), while SO_4^{2-} and CO_2 are weaker oxidants than the others, with the cheapest first, based on that order. will be refunded. This process continues until the electron donors and acceptors are completely consumed. Since NO_3^- is very stable under oxygen conditions, NO_3^- reduction processes may occur in groundwater. Very low concentrations of NO_3^- were found in samples from the study area. Nitrates and sulphates show weak positive correlations with Fe and Mn (Table 5.10), and the reduction itself is unlikely to oxidize Fe^{2+} and Mn^{2+}. Fe and Mn can combine as SO_4^{2-} and Fe and Mn sulphides and can be precipitated by sulphate reduction. Instead, the high SO_4^{2-} concentration resulted in high salinity, which could have facilitated the leaching of Fe and Mn from soil to groundwater through ion exchange processes. High salinity may facilitate the leaching process of Fe and Mn from minerals into groundwater, possibly through ion exchange processes.

The sources of CO_2 in groundwater samples are atmospheric CO_2 leaching, carbonate mineral dissolution, bacteriological oxidation of organic carbon under oxygen conditions, and plant root zone respiration. Carbon dioxide has a very low oxidation efficiency and does not participate in the redox reactions of the metal dissolution process. In general, CO_2 controls the equilibrium state of Fe and Mn in aqueous media through its acid-balancing capacity. Infiltration water plays an active role in rock weathering, especially when saturated with CO_2. In the present study area, $\log_{10}(pCO_2)$ values for groundwater sample ranged from -2.461 to -0.880 in pre-monsoon (PRM) with an average of -1.953 (±0.261) and -3.512 to -1.259 PRM period, which averages -2.272 (±0.446) during the POM period, which is higher than the atmospheric CO_2 partial pressure under equilibrium conditions ($pCO_2 = 10$–3.414 or $\log_{10}(pCO_2) = -3.414$ or 385 ppm) [27]. Therefore, the $\log_{10}(pCO_2)$ value of groundwater is sufficiently high. This saturated CO_2 reacts with water to produce H^+ through an equilibrium reaction, which is used in Fe and Mn dissolution reactions.

(iv) *Salinity and Total Hardness (TH)*

Both correlation and factor analyses show strong positive associations and loadings ($p < 0.01$) of EC with Fe and Mn (Table 5.10). The EC value represents the total salinity of the water and the higher EC values of the samples found in the study area. Possibly, high salinity due to ion exchange processes might support the leaching process of Fe and Mn from soil minerals into groundwater. Due to the salt effect, an increase in EC leads to an increase in ionic strength and a decrease in the activity coefficient, thus facilitating the weathering of Fe and Mn in inorganic complexes. Higher EC values of POM season indicate higher Fe and Mn contents over the same period. Studies suggest that Fe and Mn ions in water also form inorganic complexes with anions such as SO_4^{2-} and HCO_3^-, thereby increasing their concentrations.

Especially in his POM season, total hardness (TH) showed a strong and significant correlation with Fe and Mn in this study (Table 5.10). This is primarily measured by Ca^{2+}, Mg^{2+}, HCO_3^-, and SO_4^{2-} concentrations. Hydrochemical classification of the region indicates that samples from alluvial aquifers frequently exhibit a Ca–HCO_3 class with slightly enriched in Mg and SO_4^{2-}. These facts suggest the influence of HCO_3^--related inorganic complex formation, since water hardness is mainly determined by $CaHCO_3^-$-type hardness.

5.5.8 *Natural and Anthropogenic Impacts on Iron and Manganese Concentrations*

The log-normal graph in Fig. 5.11 helps evaluate the high amount of Fe and Mn distribution related to the aquifer system. In this figure, the transition of the curve's slope distinguishes whether the metals are present in the groundwater under natural conditions or have been impacted by external disturbers. In the graph of log-normal distribution of Fe, samples illustration high probabilities of naturally occurring in disturbed groundwater at concentrations of <9.63 and <10.74 mg/L in the PRM and POM seasons, respectively. In addition, for Mn, these concentrations are <3.44 and <3.76 mg/L in the same seasons (Fig. 5.11). Statistically, nearly half (50%) of the samples are reported in the externally affected aquifer region of the graph (above the line). Undisturbed confined groundwater is characterized by relatively low salinity (EC and TH), lower clay thickness, lower DOC, and higher DO and is located in the southern region. The disturbance criteria may be either anthropogenetic or geogenic or both. The field study presented that no human events such as massive industrial activities, no use of iron–manganese-rich chemical fertilizers in agriculture, no waste disposal inland, and no landfill were observed, but overexploitation of groundwater for irrigation has occurred. Therefore, overexploitation, substantial fluctuation of river water flow, and some soil erosions are the major causes of disturbance in aquifer systems. Therefore, along with the physicochemical influence on higher level

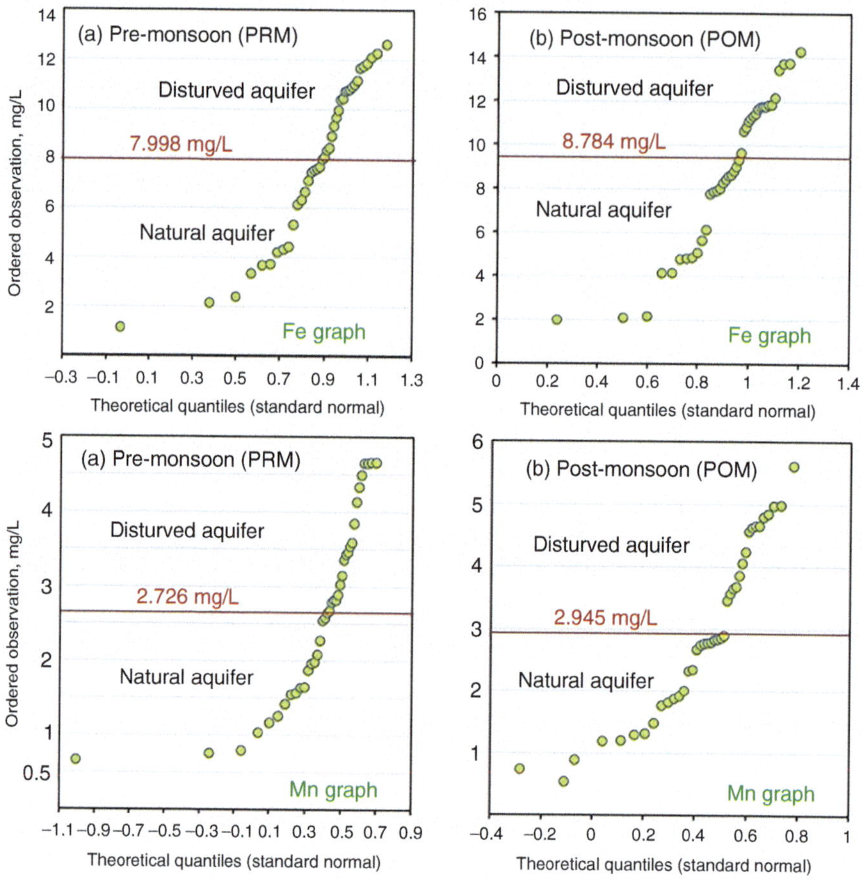

Fig. 5.11 Log-normal probability plots of Fe (**a**, **b**) and Mn (**c**, **d**). Proof of individuality of the slope changes consistently to the natural background level, NBL (red line)

of Fe and Mn concentrations in groundwater, these natural and human actions are added causes for this factor.

The findings of the present and previous studies confirmed that except for any special causes, the higher loading of Fe and Mn may decrease the As dissolution rate into groundwater. Most recently, Podgorski et al. [28] studied the geological release of As, Mn, and Fe in groundwater in Southeast Asia. They found that drier climatic conditions tended to be associated with increased manganese concentrations, while wetter climates with higher soil organic carbon levels tended to produce higher iron levels. In addition, the results confirmed that high Fe values are often related with high As content. Although this contradicts the results of the present study, regions with high concentrations of Mn and As are often adjacent. Similarly, the greatest threat from manganese and iron lies in the densely populated alluvial deposits in north-western Bangladesh. Another study of quaternary alluvial sediments in north-eastern China [29] found that sources of Fe and Mn in groundwater were Fe and Mn nodules in clay beds, whereas they were found in undecorated soils and river valleys. Another result of this study is that TDS, groundwater level variation, and residence time are important factors affecting the Fe and Mn content in groundwater. Weathering of Fe and Mn rocks results in the penetration of As adsorbed on them into groundwater.

5.6 Impact of Elevated Fe and Mn on Aquaculture and Irrigation Water Suitability

About 80% of the arable land in the study zone is irrigated with shallow groundwater by motor pumps. High levels of iron and manganese in groundwater can adversely affect irrigation water and soil environment, affecting plant growth and production. However, Fe and Mn above 5 mg/L and 2 mg/L, respectively, in irrigation water are undesirable. Potential problems due to high iron content in irrigation water are mainly fouling of concrete surfaces, clogging of irrigation systems and drainage tiles, and deposition of an unsightly rust-coloured material on ornamental plants (Fig. 5.12). High concentrations of iron in spray tank water can clog screens and nozzles and adversely affect the performance of some pesticides. Frequent and continuous use of water containing iron in very high concentrations (above 10 mg/L) can lead to nutritional deficiencies. Most of the water samples in the study area are highly mineralized and contain large amounts of soluble salts. Increased levels of dissolved salts in soil increase the osmotic pressure of soil solutions, making it difficult for plants to adapt to such conditions. On the other hand, iron in aerated soil is generally not toxic to plants, but excess iron in soil can make it difficult for plants to absorb other essential minerals such as phosphorus, magnesium, and molybdenum. Plants require phosphorus for cell division and photosynthesis. Vegetation needs magnesium to produce chlorophyll, which gives them their green colour. When there is too much iron in the soil, plants lose these important minerals and face discoloration, photosynthetic control problems, and weak roots. Excessive iron

Fig. 5.12 Rust-coloured leaves due to high iron in the soil

can lead to imbalances in some other important elements, such as zinc and copper. As for manganese, it is toxic to some crops at concentrations of 1 mg/L or less but usually only in acidic soils. Especially in the pre-monsoon season, samples from the study area exhibit lower pH values and increased toxicity of iron and manganese to plants.

In aquaculture water, when the sediment becomes oxygen-deficient, the Fe^{3+} of the sediment is reduced to Fe^{2+}, which darkens the colour of the layered sediment. However, Fe also helps reduce the toxicity of H_2S produced by anaerobic bacteria by precipitating it onto insoluble Fe_2S_3. Ferric hydroxide, $Fe(OH)_3$, formation can be a problem when water with a high Fe^{2+} content is used in hatcheries. This sediment can stick to and damage the eggs and can also clog the fish's gills. In many aquaculture applications, the presence of Fe in concentrations above 0.1 mg/L has been shown to damage fish gills. This value is much lower than the concentrations typically reported in this literature (see Table 5.1). The harmful effects of iron depend on the type and size of the fish. Fish gills act as a mechanical filter, with tiny iron particles of a few microns getting stuck in the gill lamellae. The presence of these particles irritates the gill tissue, causing damage to the gills and secondary bacterial and fungal infections. Iron acts as a catalyst in water, promoting the dissociation of the O_2 in water to form highly reactive free radicals. However, on the gill surface, iron-generated free radicals cause oxidation of the surrounding tissue, resulting in extensive destruction of the gill tissue and anaemia. Mn is less toxic than iron and has a very low bioaccumulation rate in fish.

5.7 Concentration and Mobilization of Arsenic

5.7.1 Distribution of As in Groundwater of Bangladesh

Arsenic (As), a category 1 carcinogen, occurs naturally in groundwater supplies throughout the Southeast Asia. In the Ganges plains of Bangladesh and northern part of India, 25% of hand pump wells in two shallow aquifers in the region suffer from severe groundwater pollution from naturally occurring asphalt. It was first recognized as a serious problem in Bangladesh in the 1990s, and concentration in most places exceeded the maximum level recommended by the WHO. In Bangladesh, more than 75 million people in 59 out of 64 districts (Fig. 5.13) were at risk of drinking water contamination. An estimated 43,000 people are affected by As intoxication each year in this country. The agency has taken some measures and formulated approaches to deal with the disaster. However, national campaigns and social mobilization efforts by the government and NGOs have fallen far short of expectations. Including in Bangladesh, several investigations have been conducted in Southeast Asia and other countries around the world, but the situation is still not fully controlled.

In 1992–1993, groundwater samples from several shallow pipe wells in northern Bangladesh, bordering India, were analysed for As pollution, and arsenic was found

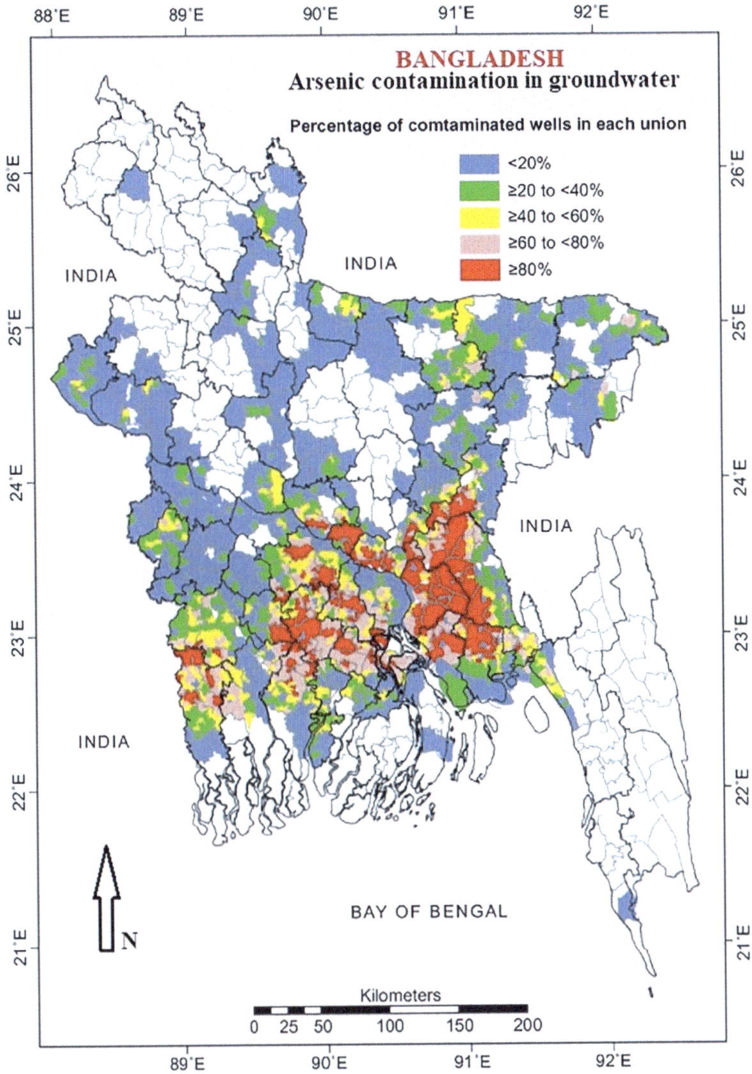

Fig. 5.13 Distribution of arsenic in groundwater of Bangladesh

in the variation of 59 to 388 ppb (parts per billion). The WHO and Bangladesh drinking water standards are 10 ppb (0.01 ppm) and 50 ppb (0.05 ppm), respectively. Twelve years after the first detection in groundwater, arsenic levels were detected in 59 of his 64 districts in Bangladesh. Due to irregular findings and reports of groundwater contamination, a nationwide survey of arsenic concentrations in tube well water was initiated in 1996. In the year of 1998, the British Geological Survey (BGS) collected 2022 groundwater samples from 41 counties with high concentrations of arsenic. Laboratory tests showed that 35% of these samples contained

Table 5.12 Statistics of arsenic concentration in Bangladesh

Basic data	Value	Percentage (%)
The whole area of Bangladesh	147,570 km^2	–
The cross population in Bangladesh (June 2022)	169.6 million	–
Total number of districts in Bangladesh (new)	64	–
WHO standard (As in potable water)	10 ppb	–
Bangladesh standard (As in potable water)	50 ppb	–
Number of districts surveyed for As contamination	64	100
Number of districts having As above 50 ppb in groundwater	59	92.2
Total area of affected 59 districts	126,134 km^2	85.5
Population at risk condition	>80 million	>48
Potentially exposed population	>35 million	>21
Number of patients suffering from arsenicosis	>38,000	–
Number of patients who died	10	–

arsenic levels greater than 50 ppb. A preliminary study conducted in 1999 examined 51,000 tube wells and quantified the presence of arsenic noxiousness in 211 out of 460 counties (subdistricts), accounting for nearly a third of the tube wells examined. Subsequently, in 2003, a nationwide inclusive survey led by the Arsenic Mitigated Water Supply Project (AMWSP) targeted 57,482 villages in 271 counties, with 1.44 million of a total of 4.95 million pipe wells contaminated with arsenic (Table 5.12 and Fig. 5.13). Arsenic levels in water from the most severely affected tubular wells were within limits of 0.10–0.30 mg/L. The highest arsenic level found in shallow pipe well water is 470 ppb, and previous studies by experts from the Bangladesh Council for Scientific and Industrial Research (BCSIR) also observed the highest level of 1400 ppb in the same type of pipe well waste. Another study found that water from 13,423 domestic pipe wells across the country contained arsenic levels exceeding Bangladesh standards in 12.6% of the samples tested [30]. Full scenarios for arsenic content in groundwater nationwide are summarized in Tables 5.12 and 5.13.

Arsenic toxicity in aquifer water can increase through the geogenic and anthropogenic sources. This problem rises in this area because of an unfortunate amalgamation of three usual features: a source of As (arsenic present in the aquifer sediment), dissolution and mobilization (arsenic weathered from the arsenic-bearing rocks to the aquifer water), and transport (As journeyed in the groundwater). Two obstructive hypotheses to relate the dissolution of As into groundwater in Bangladesh are the oxidation of pyrite rocks and the reduction of oxyhydroxide rocks. The groundwater table has decreased regularly due to the overwater mining of groundwater for irrigation practice and household purposes, weak water management practices, and inadequate external recharge of the aquifers, which are the major anthropogenetic causes of As contamination in Bangladesh.

Arsenic toxicity is a thoughtful threat and a substantial public health apprehension in Bangladesh. Long-term overexposure to As in drinking water has been related to cancer formation in the skin (Fig. 5.14), lung, bladder, nasal channels,

Table 5.13 A complete scenario of As concentration in the groundwater of Bangladesh

Sample location (old district)	Total no. of samples	Concentration (µgm/L or ppb) distribution								% of threat	
		<10	10 to 50	51 to 99	100 to 299	300 to 499	500 to 699	700 to >1000	Max. conc.	BTL[a]	WHO
Dhaka Division	***13,597***	***8236***	***2024***	***983***	***1536***	***558***	***167***	***93***			
Dhaka District	4393	3850	169	90	214	61	09	00	**5533**	16.72	21.78
Faridpur District	3726	924	826	483	956	376	104	48	**1630**	41.86	65.63
Mymensingh District	2805	2186	411	110	61	16	08	10	**1172**	01.04	06.58
Narayanganj District	748	306	58	41	170	92	46	35	**1750**	76.70	86.89
Tangail District	1925	970	560	259	135	13	00	00	**365**	03.53	25.80
Rajshahi Division	***18,581***	***12,532***	***3365***	***1181***	***1142***	***209***	***86***	***66***			
Bogra District	1165	995	135	20	16	01	00	01	**1040**	04.43	79.41
Pabna District	5395	1762	1886	815	695	124	52	41	**2108**	33.52	68.83
Rangpur District	2700	2049	524	59	37	19	10	02	**939**	14.01	38.58
Rajshahi District	5254	3735	732	280	393	65	24	22	**1600**	08.71	18.57
Dinajpur District	4087	3991	88	07	01	00	00	00	**130**	~0.00	01.10
Chittagong Division	***6734***	***1514***	***601***	***613***	***2144***	***926***	***455***	***481***			
Chittagong District	551	504	26	13	08	00	00	00	**275**	05.74	12.84
Comilla District	1757	175	71	68	820	417	129	77	**1769**	74.50	79.27
Noakhali District	3691	367	324	471	1293	506	326	404	**4730**	95.14	99.42
Sylhet District	735	468	180	61	23	03	00	00	**302**	~0.00	15.35

(continued)

Table 5.13 (continued)

Sample location (old district)	Total no. of samples	Concentration (µgm/L or ppb) distribution								% of threat	
		<10	10 to 50	51 to 99	100 to 299	300 to 499	500 to 699	700 to >1000	Max. conc.	BTL[a]	WHO
Khulna Division	***13,570***	***7766***	***2240***	***938***	***1664***	***570***	***230***	***162***			22.65
Jessore District	5465	4751	614	420	681	136	80	38	**1120**	17.38	
Khulna District	1300	940	378	145	453	239	24	24	**3143**	24.70	48.20
Kushtia Dist.	2065	1732	1051	322	274	80	51	36	**2190**	20.63	47.60
Barisal Dist.	803	295	187	51	256	115	75	64	**1770**	63.64	77.71
Patuakhali District	15	48	10	00	00	00	00	00	**15**	00	13.33
Total	***52,202***	***29,768***	***8230***	***3714***	***6484***	***2263***	***938***	***802***	–		
Percentage (%)	–	***57***	***15.8***	***7.1***	***12.4***	***4.3***	***1.8***	***1.6***	–		

[a] *BTL* Bangladesh threshold limit

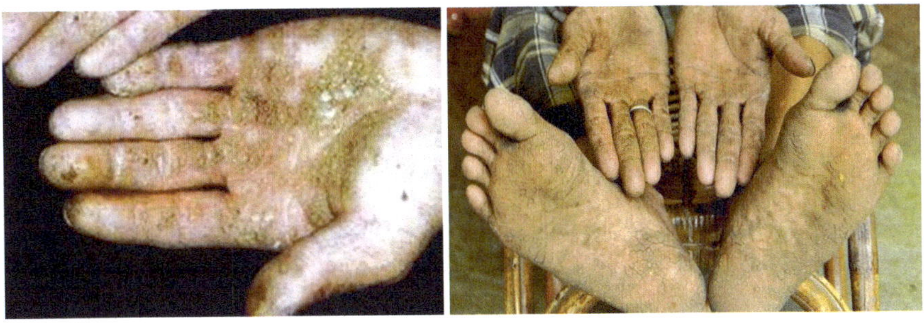

Fig. 5.14 Symptoms of arsenic-affected human skin in Bangladesh

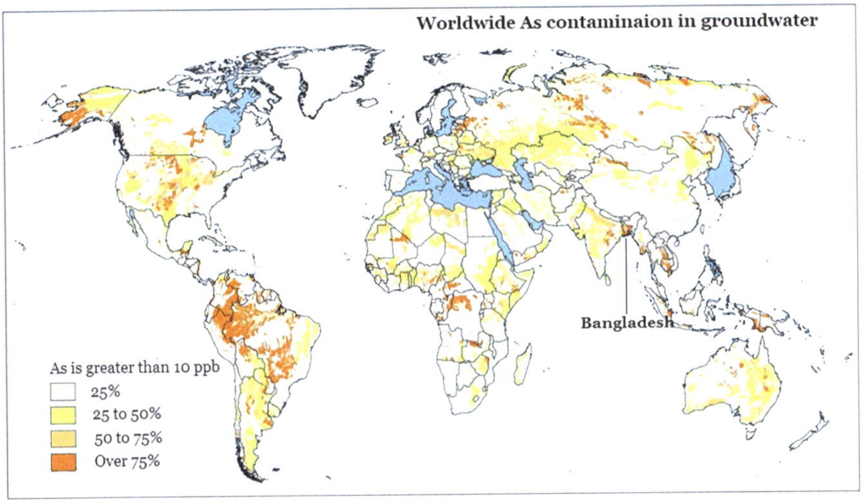

Fig. 5.15 Global picture of As contamination in groundwater

prostate, liver, and kidneys. It also relays to damage to the cardiovascular, endocrine, pulmonary, immunological, and neurological systems. Data collected by governmental authorities, NGOs, and private bodies expose that many people in Bangladesh suffer from melanosis (93.5%) and keratosis (68.3%), which are the more common manifestations between affected people. In addition, patients with hyperkeratosis (37.6%) and leucomelanosis (39.1%) have been found in several cases. Governmental organizations have analysed 11,000 urine, skin, hair, and nail samples collected from affected rural villages in Bangladesh, and the results show that approximately 90% of people have As in their urine, hair, and nails above the typical level. The usual concentrations of As in hair, nails, and urine are 0.08–0.25, 0.005–0.040, and 0.43–1.08, mg/kg, respectively.

5.7.2 *World Scenarios*

Over 300 million people worldwide use groundwater contaminated with As as a source of drinking water. The natural contamination of its groundwater has been reported worldwide, and the majority of this groundwater belongs to the South Asian and South American regions (Fig. 5.15). The severely affected countries include Bangladesh, India, China, Indonesia, Nepal, Laos, Vietnam, Myanmar, and the USA. Moreover, countries such as South Africa, Pakistan, Argentina, Hungary, Canada, Mexico, and Chile are also affected. However, the South and Southeast Asian regions are considered the most As-contaminated regions, including Bangladesh, China, India, Nepal, and Vietnam. Developed countries, such as the USA and Canada, also experience widespread levels of As in groundwater, although the concentrations are typically lower in comparison with Asian countries. Global data reveal that 107 countries are affected by As contamination in groundwater (beyond the WHO's maximum tolerable limit of 0.01 mg/L), with the highest reports from Asia (32) and Europe (31), followed by Africa (20), North America (11), South America (9), and Australia (4). Most As contamination-prone regions are located in the sedimentary basins close to modern mountain belts and deltaic areas. Areas with tropical climates are more vulnerable to As contamination, as this climate favours the release of arsenic from As-laden rocks.

5.7.3 *Sources and Mobilization of As*

The most important origin of As in the country was geogenic; meanwhile, it was contained in the sediments of the shallow Holocene aquifers of the Ganges basin. In the present study, most of the water samples contain low level of As (average 8 μg/L) related to the central portion of the country. In this area, approximately 60% of groundwater samples from ten districts exceeded the national arsenic standard level of 50 μg/L. Several studies have recognized that the shallow aquifers in this zone enriched in Fe, Mn, and Al oxides, PO_4^{3-}, NH_4^+, and NO_3^- with organic matter contain moderately high levels of As. In this present investigation, weak associations between As and influencing components in water (e.g. Fe, Mn, NO_3^-, SO_4^{2-}, DO, DOC) were found (r = <0.4 with $p \sim 0.05$) (Table 5.9). Fe and Mn are straight involved in the dissolution and distribution of As in the aquifer systems. Decreases in As concentrations are expected with increasing levels of Fe and Mn in groundwater. The oxidizing agent converts Fe^{2+} to Fe^{3+}, forming a solid phase of FeO(OH). Arsenic (+3 and +5) is adsorbed on the surface of the FeO(OH) solid, after which both are precipitated. Previous studies by the authors have confirmed that the above factors influence arsenic mobilization, but this influence mainly depends on local lithological conditions. Factors such as Fe, Mn, NO_3^-, SO_4^{2-}, DO, DOC, and Eh value of water could be adequately effective when the lithology is clayey, peaty, or silty in nature and has a rich bacterial community. However, if it is not, it will not be effective. The soil of the present study area consists of sand, silt, and silt loamy, and the impact of these influencing factors is unfavourable.

The occurrence and mobility of As in sedimentary aquifers is a complex of microbial-assisted reactions and hydrogeochemical processes that depend on site-specific hydrology and sediment types as well as on anthropogenic activities such as chemical fertilizer application. Numerous studies have been carried out in the last decades but failed to provide a clear conception of the arsenic mobilization mechanism, especially in the Ganges basin region. Four mechanisms have been proposed over time to explain this process of As contamination of the Bengal delta plain:

1. *Oxidation of pyrite*: Near the year of 2000, numerous investigators, e.g. Chowdhury et al. [30], Bhattacharya et al. [31], Nickson et al. [32], and many others, assumed that iron pyrite and arsenopyrite were oxidized by the entrance of air O_2 into aquifer sediments, and then arsenic was unconfined as arsenate into the groundwater:

$$FeS + O_2 + H_2O \rightarrow Fe(OH)_3 + SO_4{}^{2-} + H^+$$

$$FeAsS + Fe^{3+} + H_2O \rightarrow Fe^{2+} + SO_4{}^{2-} + H^+ + H_3AsO_4\ (aq.)$$

   The rationality of this supposition is not the fact; meanwhile, the groundwater in this area is (a) anoxic type, with low concentration of DO; (b) at close to neutral pH or moderate alkaline; (c) categorized by low $SO_4{}^{2-}$ concentrations with no correlation to arsenic; and (d) categorized by the very low concentration of $NO_3{}^-$. Therefore, this mechanism was postponed at that time, and the following approaches were recognized later.
2. *FeOOH reduction*: Bacterial metabolism of DOM in the aquifer basements leads to anoxic circumstances, consequently leading to the reduction of the lepidocrocite mineral FeO(OH) and then resulting in the release of adsorbed arsenic into the groundwater phase.
3. *Redox process*: As is unconfined by the reductive weathering of As-laden oxo-hydroxide rock, i.e. FeO(OH), and this is a redox procedure that happens after bacterial oxidation of organic material has used up DO, NO_3–, and $NH_4{}^+$ (as the nutrient feed of microbes).
4. *Ion exchange*: Arsenate anions adsorbed to the surface of solid phase of rocks in aquifer are evacuated into the aqueous medium by a competitive exchange of $PO_4{}^{3-}$, resulting in As pollution in groundwater.

The above-mentioned mechanisms 2 to 4 have been recognized since the 2000s. According to recently published reports, the availability of dissolved organic matters, different microbial community, an abundance of highly positive charge baring metal-minerals, high phosphate loads, and low Eh with moderately high pH in aquifers are the main criteria to increase the arsenic concentration in the shallow groundwater system. The study area is located in the north-western part of the country and has highly productive arable land compared to other parts of Bangladesh. About 72% of the total area is arable, of which about 30% is triple cropped. Almost all arable soils are loamy, sandy, or sandy–loamy, but clayey or silty clay layers do not occur in this area. A previous study [33] confirmed that the soils in these areas are low in organic matter and have relatively high soil Eh values. There is no As toxicity in this area, but chemical fertilizer consumption and water quality degradation are very high. In the study area, other factors such as soil pH, DOC, and Eh, which are the main factors for As dissolution, may be favourable for reducing As contamination. Some researchers [34–36] believe that intensive irrigation lowers the water table, allowing atmospheric O_2 to enter the aquifer and possibly promoting the release of arsenic by oxygen oxidation of pyrite. This statement does not apply to the study area, as irrigation activities are very common in these areas and the overuse of water in hot summers is probably reducing the groundwater table. Nath et al. [37] and Reza et al. [38] described the comparative function of mobilizing agents in the most and least arsenic-affected areas of Bangladesh. This study supported their works and reported that in the least affected areas, the lower level of arsenic does not depend on PO_4^{3-}, dissolved organic matters, and alkalinity of the soil, and As is released from the soil with low concentration due to less desorption of arsenic or precipitation of FeO(OH). The low level of detection suggests that sorption of arsenic on iron oxide precipitates, or other solubility controls, may limit its dissolution in groundwater.

5.8 Summary

Trace elements in groundwater can have both natural and man-made sources due to industry, industrial effluents, and agrochemicals. In this investigation, the level of trace elements was determined from 40 groundwater samples within three seasons in the study area of the middle-west part of Bangladesh. The analysis results presented that the maximum metal concentration exceeded the guideline values of the respective components. Three metal ions, viz. Fe, Mn, and Pb, exceed the limit of WHO standards in most of the water samples, representing severe human health hazards. The levels of Fe (8.43 mg/L) and Mn (2.85 mg/L) were 28-fold greater than the standard values of 0.3 and 0.1 mg/L, respectively, for drinking purposes. Another finding is that the trace metal concentrations in the MON and POM seasons are somewhat higher than those in the PRM. The investigation results revealed that mainly natural processes and few man-made activities impacted the trace metal loads in the groundwater system of the area. To measure factors affecting the level of trace metals in groundwater samples and determine their probable sources and mobilization, multivariate statistical techniques, including Pearson's correlation coefficient, principal component analysis, and dendrogram cluster analysis, are utilized. The results show that the sources of trace metals were geogenic and slightly from human actions, i.e. excess use of agrochemicals, overexploitation, and industrial wastewater. In addition, it was found that the level of trace metals in samples somewhat depends on water pH and salinity. Groundwater in this area poses a huge threat due to high concentrations of trace elements, and without any purification, it is necessary to avoid this water for drinking uses.

References and Further Study

1. Islam, M. S., & Mostafa, M. G. (2023). Identification of sources and mobilization of trace elements in shallow groundwater of the upper Ganges River basin. *Journal of Chemistry and Environment., 1*(2), 14–29. https://doi.org/10.56946/jce.v1i01.34
2. WHO. (2004). *The World Health report 2004: Changing history*. 96 p.
3. WHO. (2022). *Guidelines for drinking-water quality. 4th edn incorporating the first and second addenda*. World Health Organization.
4. Islam, M. S., & Mostafa, M. G. (2021). Influence of chemical fertilizers on arsenic mobilization in the alluvial Bengal delta plain: A critical review. *Journal of Water Supply: Research and Technology-AQUA, 70*(7), 948–970. https://doi.org/10.2166/aqua.2021.043
5. Tadiboyinaa, R., & Ptsrkb, P. R. (2016). Trace analysis of heavy metals in ground waters of Vijayawada Industrial Area. *International Journal of Environmental Science and Education, 11*(10), 3215–3229.
6. Aduck, J. N., Ma-Effeti, A. N., Ngwese, S. N., & Akoachere, R. A., II. (2022). Trace metals evaluation of groundwater in the Granito-Basaltic fractured rock Aquiferous formations in Bafoussam, West Region-Cameroon. *Open Access Library Journal, 9*, e8331. https://doi.org/10.4236/oalib.1108331
7. Islam, A. R. M. T., Shen, S., Bodrud-Doza, M., et al. (2017). Assessment of trace elements of groundwater and their spatial distribution in Rangpur district, Bangladesh. *Arabian Journal of Geosciences, 10*(4), 1–14. https://doi.org/10.1007/s12517-017-2886-3
8. Brindha, K., Paul, R., Walter, J., et al. (2020). Trace metals contamination in groundwater and implications on human health: Comprehensive assessment using hydrogeochemical and geostatistical methods. *Environmental Geochemistry and Health, 11*(14), 3819–3839. https://doi.org/10.1007/s10653-020-00637-9
9. Bakyayita, G. K., Norrström, A. C., & Kulabako, R. N. (2019). Assessment of levels, speciation, and toxicity of trace metal contaminants in selected shallow groundwater sources, surface runoff, wastewater, and surface water from designated streams in Lake Victoria Basin, Uganda. *Journal of Environmental and Public Health, 2019*(ID 6734017), 1–18. https://doi.org/10.1155/2019/6734017
10. Gu, C., Zhang, Y., Peng, Y., et al. (2020). Spatial distribution and health risk assessment of dissolved trace elements in groundwater in southern China. *Scientific Reports, 10*, 7886. https://doi.org/10.1038/s41598-020-64267-y
11. Toscano, C. A. R., Villanueva, R. A. C. V., Raúl Cortés-Martínez, R., et al. (2020). Hydrogeochemical characteristics and assessment of drinking water quality in the urban area of Zamora, Mexico. *Water, 12*(2), 556. https://doi.org/10.3390/w12020556
12. US-EPA. (2003). *Final guide for incorporating bioavailability adjustments into human health and ecological risk assessments at U. S. Department of Defense Facilities Part 1: Overview of metals bioavailability*.
13. DPHE. (2009). *Bangladesh national drinking water quality survey*. Department of Public Health and Engineering, Govt. of Bangladesh.
14. Islam, M. S., Haque, M. N., & Hasan, M. A. F. M. R. (2001). Hydrogeological conditions in the Northern Part of Kushtia Sadar Upazila of Bangladesh. *The Journal of Geo-Environment, 1*, 79–89.
15. Huang, S. B., Wang, Y. X., Ma, T., Tong, L., Wang, Y. Y., Liu, C. R., & Zhao, L. (2015). Linking groundwater dissolved organic matter to sedimentary organic matter from a fluvio-

lacustrine aquifer at Jianghan Plain, China by EEM-PARAFAC and hydrochemical analyses. *Science of the Total Environment, 529*, 131–139. https://doi.org/10.1016/j.scitotenv.2015.05.051
16. Zhang, H., Cheng, S., Li, H., Fu, K., & Xu, Y. (2020). Groundwater pollution source identification and apportionment using PMF and PCA-APCA-MLR receptor models in a typical mixed land-use area in Southwestern China. *Science of the Total Environment, 741*, 140383.
17. Rusydi, A. F., Onodera, S.-I., Saito, M., et al. (2021). Vulnerability of groundwater to iron and manganese contamination in the coastal alluvial plain of a developing Indonesian city. *SN Applied Sciences, 399*, 1–12. https://doi.org/10.1007/s42452-021-04385-y
18. Palmucci, W., Rusi, S., & Curzio, D. D. (2016). Mobilisation processes responsible for iron and manganese contamination of groundwater in Central Adriatic Italy. *Environmental Science and Pollution Research, 23*, 11790–11805. https://doi.org/10.1007/s11356-016-6371-4
19. Adeyeye, O., Xiao, C., Zhang, Z., & Liang, X. (2020). State, source and triggering mechanism of iron and manganese pollution in groundwater of Changchun, Northeastern China. *Environmental Monitoring Assessment, 192*(10), 115. https://doi.org/10.1007/s10661-020-08571-0
20. Guo, H., Wen, D., Liu, Z., Jia, Y., & Guo, Q. (2014). A review of high arsenic groundwater in Mainland and Taiwan, China: Distribution, characteristics and geochemical processes. *Applied Geochemistry, 41*, 196–217. https://doi.org/10.1016/j.apgeochem.2013.12.016
21. Rawson, J., Siade, A., Sun, J., Neidhardt, H., Berg, M., & Prommer, H. (2017). Quantifying reactive transport processes governing arsenic mobility after injection of reactive organic carbon into a Bengal Delta aquifer. *Environmental Science & Technology, 51*(15), 8471–8480. https://doi.org/10.1021/acs.est.7b02097
22. Lovely, C. J., Du, H., He, Y., & Dias, H. V. R. (2004). Oxidative rearrangement of Imidazoles with Dimethyldioxirane. *Organic Letters, 6*(5), 735–738. https://doi.org/10.1021/ol036403w
23. McMahon, P. B., & Chapelle, F. H. (2008). Redox processes and water quality of selected principal aquifer systems. *Ground Water, 46*(2), 259–271. https://doi.org/10.1111/j.1745-6584.2007.00385.x
24. Khozyem, H., Hamdan, A., Tantawy, A. A., Emam, A., & Elbadry, E. (2019). Distribution and origin of iron and manganese in groundwater: Case study, Balat-Teneida area, El-Dakhla Basin, Egypt. *Arabian Journal of Geosciences, 12*, 523. https://doi.org/10.1007/s12517-019-4689-1
25. Boyd, C. E. (2008). *Dissolved oxygen management in aquaculture*. Global sea fod allaience.
26. Schmidt-Rohr, K. (2020). Oxygen is the high-energy molecule powering complex multicellular life: Fundamental corrections to traditional bioenergetics. *ACS Omega, 5*(5), 2221–2233. https://doi.org/10.1021/acsomega.9b03352
27. Macpherson, G. L. (2009). CO_2 distribution in groundwater and the impact of groundwater extraction on the global C cycle. *Chemical Geology, 264*(1–4), 328–336. https://doi.org/10.1016/j.chemgeo.2009.03.018
28. Podgorski, J., Araya, D., & Berg, M. (2022). Geogenic manganese and iron in groundwater of Southeast Asia and Bangladesh–Machine learning spatial prediction modeling and comparison with arsenic. *Science of the Total Environment, 833*, 155131. https://doi.org/10.1016/j.scitotenv.2022.155131
29. Zhai, Y., Cao, X., Xia, X., Wang, B., Teng, Y., & Li, X. (2021). Elevated Fe and Mn concentrations in groundwater in the Songnen Plain, Northeast China, and the factors and mechanisms involved. *Agronomy, 11*, 2392. https://doi.org/10.3390/agronomy11122392
30. Chowdhury, T. R., Basu, G. K., Mandal, B. K., et al. (1999). Arsenic poisoning in the Ganges delta. *Nature, 401*, 545–546. https://doi.org/10.1038/44056
31. Bhattacharya, P., Jacks, G., Larsson, M., et al. (2000). Suitability of low-cost remediation techniques for the amendment of high-arsenic groundwater in West Bengal and Bangladesh. In *Abstract, international workshop on control of arsenic concentration in groundwater PHED, Calcutta, India*.
32. Nickson, R. T., McArthur, J. M., Ravenscroft, P., Burgess, W. G., & Ahmed, K. M. (2000). Mechanism of arsenic release to groundwater, Bangladesh and West Bengal. *Applied Geochemistry, l5*(4), 403–413.

33. Ravenscroft, P., McArthur, J. M., & Hoque, B. A. (2001). Geochemical and palaeohydrological controls on pollution of groundwater by arsenic. In W. R. Chappell, C. O. Abernathy, & R. Calderon (Eds.), *Arsenic exposure and health effects, IV* (pp. 53–77). Elsevier Science Ltd, Oxford, UK.
34. Chakraborty, T. K., Ghosh, G. C., Ghosh, P., et al. (2022). Arsenic, iron, and manganese in groundwater and its associated human health risk assessment in the rural area of Jeshore, Bangladesh. *Journal of Water and Health, 20*(6), 888–902.
35. Das, K., Roy, D., & Mostafa, M. G. (2015). Sustainable management for arsenic (As) free safe drinking water in Bangladesh: A review. *Journal of Petroleum and Environmental Biotechnology, 6*, 228.
36. Chowdhury, T. R., Basu, G. K., & Mandal, B. K. (1999). Arsenic poisoning in the Ganges delta. *Nature, 401*, 545–546.
37. Nath, B., Stüben, D., & Mallik, S. B. (2008). Mobility of arsenic in West Bengal aquifers conducting low and high groundwater arsenic. Part I: Comparative hydrochemical and hydrogeological characteristics. *Applied Geochemistry, 23*(5), 977–995.
38. Reza, A. H. M., Selim, J., Jiin-Shuh, Y., et al. (2010). Occurrence of arsenic in core sediments and groundwater in the Chapai-Nawabganj District, northwestern Bangladesh. *Water Research, 44*(6), 2021–2037.

Chapter 6
Drinking Water Quality

Abbreviations

BOD	Biological oxygen demand
Cd	Degree of contamination
CDI	Chronic daily intake
COD	Chemical oxygen demand
CSF	Cancer slope factor
CWQI	Canadian Water Quality Index
DO	Dissolved oxygen
HI	Hazard index
HMEI	Heavy metal evaluation index
HMPI	Metal pollution index
HQ	Hazard quotient
HRA	Health risk assessment
ILCR	Incremental lifetime cancer risk
ISF	Incremental slope factor
MTL	Maximum threshold levels

M. S. Islam, *Hydrogeochemical Evaluation and Groundwater Quality*,
https://doi.org/10.1007/978-3-031-44304-6_6

NSF	National Sanitation Foundation
PHED	Public Health and Engineering Department
RAIS	Risk Assessment Information System
US-DOE	United States Department of Energy
WHO	World Health Organization
WQI	Water quality index
WWQI	Weighted average water quality index

6.1 Drinking Water Quality: *An Overview*

Presently, in Bangladesh, a densely populated rural country, water quality problems have become a major problem due to excessive water withdrawal for irrigation, agricultural diversity, urban sprawl, industrial development, and inadequate water management. As said by the World Health Organization (WHO), more than three million people (90% of whom are under the age of 5) die each year in developing countries from water-borne diseases. More than 12 million people in rural Bangladesh drink raw groundwater without any purification. Moreover, 90% of the country's irrigation water comes from this source. In the study area, large-scale pesticide leakage, topsoil contamination, salinization, landfills, and large-scale flooding affect groundwater quality. In Bangladesh, 40,000 tonnes of pesticides was used in 2018, with the rest seeping into the subsoil and causing groundwater contamination [1]. Severe floods occur frequently every year, contaminating the country's topsoil and groundwater. Seawater intrusion due to climate change and groundwater overmining is a major salinization problem in the country's coastal areas. In addition, geological springs have significantly degraded groundwater quality in some parts of the country over the last 30 years.

Natural surface water quality in rivers, lakes, and reservoirs is used for drinking, domestic water, irrigation, and aquatic life and could play an important role in the social and economic structure of developing countries like Bangladesh. Several investigations have been conducted to assess the suitability of surface water in this country. These studies presented that the quality of this water resource continues to be polluted by heavy industrialization, increased pesticide-based crops, and municipal waste. Numerous studies have been conducted in coastal areas where groundwater quality in Bangladesh is a major problem, especially sodicity problem. However, in the upper delta plains, where heavy mineralization and water hardness are the main water quality problems, there is insufficient information on which to base geochemical studies to date. Continuous monitoring is therefore important to assess the suitability of water for various purposes and to prevent further deterioration of groundwater quality in the study area.

Himalayan and non-Himalayan rivers flow into the Bay of Bengal as a common river, carrying the largest alluvial sediment and forming the world's largest agricultural Bengal delta basin. This study considered the western upper part of this basin as the study area of the Kushtia District of Bangladesh. Previous studies have identified frequent fluctuations in hydrogeological and aquifer conditions in the study area. The flow of rivers across the Bangladesh border on the Indian side has been cut off, severely obstructing waterways within Bangladeshi territory. Rather, these rivers are mostly submerged during the dry season, greatly affecting groundwater composition in the area. Therefore, the morphology and banks of river basins are constantly changing, adversely affecting the ecosystem services of river basins. Recently, the amount of water in transboundary rivers in the study area decreased significantly during winter and pre-monsoon season (PRM), causing surface water shortage. Residents of the study area therefore rely entirely on raw groundwater for drinking and other domestic purposes. Proper assessment of groundwater quality is therefore essential to ensure safe water for all purposes, justifying the widely used water quality index techniques to accurately assess water quality.

The water quality index (WQI) is intended to be scored by an arithmetic number calculated based on a system of fitting all the different parameters and their concentrations present in a sample into one number. This is a method of operation that allows us to assess the quality of various water samples based not only on the parameter values of each sample but on the basis of a single arithmetic value. Though there is no universally accepted composite water quality index model, some countries and regions have used or still use composite water quality datasets to develop WQIs. The method of indexing maximum water quality depends on the normalized dataset of water quality parameters. Parameters are often weighted according to their apparent importance to overall water quality, and indices are calculated as weighted averages of all observations of interest. Frequent fluctuations in WQI have been documented in the literature over the past 50 years. There is an urgent need for users around the world to develop a widely accepted WQI that is flexible enough to represent water suitable for drinking and other purposes. The use of indicators to assess water suitability has recently been innovated by several researchers. Recently, Najafzadeh et al. [2, 3] evaluated the dependability of the groundwater quality index utilizing remote sensing and data-driven models. In these two studies, four robust data-driven methods such as evolutionary polynomial regression (EPR), M5 model tree (MT), gene-expression programming (GEP), and multivariate adaptive regression spline (MARS) – based on evolutionary algorithms and classification ideas – have been applied to present formulations for the calculation of ground and surface WQI values.

The largest quantities of WQIs have been developed for surface waters, especially river waters, and a limited number of index models are designed for groundwaters. Horton [4] was the first methodologically proposed indicator to measure water quality using his ten most commonly used water parameters. This method is widely used and accepted in European, Asian, and African countries. One advantage of this method is that fewer parameters are required to compare water quality for a given application. However, this model was later modified by Brown et al. [5]. In the mid-1990s, a new WQI, the Canadian Council of Environment Ministers – Water Quality Index (CCEM-WQI), was introduced. This became known as the Canadian Water Quality Index (CWQI) in 2001. In this model, WQI was established based on frequency of sampled variables, failed variables, and deviations from default values. The CWQI model was later accepted by the United Nations Environment Programme as a suitable model for measuring the quality of drinking water worldwide. Later, the sensitivity test of the Canadian WQI was conducted by altering some input variable quantity and the number of sampling sites with sampling seasons. Each WQI method has a different scale. For example, good water quality had an index value between 71 and 90 according to the US NSF-WQI model (National Sanitation Foundation WQI) and an index value between 80 and 94 according to the Canadian WQI model. In addition, a number of selected parameters were used in certain models that were related to water suitability in selected regions but not in others. Dissolved oxygen (DO), chemical oxygen demand (COD), biological oxygen demand (BOD), and/or microbial parameters are not relevant for groundwater testing but are very important for surface water indexing. This study used different methods of indexing water quality to evaluate drinking purposes and showed comparisons between the methods. For this purpose, annual spatiotemporal and seasonal variations in the WQIs of the groundwater around the Ganges River basin area in Kushtia district, Bangladesh, were investigated using the CWQI and weight average WQI. In this investigation, a comparative study was conducted using several statistical methods (correlation and t-test) to select the appropriate model for determining the suitability of water quality for drinking water and irrigation purposes.

Metal-contaminated groundwater has delirious impact on the human physiology, causing acute and chronic toxicities. This water is a significant natural resource for domestic, industrial, and irrigation purposes. Harmless drinking water is vital for healthy living. Approximately 90% of the population of a developing country, such as Bangladesh, has access to drinking water from shallow tube wells. The metal pollution in groundwater has concerned global consideration due to its abundance, serious toxicity, and persistence. Several pollution sources, including industrial effluents, mining and smelting, urban waste, atmospheric depositions in soil, and agrochemicals, degrade groundwater through their growing activities. The presence and circulation of trace metals in soils are influenced mostly by the parent material, soil properties, metal speciation, and climatic conditions. Toxic trace elements can influence the mental and central nervous system, lungs, kidneys, liver, blood composition, and other important organs in the human body. To maintain good human health, the US-EPA, WHO, and PHED have set standard values for adapting the maximum threshold levels (MTLs) in drinking water for trace metals. Also, health risk assessment (HRA) and several models were recognized by local and worldwide authorities for the intake of trace metals through water consumption.

Several studies have explored the incidence of heavy metals in the groundwater of Bangladesh. Sarkar et al. [6], Saha and Zaman [7], and Islam and Mostafa [8] measured excess concentration of lead in groundwater in the northern region of the country. Except for coastal areas, a higher level (over 5 mg/L) of iron (Fe) in drinking and irrigation water is the greatest environmental threat to almost the entire country. Other trace metals, such as manganese (Mn) and arsenic (As), were observed at higher concentrations than the standard level in the selected area's groundwater. Except for those, other trace metal concentrations were practically within the safe ranges in the countrywide groundwater.

Numerous water suitability indices, such as the heavy metal pollution index (HMPI), degree of contamination (C_d), and heavy metal evaluation index (HMEI), have been expected for the evaluation of water quality in view of trace element parameter. This study considered those indices as supporting influences for the assessment of the health risk assessment (HRA) of groundwater in the study area. The HRA is well defined as a procedure used to estimate health effects over time that might result from the intake of toxic materials. It is characteristically attained in four steps: risk findings, assessment of toxicity (dose–response), exposure evaluation, and risk categorizations. The major routes for heavy metal consumption in the human body are oral, dermal, and nasal through drinking water, foodstuff, and dust, in which ingestion and dermal absorption are common pathways. The stage of hazard identification is used to establish a connection between the identification of carcinogenic and noncarcinogenic substances and their health impacts on inhabitants in the investigation area. Exposure calculation is used to evaluate the category and degree of exposure from substances of probable concern that are permanently present or migrated from other locations. To measure exposure, it is essential to estimate the chronic daily intake (CDI) of toxic chemicals by drinking groundwater paths. Toxicity evaluation (dose–response) offers the connection between the degree of exposure and opposing health effects. For noncarcinogens, the dose–response calculation and reference doses (RfDs) were computed, and for carcinogens, incremental slope factors (ISFs) were determined by the Risk Assessment Information System (RAIS). The risk classification step creates all the information met in the three prior steps to calculate the likelihood that a theoretical exposure may harmfully influence human health. After calculation of the HRA, the results were then related with the output of numerous recent studies on that topic that were conducted in various regions of the world.

The present study zone was part of the Ganges River basin, located in the middle-west zone of Bangladesh. The source of domestic water in this area is mostly raw groundwater. In the Ganges River basin, the groundwater suitability is poor in some regions due to contamination with high mineralization and the presence of some toxic elements. Therefore, suitable assessment and reporting of groundwater quality are vigorous issues in the study area. The study delivers important information for groundwater suitability indices and HRA that support the sustainability of drinking water management considering water demand and the safe environment.

The study used different water quality indexing models to evaluate drinking purposes and presented a judgement between the methods. For this purpose, yearly spatiotemporal and periodic variations in the WQIs of the groundwater around the Ganges River basin area, Kushtia, Bangladesh, are explored utilizing the CWQI and WWQI. Using some statistical techniques (correlation and t-test), the study made a relative study between them and designated a suitable method to determine the suitability of water quality for domestic and irrigation purposes.

6.2 Assessment of Drinking Water Quality

Various methods have been used to assess groundwater suitability for drinking purposes. For this reason, three water quality index (WQI) methods, heavy metal contamination indices, and health risk assessment (HRA) for trace metals were used. Here, the computing processes are described below:

6.2.1 *Indexing Methods*

6.2.1.1 Canadian Water Quality Index (CWQI)

The CWQI has been recognized by the Canadian Council of Ministers of the Environment based on the water quality index equation (WQI = $[F_1^2 + F_2^2 + F_3^2]^{1/2}$) announced by the British Columbia Ministry of Environment. The Canadian WQI does not select any water quality parameters or variables and sampling seasons since the value of parameters varies from sampling location to location and is subject to environmental conditions. A minimum of four water variables and a minimum of four measurements of these variables are obligatory for the calculation of this index [9, 10].

The CWQI comprised three (3) aspects, each fluctuating from 0 to 100. The theoretical model for the index is presented in Fig. 6.1. The values of three (3) variances, viz. *scope, amplitude, and frequency*, have produced a vector in an imaginary 'objective exceedance' space. The length of the vector is then scaled to vary from 0 to 100 and subtracted from 100 to yield an index value that is zero or close to zero for 'very poor' water suitability and close to 100 for 'excellent' water quality for any purpose.

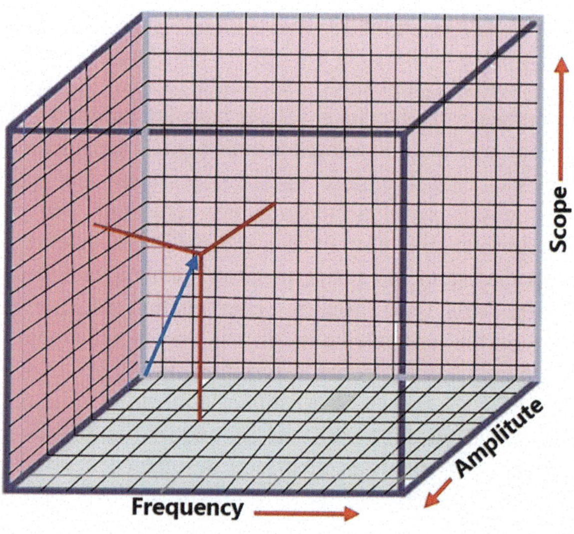

Fig. 6.1 Theoretical model of the Canadian Water Quality Index

The index is made up of three factors:

Factor 1 *F*1 is denoted as factor 1 and is represented as ***scope*** (Fig. 6.1a). It gives the % of variables that surpass the objective or standard value in the documented guidelines comparative to the entire number of variables. Therefore,

$$F_1 = \left(\frac{\text{Number of failed variables}}{\text{Total number of variables}}\right) \times 100 \tag{6.1a}$$

Factor 2 F_2 is factor 2 and is signified as ***frequency*** (Fig. 6.1b). It gives the % failed tests comparative to the whole number of tests carried out throughout the monitoring course:

$$F_2 = \left(\frac{\text{Number of failed test}}{\text{Total number of tests}}\right) \times 100 \tag{6.1b}$$

Factor 3 F_3 is factor 3 and is named ***amplitude***. F_3 gives the number of failed test values that crossed the objective value in the guidelines. F_3 can be computed in three steps:

Step a. The quantity of times that the value of the variable or parameter does not encounter the objective is denoted as 'excursion' and can be computed as follows:

(i) For the situation in which the value of the parameter should not be greater than the objective/guideline value:

$$\text{excursion}_i = \left(\frac{\text{Failed test value}_i}{\text{Objective}_i}\right) - 1 \tag{6.1c}$$

(ii) For the condition in which the value of the parameter should not be fewer than the objective/standard value:

$$\text{excursion}_i = \left(\frac{\text{Objective}_i}{\text{Failed test value}_i}\right) - 1 \tag{6.1d}$$

(iii) For the situation in which the objective (standard) value of parameters is zero:

$$\text{excursion}_i = \text{Failed test value}_i \tag{6.1e}$$

Step b. The next phase is calculating the ratio of the sum of *excursions* obtained in Step 1 to the total number of investigations. This ratio represents the normalized sum of *excursions* (nse):

$$\text{nse} = \frac{\sum_{i=1}^{n} \text{excursion}i_i}{\text{Total number of tests}} \tag{6.1f}$$

Step c. The concluding step is the computation of F_3 with the scaling of the 'nse' value from the objectives or standard to the range among 0 and 100. Hence,

$$F_3 = \left(\frac{\text{nse}}{0.01\text{nse} + 0.01}\right) \tag{6.1g}$$

Finally, the Canadian WQI is then represented as Eq. 6.1h through the combination method:

$$\text{CWQI} = 100 - \left(\frac{\sqrt{F_1^2 + F_2^2 + F_3^3}}{1.732}\right) \tag{6.1h}$$

The value of 1.732 increases because each of the three (3) individual index factors can range as high as 100. This means that the vector length can reach $\sqrt{100^2 + 100^2 + 100^2} = 1.732$ as a maximum value. The vector length down to 100 as a maximum value is obtained through dividing by 1.732. The calculation of all steps should be performed by SPSS statistics or MS Excel software.

6.2.1.2 Classical Water Quality Index (WQI)

The water quality index (WQI) has been adopted in various investigations for such purpose. WQI was first anticipated by Horton [4]. The same was adopted in this study, with a modified weighted arithmetic index approach from Brown et al. [5], to evaluate the suitability of groundwater for drinking. This indexing model was calculated in three steps, based on 16 physicochemical parameters ($n = 16$). In this method, weights (w_i) are assigned to water quality parameters based on the health impact of drinking water. A maximum value of 5 was assigned to parameters such as Pb, As, Cr, Cd, NO_3^-, and pH, which are considered to be of extreme health importance and have a significant impact on groundwater quality. TDS, TH, Ca^{2+}, Mg^{2+}, Na^+, K^+, Cl^-, SO_4^{2-}, PO_4^{3-}, and Fe have a scale of 2–4, depending on their perceived significance for drinking water quality and health effects (Table 6.1). The method also includes calculation of comparative weights (*Wi*) and quality rating scales (*qi*) and determination of subindexes (SI) and water quality indices (WQI). The calculated WQIs were then classified according to Sahu and Sikdar [11] (Table 6.2):

Table 6.1 Standards (S_i), weights (w_i), and relative weights (W_i) used for WQI calculation

Chemical parameter	Unit	Objective value (S_i)	Weight (w_i)	Relative weight (W_i)
pH	-	7.5	5	0.0847
TDS	mg/L	500	4	0.0678
Total hardness, TH	mg/L	200	4	0.0678
Na^+	mg/L	200	2	0.0339
K^+	mg/L	30	2	0.0339
Ca^{2+}	mg/L	200	2	0.0339
Mg^{2+}	mg/L	150	2	0.0339
Cl^-	mg/L	250	3	0.0508
SO_4^{2-}	mg/L	250	3	0.0508
NO_3^-	mg/L	50	5	0.0847
PO_4^{3-}	mg/L	0.7	4	0.0678
Cr	mg/L	0.05	5	0.0847
As	mg/L	0.01	5	0.0847
Pb	mg/L	0.01	5	0.0847
Fe	mg/L	0.3	3	0.0508
Cd	mg/L	0.003	5	0.0847
			Total weight = 59	$\sum W_i = 1$

Table 6.2 Drinking water suitability classification based on the classical WQI method

WQI range	Category	Type of water
<50	Class I	Excellent water quality
50–100	Class II	Good water quality
<100–200	Class III	Poor water quality
<200–300	Class IV	Very poor water quality
<300	Class V	Rejected water for drinking

$$W_i = \frac{w_i}{\sum w_i} \tag{6.2}$$

$$q_i = \frac{C_i}{S_i} \times 100 \tag{6.3}$$

where *Ci* and *Si* are the concentration of water parameter and objective value to be met, respectively:

$$\mathrm{SI} = W_i \times q_i \tag{6.4}$$

$$\mathrm{WQI} = \sum_{n=0}^{n} \mathrm{SI} \tag{6.5}$$

6.2.1.3 Weighted Average Water Quality Index (WWQI)

The value of the weighted average water quality index was computed from Eq. 6.8. The WWQI was calculated utilizing the weighted arithmetic/average water quality parameters, which was predicted by Horton [4]. Brown et al. [5] and Cude [12] then developed it in which water quality parameters are multiplied by a weighting factor (w_i) and are then aggregated using the following three equations of 6.6, 6.7, and 6.8:

$$W_i = \frac{K}{S_i} = \frac{1}{\sum \frac{1}{S_i}} / S_i \tag{6.6}$$

$$Q_i = \frac{(V_n - V_0)}{(S_n - V_0)} \times 100 \tag{6.7}$$

$$\text{WWQI} = \sum_{i=1}^{n} \frac{W_i Q_i}{\sum_{i=1}^{n} W_i} \tag{6.8}$$

where Q_i is the subindex of the ith water parameter, W_i is the unit weightage of the ith water parameter, n is the number of parameters comprised, V_n is the experimental value of the parameter, V_O is the perfect value of fully clean water, and S_i is the objective or guideline value of the ith parameter.

The perfect value for pH is equal to 7, DO is equal to 14.6 mg/L, and for other water parameters, it is usually equal to zero for maximum parameters excluding pH.

That is,

$$Q_{\text{pH}} = \frac{(V_{\text{pH}}\text{–}7)}{8.5\text{–}6.5} \times 100 \tag{6.9}$$

The weightage unit (W_i) of each parameter has computed a value in inverse proportional to the recommendation of the World Health Organization (S_i).

6.2.2 *Heavy Metal Pollution Indices*

Three recognized methods, the heavy metal evaluation index (HMEI), heavy metal potential index (HMPI), and degree of contamination (C_d), were assessed for their suitability for pollution monitoring of groundwater in middle- east parts of Bangladesh.

6.2.2.1 Heavy Metal Evaluation Index (HEI)

The heavy metal evaluation index (HEI) model delivers insight into the complete quality of groundwater concerning heavy and trace metals, and it was calculated by Eq. 6.10:

$$\mathrm{HEI} = \sum_{i=1}^{n} \frac{H_c}{H_{\mathrm{MAC}}} \tag{6.10}$$

where H_c is the measured value and H_{MAC} is defined as maximum admissible concentration (MAC) of the ith metal parameter.

6.2.2.2 Heavy Metal Pollution Index (HPI)

The heavy metal pollution index (HPI) technique was recognized by assigning the weightage (W_i) for the designated parameter and choosing the groundwater parameter on which the index must be based. The score is close 0 to 1, and its selection exposes the consequence of each water suitability parameter. HPI is calculated by Eq. 6.11 as follows:

$$\mathrm{HPI} = \frac{\sum_{i=1}^{n} W_i Q_i}{\sum_{i=1}^{n} W_i} \tag{6.11}$$

where W_i is the unit weight of the ith variable (Table 6.3) and n is the number of variables.

The subindex Q_i is calculated by Eq. 6.12:

$$Q_i = \sum_{i=1}^{n} \frac{|M_i - I_i|}{(S_i - I_i)} \times 100 \tag{6.12}$$

Here, S_i, M_i, and I_i provide the standard values, monitored value, and perfect value of the ith parameter, respectively (Table 6.3).

6.2.2.3 The Degree of Contamination (C_d)

The degree of contamination (C_d) was implemented by Backman et al. [13], and C_d was calculated by Eq. 6.13:

$$C_d = \sum_{i=1}^{n} \frac{C_{ai}}{C_{ni}} - 1 \tag{6.13}$$

Table 6.3 Standard values (S_i), ideal values (I_i), and maximum admissible concentrations (all are in mg/L unit) for the analysed metal elements

Metal elements	S_i	I_i	MAC	W_i
Fe	1.0	0.3	1.0	0.001855
Mn	0.3	0.1	0.3	0.005958
B	2.4	1.0	2.4	0.000852
Cr	0.05	0	0.05	0.039977
Pb	0.01	0	0.01	0.202634
Co	0.05	0	0.05	0.045677
Ni	0.1	0.07	0.1	0.022053
Cd	0.003	0	0.003	0.674379
Cu	2.0	0.5	2.0	0.001023
Zn	3.0	0.5	3.0	0.000682
As	0.01	0	0.01	0.005260
				$\sum W_i = 1$

where $[(C_{ai}/C_{ni}) - 1]$ is the pollution factor, which is denoted as C_f; C_{ai} is the analysed value, C_{ni} is the maximum permitted level for the ith parameter, and n is the normalizing value. Here, C_{ni} is denoted as the maximum admissible concentration (MAC).

6.2.3 *Human Health Risk Assessment (HRA)*

HRA is a multistep procedure that includes the assembly and assessment of data, valuation of exposure, approximation of toxicity, and risk characterization. Experience with trace metals in the human body can occur through oral and dermal absorption, which are general routes for water intake. Both carcinogenic and noncarcinogenic hazards correlated to these contact routes were distinctly evaluated. The dose received (chronic daily intake, $\text{CDI}_{nc\text{-}ca}$) through two exposure pathways was calculated using Eqs. 6.14 to 6.17 adapted from the US-DOE Risk Assessment Information System [14] and US-EPA [15, 16] for two sub-resident groups of children and adults.

6.2.3.1 For Noncarcinogenic (nc) Risk Calculation

$$\text{CDI}_{\text{oral}-\text{nc}} = \frac{C_w \times \text{IRW} \times \text{EF} \times \text{ED}}{\text{BW} \times \text{AT}} \tag{6.14}$$

$$= C_w \times \text{Factor A}_{\text{adult/child}} \text{ mg/kg/day} \tag{6.15}$$

Table 6.4 Crucial factors for calculating the health risks for drinking water through oral and dermal pathways

Symbol	Exposure route/exposure factor	Value	References/sources
IRW_a	Water ingestion rate for adult	2 L/day	Rani et al. [17]
BW_a	Body weight for adult	70 kg	Rani et al. [17]
IRW_c	Water ingestion rate for child	1 L/day	Harries and Harper [18]
IRW_{adj}	Age-adjusted water ingestion rate	2.229 L-year/kg-day	Equation (6.10)
C_w	Contaminant conc. of each metal	(mg/L)	–
ED_a	Exposure duration for adult	70 years	Harries and Harper [18]
BW_c	Body weight for a child	15 kg	Harries and Harper [18]
ED_c	Exposure duration for child	6 years	Harries and Harper [18]
EF	Exposure frequency	365 days/year	Duggal et al. [19]
ET_a	Absorbed dose per event	0.58 hours/day	Equation (6.12)
AT_a	Average time for adult	25,550 days	AT = EF × ED
AT_c	Average time for a child	2190 days	AT = EF × ED
CF	Conversion factor	0.001 L/cm^3	1 L = 1000 cm^3
ET_c	Exposure time for a child	1 hour/day	US-EPA [20]
ET_{adj}	Age-adjusted exposure time	0.616 hours/day	Equation (6.16)
EV_a	Event frequency for an adult	1 event/day	US-EPA [20]
EV_c	Event frequency for child	1 event/day	US-EPA [20]
SA_a	Skin surface area for adult	18,000 cm^2	US-EPA [20]
SA_c	Skin surface area of the child	6600 cm^2	US-EPA [20]
SA_{adj}	Age-adjusted skin surface area	19,097 cm^2-year/kg-day	Equation (6.16)
K_p	Permeability coefficient for each metal	COPC-specified (see Table 6.5)	US-DOE [14]

where the measured values of Factor A_{adult} and Factor A_{child} are 0.0285 and 0.0666, respectively (based on Table 6.4):

$$\mathrm{CDI}_{\mathrm{derm-nc}} = \frac{K_p \times C_W \times \mathrm{ET_a} \times \mathrm{CF} \times \mathrm{SA} \times \mathrm{EF} \times \mathrm{ED} \times \mathrm{EV}}{\mathrm{BW} \times \mathrm{AT}} \quad (6.16)$$

$$= K_p \times C_W \times \text{Factor } \mathrm{B}_{\mathrm{adult/child}} \ \mathrm{mg/kg/day} \quad (6.17)$$

where the intended values of Factor B_{adult} and Factor B_{child} are 0.000149 and 0.00044, respectively (based on Table 6.4). The K_p values of separate metals are listed in Table 6.5.

6.2.3.2 For Carcinogenic (ca) Risk Calculation

$$\mathrm{CDI}_{\mathrm{oral-ca}} = \frac{C_w \times \mathrm{IRW}_{\mathrm{adj}} \times \mathrm{EF}}{\mathrm{AT}} \tag{6.18}$$

$$= C_w \times \text{Factor } \mathrm{C}_{\frac{\mathrm{adult}}{\mathrm{child}}} \mathrm{mg/kg/day} \tag{6.19}$$

where Age – adjusted water ingestion rate, $\mathrm{IRW}_{\mathrm{adj}} = 2229\ \mathrm{L}.\frac{\mathrm{year}}{\mathrm{kg}}.\mathrm{day}$

and the calculated values of Factor $\mathrm{C}_{\mathrm{adult}}$ and Factor $\mathrm{C}_{\mathrm{child}}$ are 0.0317 and 0.2049, respectively (based on Table 6.4).

$$\mathrm{CDI}_{\mathrm{derm-ca}} = \frac{K_p \times C_W \times \mathrm{ET}_{\mathrm{adj}} \times \mathrm{CF} \times \mathrm{SA}_{\mathrm{adj}} \times \mathrm{EF}}{\mathrm{AT}} \tag{6.20}$$

$$= K_p \times C_W \times \text{Factor } \mathrm{D}_{\frac{\mathrm{adult}}{\mathrm{child}}} \mathrm{mg/kg/day} \tag{6.21}$$

where Age – adjusted skin surface area, $\mathrm{SA}_{\mathrm{adj}} = 19097\ \mathrm{cm}^2.\frac{\mathrm{year}}{\mathrm{kg}}.\mathrm{day}$

and Age – adjusted exposure time, $\mathrm{ET}_{\mathrm{adj}} = 0.616\ \frac{\mathrm{h}}{\mathrm{day}}$

where the measured values of Factor $\mathrm{D}_{\mathrm{adult}}$ and Factor $\mathrm{D}_{\mathrm{child}}$ are 0.0002 and 0.002, respectively (based on Table 6.4). The K_p values of separate metals are listed in Table 6.5. All factor/parameter values (Eqs. 6.18 to 6.21) are collected from the US-EPA [20], Harries and Harper [18], Rani et al. [17], and Duggal et al. [19].

Table 6.5 Values of $\mathrm{RfD}_{\mathrm{oral}}$, $\mathrm{CSF}_{\mathrm{oral}}$, and $\mathrm{GI}_{\mathrm{ABS}}$ for analysed metals (US-DOE [14])

Element	K_p (cm/hour)	$\mathrm{CSF}_{\mathrm{oral}}$ (mg/kg/day)	$\mathrm{RfD}_{\mathrm{oral}}$ (mg/kg/day)	Factor for gastrointestinal absorption ($\mathrm{GI}_{\mathrm{ABS}}$)
Fe	0.005	–	7.00E−01	1
Mn	0.003	–	4.60E−02	4.00E−02
Cr	0.002	4.10E−02	1.50	1.30E−02
B	0.002	–	2.00E−01	1
Pb	0.001	8.50E−03	3.50E−03	1
Co	0.001	–	3.00E−04	1
Ni	0.002	8.60E−04	2.00E−02	4.00E−02
Cd	0.001	6.10E−03	5.00E−04	5.00E−02
Cu	0.001	–	4.00E−02	1
Zn	0.006	–	3.00E−01	1
As	0.002	–	3.00E−04	5.00E−01

Assessing the hazard quotient (HQ) was utilized for the assessment of noncarcinogenic risks. To obtain the HQ_{oral} and HQ_{derm} values, the assessed CDIs for individual metals were divided by their reference dose (*RfD*) and *RfD* with gastrointestinal absorption factor (GI_{ABS}), respectively (Table 6.5):

$$HQ_{oral} = \frac{CDI_{oral-nc}}{RfD_{oral}} \tag{6.22}$$

$$HQ_{derm} = \frac{CDI_{derm-nc}}{RfD_{oral} \times GI_{ABS}} \tag{6.23}$$

where GI_{ABS} is a gastrointestinal absorption factor (dimensionless).

In the human body, a combination of noncarcinogenic risks across varied intake paths can be obtained from the total HQ value of an individual contact pathway for a trace element to produce the screening level hazard index (HI) (Eq. 6.24). If the HI value is less than 1 ($HI < 1$), noncarcinogenic risks are not predictable to occur from any metal. If the HI crossed the unit value ($HI > 1$), there may be a concern for potential noncarcinogenic impact: [21]

$$HI = \sum_{i=1}^{n} HQ_i \tag{6.24}$$

The potential incremental lifetime cancer risks (ILCRs) for oral and dermal water exposure to Cr, Ni, Pb, and Cd were calculated by Eqs. 6.25 and 6.26 with corresponding cancer slope factors (CSFs) at a 95% confidence limit [14, 20]. The prescribed *CSF* values of these metals are included in Table 6.5. Lastly, the total incremental lifetime cancer risks ($ILCR_{total}$) were measured by addition of the ILCRs through oral and dermal absorption routes. As a whole, ILCRs less than 10^{-6} (a possibility of 1 chance in 1,000,000 of discretely developing cancer) are well thought-out to be insignificant, though *ILCRs* above 10^{-4} are measured unsafe for human consumption, as declared by most international regulatory agencies:

$$ILCR_{oral} = CDI_{oral-ca} \times CSF_{oral} \tag{6.25}$$

$$ILCR_{derm} = CDI_{derm-ca} \times \frac{CSF_{oral}}{GI_{ABS}} \tag{6.26}$$

$$ILCR_{oral/derm} = \sum ILCR_{oral/derm} \tag{6.27}$$

$$ILCR_{total} = ILCR_{oral} + ILCR_{derm} \tag{6.28}$$

6.3 Drinking Water Quality Evaluation

Water quality assessment can be evaluated for drinking by physicochemical and biological aspects. Water pollution by organic and inorganic contaminants harms the environment and human health. Heavy mineralization of groundwater, excess use of agrochemicals and petroleum substances, municipal solid and liquid waste, landfills, etc., can deteriorate drinking groundwater. In this section, numerous water quality indexing methods and various drinking water quality indicators, such as heavy metal pollution indices (HMPI), water quality indices (WQI), and human health risk assessment (HRA), are used to assess the suitability of drinking water.

6.3.1 Application of WQI Methods

The classical water quality indexing method was first approved (1965) to evaluate the quality of drinking water through mathematical equations using some physiochemical water parameters. It is considered an indicator for the cataloguing of water into five categories: excellent, good, poor, very poor, and water rejected for drinking (Table 6.2). Index calculations were made for 40 groundwater samples in each of the three sampling periods in the study area. Due to the higher concentration of total hardness, Ca, Fe, and Pd, the index value was very high in several samples. The WQI values fluctuating from 88.8 to 441 designated that only 29.5% of the groundwater is 'good' for drinking purposes, and 38.7%, 17.3%, and 14.5% of the groundwater are categorized as 'poor', 'very poor', and 'unsuitable' for drinking purposes, respectively. These data are the average value for three sampling seasons.

The study followed another two generally used water quality index models, which were then improved by numerous researchers or water managers. The weighted average indexing method, whereby subindices are created and further adding to an overall index value. Subindices are value functions used to change the different units of water quality parameters to a common scale. The second method is the amplitude procedure (objective-based), where the overall water quality index value is invented by measuring the extent to which water quality parameters diverge from the objectives or standard values. In this study, the weighted water quality index model (WWQI) and Meireles water quality index model (MWQI) as the first category and the Canadian Water Quality Index (CWQI) as the second category methods were used to investigate the groundwater suitability of the study zone (Appendix X). Using the numerous statistical methods, the results of those techniques were compared with each other, and select one to determine the groundwater suitability for both drinking and irrigation prepossess properly.

Canadian Water Quality Index (CWQI)

The complete dataset for calculating the results of the CWQI and CWQI for drinking purposes is presented in the Appendix section. The summary statistics of these two indices are shown in Table 6.6 and Fig. 6.2. The results presented that the mean value of the CWQI was 54.46 (38.3–70.5), with a standard deviation (SD) of ±11.57

Table 6.6 Statistical summary of WQI values of collected water samples (Appendix X)

Value	CWQI (drinking)				WWQI (drinking)			
	PRM	MON	POM	Average	PRM	MON	POM	Average
Min.	40.21	38.35	37.09	38.55	88.8	93.34	95.51	92.66
Max.	73.09	70.57	70.32	71.33	488.62	531.17	516.56	478.21
Mean	**56.24**	**54.19**	**52.94**	**54.46**	**258.22**	**295.18**	**312.52**	**289.07**
Variance	134.86	133.68	134.17	133.89	15057.83	17104.54	16808.02	15542.09
SD (±)	11.61	11.56	11.58	11.57	122.71	130.78	129.65	124.67
Skewness	0.046	0.068	0.087	0.065	−0.038	−0.035	−0.228	−0.180

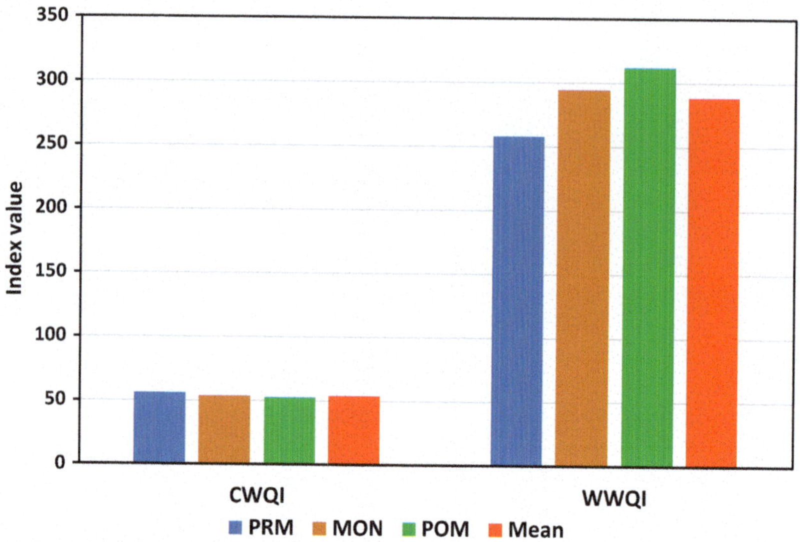

Fig. 6.2 Index values of CWQI and WWQI of three sampling seasons with the mean values

and a variance of 133.89. The water suitability classification for sampling sites was obtained to be 'fair' (C class) for 37.5%, 'marginal' (D class) for 22.5%, and 'poor' (E class) for 40% (Table 6.6). The EC, total hardness (TH), turbidity, Ca^{2+}, HCO_3^-, Fe, and Mn values were very high regarding national and WHO guideline values that seriously impacted the lowering values of the CWQI. The groundwater in the seaside zone of southern part of the country contained elevated level of Na salinity with higher value of EC compared to the present study zone. But the total hardness (Ca^{2+}, Mg^{2+}, and HCO_3^- level) is very low in the groundwater of that coastal area [22, 23]. Thus, the study predicted that the Canadian WQI value of this groundwater presented the similar results as the present study. In addition, the values of the most water parameters of groundwater in the northwest region of Bangladesh were significantly less than those in the seaside areas as well as the study zone. A literature report established that the water quality index value of the northern area of the country was in very good condition [24].

The Canadian Council of Environment Ministers investigated the sensitivity of this model to changes in some inputs and conducted a study on large volumes of water quality data. One of the results of these tests is that the performance of the Canadian WQI model is reasonable if at least ten parameters are included and at least 30 observations over at least 3 years are included in the index calculation. In this regard, the study indicated that the number of inputs (parameters and sampling sites) was sufficient and that the study complied with WHO and national guidelines on water quality standards (whichever is appropriate), so that the present dataset provided enough information. The model is flexible in choosing input parameters and targets and tolerates missing data. The CWQI model has great success to measure the water quality but some limitations are identified such as the following: (a) all variables were given equal importance, (b) there was lack of guidance on site-specific targets and specific water uses, (c) there was only partial diagnosis of water quality, (d) it was easy to manipulate, and (e) *F*1 did not perform well when too few variables were measured.

Weighted Average Water Quality Index (WWQI)

The WWQI value and water type of the samples are presented in Table 6.6, and the classification of that method is presented in Table 6.7. This index was calculated for the samples using the concentration/value of 14 parameters such as pH, EC, TDS, TH, turbidity, Ca^{2+}, Mg^{2+}, Na^{+}, K^{+}, Cl^{-}, HCO_3^{-}, SO_4^{2-}, NO_3^{-}, and PO_4^{3-}. The WWQI ranges from 88.8 to 551.6 and 95.5 to 543.3 for the pre- and post-monsoon seasons, respectively. The results showed that the average index value of the samples ranged from 92.3 to 524.7. The mean value was 287.6 with a standard deviation and

Table 6.7 WQI designations and summarized results of samples

WQI	Parameter uses	Scale	Index value	Category	Description	Av. result (% sample)
CWQI (drinking)	Detected all parameters	0–100	95–100	A	All quantities are within objectives almost all the time: excellent	–
			80–94	B	Conditions infrequently proceed from natural or required levels: good	–
			65–79	C	Conditions occasionally depart from required levels: fair	37.5
			45–64	D	Conditions often depart from natural or wanted levels: marginal	22.5
			0–44	E	Conditions typically depart from natural or necessary levels: poor	40
WWQI (drinking)	pH, EC, TDS, TH, turbidity, Ca^{2+}, Mg^{2+}, Na^{+}, Cl^{-}, SO_4^{2-}, NO_3^{-}, PO_4^{3-}, and HCO_3^{-}	0–100	0–25	A	Excellent	–
			26–50	B	Good	–
			51–75	C	Poor	–
			76–100	D	Very poor	5
			>100	E	Undesirable	95

variance of 128.95 and 16627.40, respectively (Table 6.6). These figures are the average of two seasons, i.e. the pre- and post-monsoon seasons. According to the computed value, 95% of the tested samples are 'undesirable (E category)', and 5% of the samples fall under the 'very poor (D category)' quality type for drinking purposes (Table 6.7). Howladar et al. [24] used the same groundwater parameters to calculate the WWQI in the Dinajpur District in the northern part of Bangladesh. They found that 70.97%, 22.5%, and 6.45% of water samples fell into excellent, good, and poor ranges, respectively. This is the big difference between both results. The high EC value of the present samples provided a higher WWQI value, indicating unsuitability for potable uses. Besides, if we considered the concentrations of Fe and Mn, then these index values exceeded 500 in the case of all samples. High values of EC, TDS, turbidity, and TH (Ca, Mg, and HCO_3^-) also contributed to high index values, typically during the POM rather than the PRM period. The study areas are situated in the upper Ganges floodplain and river delta basin areas, which makes it rich in aquifers with excess mineralization by the mineral/rock weathering process, which is the main feature of geochemical processes in this area [25]. For this reason, the groundwater became highly mineralized and contained large mineral ions that could enhance the WWQI values.

The weighted average WQI values and water categories of the samples are shown in Table 6.6, and classification of that method is revealed in Table 6.7. This index uses concentrations/values of 14 parameters including pH, EC, TDS, TH, turbidity, Ca^{2+}, Mg^{2+}, Na^+, K^+, Cl^-, HCO_3^-, SO_4^{2-}, NO_3^-, and PO_4^{3-}. The calculated values of WWQI vary from 88.8 to 551.6 and 95.5 to 543.3 before and after the monsoon, respectively. The results displayed that the mean exponent values for the samples ranged from 92.3 to 524.7. The mean was 287.6 with standard deviation and variance of 128.95 and 16627.40, respectively (Table 6.6). These numbers are averages of two sampling periods before and after the monsoon season. Based on the calculated values of WWQI, 95% of the tested samples are of 'unwanted' (category E) for potable use and 5% of the samples are of quality type 'very poor' (category D) (Table 6.7). Howlader et al. [24] calculated the WWQI for Dinajpur district in northern Bangladesh using the same groundwater parameters. They found that 70.97%, 22.5%, and 6.45% of the water samples fell into the excellent, good, and poor ranges, respectively. This is the big difference between these two results. The high EC values of the presented samples lead to higher WWQI values, indicating their unsuitability for drinking water purposes. Furthermore, considering the concentrations of Fe and Mn, these index values were above 500 for all samples, and high EC, TDS, turbidity, and TH (Ca, Mg, HCO_3^-) values also contributed to the high index values. In addition, it was found that usually index value obtained higher ranges in the POM period rather than the PRM period. The study area is located in the upper Ganges floodplain and delta region, an area rich in aquifers with excessive mineralization due to mineral/rock weathering processes, which is a major feature of geochemical processes in the region [25]. For this reason, groundwater is highly mineralized and contains large amounts of mineral ions that can increase WWQI values.

This WWQI model has some limitations and doubts. It may not get enough information about the actual water suitability. Various important water quality parameters (such as trace elements) are not included in the calculation of this index. Hiding or overemphasizing a single erroneous parameter value can affect this model. No single parameter could represent the entire history of water quality, and many other water quality parameters were missing from the index. However, the method is simple and easy to calculate, and several very important water quality parameters are used to assess drinking water quality.

Judgement of WQIs and Their Use in Practice

Concerning water suitability parameters, in actual fact, there are somewhat differences between groundwater and surface water. Most of the indexing methods were computed using parameters of surface water, such as bacteriological contents and biochemical parameter loads, which are unimportant for groundwater quality evaluation. Because of the inadequate index method, it is very problematic to accurately evaluate the suitability of groundwater for drinking, industrial, and irrigation use. The present study applied two methods for each drinking and irrigation water quality evaluation. It made a judgement between the CWQI (drinking) vs. WWQI and CWQI (irrigation) vs. MWQI and determined which one is better than the other for the present investigation.

Judgement Between CWQI and WWQI

There is a substantial variance between the results of the two models at a certain substantial limit, and the paired *t*-test was implemented, which is a statistical procedure to determine whether the mean difference between the two sets of observations is zero. The results of running paired *t*-tests by the SPSS computer programme are quantified in Table 6.8.

Table 6.8 shows that there is a large difference between the average Canadian WQI (54.17) and the weighted average WQI (287.6). Similarly, the standard deviation, variance, and standard error of both models are arithmetically different and do not yield statistically significant signs. Again, Table 6.8 illustrates that it has obtained a very high negative correlation ($r = -0.91$) and a population correlation coefficient of 0 ($p = 0$) and that the correlation could be zero (significance = 0.000). The mean difference was −233.46, the standard deviation of the difference was 141.299, and the mean standard error was 22.341, which were statistically abnormal values. Furthermore, the approximately 95% confidence interval (CI) for the mean difference was −278.652 to −188.273. The calculated *t*-value (T) was −10.450, the degree of freedom (*df*) was 39, and the *p*-value labelled as significant (two-tailed) was 0.000. This statistic showed that there was no statistical significance between the two indices in assessing drinking water quality.

Of these three index methods, the Canadian WQI is the only method that allows the use of all available parameters in calculating the overall index value. Using this method, the water quality data for all sampling stations were rated from fair to poor (Table 6.7). EC, TH (Ca^{2+}, Mg^{2+}, and HCO_3^-), turbidity, NH_3, Fe, and Mn were the main parameters that lowered the total CWQI values at all sampling stations.

Table 6.8 Paired sample *t*-test between CWQI–WWQI methods for comparative study

Pair CWQI–WWQI (drinking)	*Paired differences*							*T*	*df*	*Significance (two-tailed)*
	Mean	Std. deviation	Std. error mean	Paired sample correlations		95% confidence interval (CI) of the difference				
				Correlation	Sig.	Lower	Upper			
	−233.46	141.299	22.341	**−0.910**	0.000	−278.652	−188.273	−10.45	39	0.000

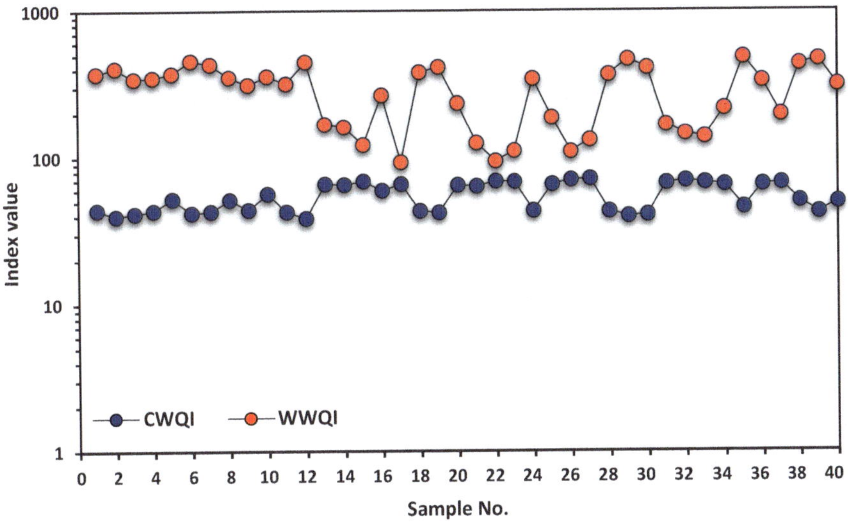

Fig. 6.3 The pair difference statistics between CWQI (drinking) and WWQI values for all water samples

Subindex models (such as WWQI) had a limited number of parameters, and subindex equations were provided for each parameter. The WWQI model contains 14 key parameters for assessing water quality. The indicator values obtained at all stations were classified as 'undesirable'. In this method, the combination of subindex values eliminated the eclipse problem, and the use of weighting factors for each parameter yielded reasonable results. As a result, there was a large difference between the index values of CWQI and WWQI for beverage purposes at all stations (Fig. 6.3). For CWQI, samples were classified into three categories: fair, borderline, and poor, whereas most samples were undesired in the WWQI model for drinking uses (Table 6.7). The largest samples (mainly S13–S40) showed completely opposite results of these two methods. Samples had lower index values, but more different index values were obtained from the same sample (Fig. 6.3). CWQI included all investigated parameters (including trace elements), while WWQI included some common physicochemical parameters. However, if only one parameter exceeds the target or default value, the WWQI subindex value will increase exponentially. Moreover, the results of a sensitivity test study of CWQI models [10] showed that variations in input factors can affect the overall indicator value very slowly. For example, studies in which the WWQI method took into account the Fe content of the samples increased the index value by more than 100 points, whereas for the CWQI a small change in the overall index value occurred. Therefore, in this study, we hypothesized that the results of the CWQI method are superior to those of the WWQI method.

6.3.2 Metal Pollution Indices

Several water quality indices were derived from the heavy metal evaluation index (HMEI), heavy metal pollution index (HMPI), and degree of contamination (C_d) using 40 groundwater samples, and the calculated values of these indices are presented in Table 6.9. The study measured the mean value of metal concentration in both pre-monsoon (PRM) and post-monsoon (POM) sampling seasons. Concerning these documentation approaches, the findings of Table 6.9 show that an average of 60% of samples fell in a 'high degree' of contamination caused by trace and heavy elements in groundwater samples. It is an excessive threat to the public health of the study area. The values of C_d presented an important correlation with HMEI, but the values of HMPI showed abnormal results regarding C_d and HMEI. Edet and Offiong [26] demonstrated the same connection among these three indices for the surface water sources of Nigeria. The elevated concentrations of Fe, Mn, Pb, and Zn in the groundwater samples were accountable for this result. The mean values of HMPI, HMEI, and C_d were 554.0, 32.30, and 31.27, respectively, in the present study. Also, Bodrud-Doza et al. [27] and Bhuiyan et al. [28] conducted two separate investigations in different districts, such as Faridpur and Lakhimpur, in Bangladesh. The values of the first investigation for the HMPI, HMEI, and C_d were 46, 8.55, and 7.52, respectively, and the second ones for the same indices were 26.12, 7.44, and 11.2, respectively, achieved in the groundwater samples. These results were found to be good enough compared to the present study. The study expected that trace elements are measured to cause toxic properties by changing the

Table 6.9 Summarized results of heavy metal contamination indices

Parameter (1)	Average (2)	Minimum (3)	Maximum (4)	Category/degree of pollution (5)	References
HMPI	554.0	15.44	1841.42	<45: low (30%) 45–90: medium (10%) >90: high (60%)	Raja et al. [29]
HMEI	32.30	4.61	69.51	<10: low (10%) 10–20: medium (17.5%) >20: high (72.5%)	Wagh et al. [30]
C_d	31.27	1.23	68.51	<10: low (22.5%) 10–20: medium (30%) >20: high (47.5%)	Edet and Offiong [26]
Average (column 5)	***Low contamination level in samples: (30 + 10 + 22.5) % = 62.5; average of 20.84%*** ***Medium contamination level in samples: (10 + 17.5 + 30) % = 57.5; average of 19.16%*** ***High contamination risk of the samples: (60 + 72.5 + 47.5) % = 180; average of 60.0%***				

mechanisms of the biochemical purposes of the exposed people in the study zone. As stated by several reports, heavy metals are known to fix protein sites by fluctuating the required metal complexes or causing the oxidative deterioration, leading to the out of order of human body cells. Alternatively, toxic elements can substitute the hydrogen atom from S–H bonds as H^+ of enzyme groups, and metal ions form a complex compound with enzymes, resulting in the breaking of the essential enzymes of the human body. Therefore, without the elimination of raised loaded metals in groundwater, they should not be used for drinking or other household purposes.

6.3.3 *Human Health Risk Assessment (HRA)*

Human HRA is well defined as the determination of the features and degree of opposing health effects in humans who may be exposed to toxic chemicals in a polluted situation. The health risk assessment model is the most operative tool for computing the carcinogenic and noncarcinogenic health risks for distinct age groups. These kinds of risks were calculated for the trace elements Fe, Mn, Cr, Co, Cd, Ni, Pb, B, As, Cu, and Zn. The concentration of metals in water samples was used to measure human exposure through oral consumption and dermal absorption routes. Two age groups, i.e. adults and children, were considered. Table 6.6 and Fig. 6.4a–c show the HQ and HI values for three sampling periods, viz. pre-monsoon, monsoon, and post-monsoon. On the other hand, ILCR values were assessed for local people based on the oral intake and dermal absorption of water and are represented in Tables 6.10a, 6.10b, and 6.10c.

6.3.3.1 Noncarcinogenic Risk Analysis (HQs and HIs)

The investigation carried out health exposure and risk assessment based on the US-EPA process. Human intake of trace metals primarily occurs via paths of potable water, foodstuffs, inhaled aerosols, and dust particles. The potential toxicity of trace elements to human physiology is straight connected with their everyday consumption. Two consumption processes viz. oral ingestion via drinking water and dermal adsorption were taken into account for risk calculation. The noncarcinogenic risk analysis was based on the values of chronic daily intake (CDI) (Eqs. 6.14–6.21). Calculated oral and dermal hazard quittance ($HQ_{oral/dermal}$) and hazard index ($HI_{oral/dermal}$) values for groundwater samples of three sampling seasons are presented in Tables 6.10a, 6.10b, and 6.10c and Fig. 6.4a–c. The table and figures show that almost the same results were found in this study in all sampling periods. Additionally, the result demonstrates the calculated average values of HQ_{oral} and HQ_{derm} (Eqs. 6.22 and 6.23) in the study areas were detected in the order of Co > Mn > Cd > Cu > Pb > Ni > Fe > Zn > B > Cr and Co > Mn > Cu > Cd > Pb > Fe > Ni > Zn > B > Cr, respectively, for the adult and child groups. The tables also show that the total HQs of the trace metals

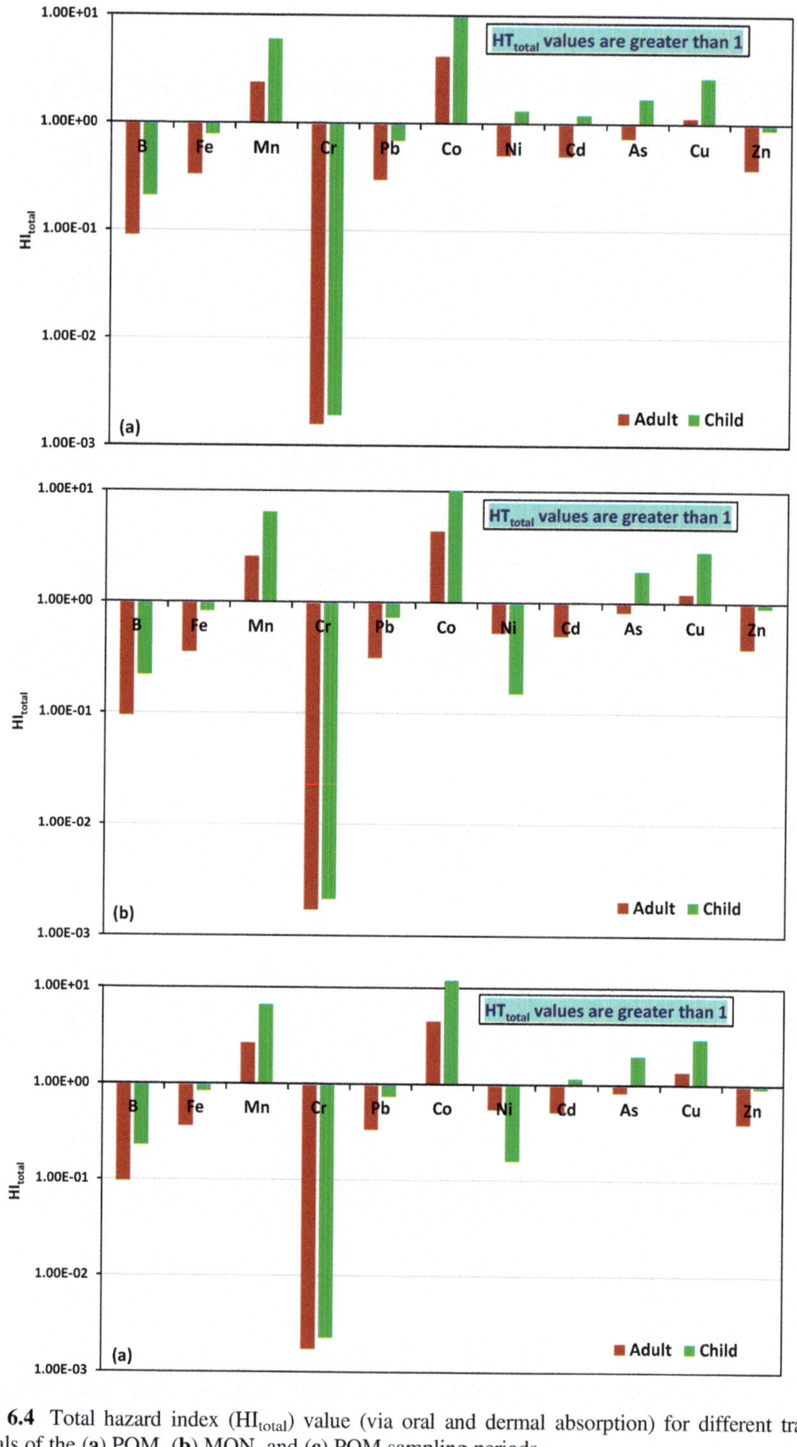

Fig. 6.4 Total hazard index (HI_{total}) value (via oral and dermal absorption) for different trace metals of the (**a**) POM, (**b**) MON, and (**c**) POM sampling periods

Table 6.10a Noncarcinogenic health risks of trace elements in samples by oral and dermal pathways in the PRM period

	HQ_{oral}		HQ_{derm}		$HI = \sum HQ_{oral/derm}$	
Element	Adult	Child	Adult	Child	Adult	Child
B	9.03E−02	2.11E−01	3.17E−04	9.35E−04	9.06E−02	2.12E−01
Fe	3.27E−01	7.62E−01	8.14E−03	2.40E−02	3.35E−01	7.86E−01
Mn	1.69E+00	3.95E+00	7.01E−01	2.07E+00	2.39E+00	6.02E+00
Cr	8.31E−04	1.94E−03	7.64E−04	3.81E−07	1.60E−03	1.94E−03
Pb	2.96E−01	6.91E−01	3.19E−03	5.15E−04	2.99E−01	6.92E−01
Co	4.16E+00	9.69E+00	2.73E−02	8.07E−02	4.19E+00	9.77E+00
Ni	4.37E−01	1.22E+00	6.97E−02	3.39E−04	5.07E−01	1.32E+00
Cd	4.35E−01	1.11E+00	6.56E−02	4.94E−04	5.01E−01	1.21E+00
As	7.36E−01	1.72E+00	3.08E−05	2.27E−05	7.36E−01	1.72E+00
Cu	1.14E+00	2.66E+00	3.33E−03	9.85E−03	1.14E+00	2.67E+00
Zn	3.76E−01	8.77E−01	5.14E−03	1.52E−02	3.81E−01	8.92E−01
Average value for all metals					**0.9610**	**2.2994**

Table 6.10b Noncarcinogenic health risks of trace elements in samples by oral and dermal pathways in the MON period

	HQ_{oral}		HQ_{derm}		$HI = \sum HQ_{oral/derm}$	
Element	Adult	Child	Adult	Child	Adult	Child
B	9.46E−02	2.21E−01	3.88E−04	9.93E−04	9.50E−02	2.22E−01
Fe	3.48E−01	8.12E−01	9.24E−03	2.40E−02	3.57E−01	8.36E−01
Mn	1.79E+00	4.17E+00	7.91E−01	2.34E+00	2.58E+00	6.51E+00
Cr	9.23E−04	2.15E−03	8.19E−04	3.97E−07	1.74E−03	2.15E−03
Pb	3.15E−01	7.36E−01	3.66E−03	5.35E−04	3.19E−01	7.37E−01
Co	4.38E+00	1.02E+01	2.92E−02	8.37E−02	4.41E+00	1.03E+01
Ni	4.65E−01	1.53E−01	7.27E−02	3.56E−04	5.38E−01	1.53E−01
Cd	4.40E−01	1.13E+00	6.88E−02	4.99E−04	5.09E−01	1.03E+00
As	8.27E−01	1.93E+00	3.38E−05	2.54E−05	8.27E−01	1.93E+00
Cu	1.21E+00	2.89E+00	3.59E−03	9.99E−03	1.21E+00	2.90E+00
Zn	3.86E−01	9.00E−01	5.71E−03	1.65E−02	3.92E−01	9.17E−01
Average value for all metals					**1.0217**	**2.3216**

were below 1 in the adult group apart from Mn and Co. In addition, the HQ_{total} (Eq. 6.24) values for Mn, Co, Cd, Pb, and Cu were higher than 1 ($HI_{total} > 1$) for the child group. The results presented that the HI_{total} value of all metals for the child group was almost double that of the adult group. Therefore, children are more susceptible to noncarcinogenic health risks than adults. The studies by Duggal et al. [19], Ukah et al. [31], and Tian et al. [32] in different areas of the world yielded the same findings as the present study. On the other hand, the HI_{total} values were slightly different in the three sampling periods. Due to the variation in trace metal concentration, the samples of MON and POM have a somewhat higher value of HI_{total} for both age groups.

Table 6.10c Noncarcinogenic health risks of trace elements in samples by oral and dermal pathways in the POM period

	HQ_{oral}		HQ_{derm}		$HI = \sum HQ_{oral/derm}$	
Element	Adult	Child	Adult	Child	Adult	Child
B	9.67E−02	2.29E−01	3.94E−04	10.05E−04	9.71E−02	2.30E−01
Fe	3.56E−01	8.25E−01	9.29E−03	2.31E−02	3.65E−01	8.48E−01
Mn	1.88E+00	4.29E+00	7.98E−01	2.42E+00	2.68E+00	6.71E+00
Cr	9.29E−04	2.31E−03	8.28E−04	4.11E−07	1.76E−03	2.31E−03
Pb	3.35E−01	7.47E−01	3.77E−03	5.44E−04	3.39E−01	7.48E−01
Co	4.49E+00	1.20E+01	3.07E−02	8.41E−02	4.52E+00	1.21E+01
Ni	4.78E−01	1.61E−01	7.42E−02	3.66E−04	5.52E−01	1.61E−01
Cd	4.52E−01	1.17E+00	6.80E−02	5.09E−04	5.20E−01	1.17E+00
As	8.36E−01	1.99E+00	3.36E−05	2.69E−05	8.36E−01	1.99E+00
Cu	1.37E+00	2.99E+00	3.61E−03	10.12E−03	1.37E+00	3.00E+00
Zn	3.90E−01	9.15E−01	5.75E−03	1.70E−02	3.96E−01	9.32E−01
Average value for all metals					**1.0615**	**2.5356**

Table 6.11 Summary results of noncarcinogenic cataloguing based on HI_{total} values (Ukah et al. [31]; US-EPA [21])

			% Samples in the category					
		Chronic risk	Pre-monsoon		Monsoon		Post-monsoon	
Risk level	HI_{total}	category	Adult	Child	Adult	Child	Adult	Child
1	>0.1	Negligible	0	0	0	0	0	0
2	≥ 0.1 < 1	Low	82.5	15	77.5	15	72.5	7.5
3	≥ 1 < 4	Medium	17.5	80	22.5	75	27.5	80
4	≥4	High	0	5	0	10	0	12.5
Average value of HI_{total} for all metals			***0.9610***	***2.2994***	***1.0217***	***2.3216***	***1.0615***	***2.5356***

The hazard index via oral (HI_{oral}) and dermal absorption (HI_{derm}) was measured at each sampling spot to evaluate the complete noncarcinogenic risk through toxic metals. The dataset of the calculated HI_{total}, the summation of HI_{oral} and HI_{derm}, for the adults and children of each water sample, is illustrated in Fig. 6.4, and a summary of the results is revealed in Table 6.11. For adults, the sample-by-sample results of the HI_{total} fluctuated from 0.7341 to 1.554 with an average value of 1.0147 in the three sampling periods, indicating low chronic risk (Fig. 6.4). Out of the total 40 groundwater samples, the HI_{total} value of the 11 samples was higher than 1 (HI > 1), and nine samples were close to 1 (HI > 0.8), illustrating that heavy metals might cause opposing health effects and noncarcinogenic health risks to the respective inhabitants. However, the same result revealed that the value of HI_{total} of all the samples for the child group was higher than 1 (HI > 1), except for the maximum of six samples among the three sampling seasons. Estimation showed that the HI_{total} values in the child group varied from 0.8668 to 4.8159 with an average of 2.6170 for three sampling periods. Therefore, the child group was more vulnerable to noncarcinogenic health risks in the study area (Table 6.11). Calculation showed that the mean values of HI_{derm} at all the sampling stations for both groups were much below 1, representing that the metal elements would not illustrate at all health risks to the consumers over dermal absorption. The computed results showed that HI_{total} was mostly attributed to the oral route.

6.3.3.2 Carcinogenic Risk Analysis

Trace metals can enhance the risk of cancer in human body part. According to the International Agency for Research on Cancer [33, 34], Cd, Cr, Mn, Cu, Zn, Pb, etc., were observed as noncancer consequence metals, whereas Cr, Co, Ni, and Cd were considered to have possible cancer effects. Continuing exposure to lower concentrations of toxic trace metals could, then, result in many kinds of carcinogens. Next, stated Eqs. 6.21 to 6.24 were used to calculate the incremental lifetime cancer risk (ILCR) by utilizing the values of K_p, CSF, and GIABS, which are itemized in Table 6.5. Cancer slope factor (CSF) values do not exist for all toxic metals, which is a major problem in calculating the total carcinogenic risk. In this study, using only Cr, Ni, Cd, and Pb as carcinogenic metals, the total intake of the inhabitants was measured in the ILCR based on the calculated CDIs values. Table 6.12 shows that the Cr cancer risk was higher than that of Ni, Cd, and Pb for water consumption through the oral and dermal absorption pathways. The carcinogenic risk assessment (ILCR) for the adult group is given away in Fig. 6.5.

Table 6.12 Carcinogenic health risks ($ILCR_{total}$) of heavy metals by oral and dermal pathways in the different sampling seasons

Element	$ILCR_{oral}$		$ILCR_{dermal}$		$ILCR_{total}$	
	Adult	Child	Adult	Child	Adult	Child
	Pre-monsoon (PRM)					
Cr	5.68E−05	3.66E−04	2.75E−05	2.75E−04	8.43E−05	**6.41E−04**
Co	9.81E−06	6.31E−05	6.17E−08	6.17E−07	9.87E−06	6.37E−05
Ni	8.35E−06	5.38E−05	1.31E−05	1.31E−04	2.15E−05	**1.85E−04**
Cd	2.42E−04	1.56E−03	1.85E−06	1.85E−05	**2.44E−04**	**1.58E−03**
	Monsoon (MON)					
Cr	5.94E−05	3.81E−04	2.94E−05	2.92E−04	8.88E−05	**6.73E−04**
Co	10.10E−06	6.49E−05	6.42E−08	6.42E−07	1.02E−05	6.55E−05
Ni	8.53E−06	5.61E−05	1.58E−05	1.58E−04	2.43E−05	**2.14E−04**
Cd	2.62E−04	1.77E−03	1.99E−06	1.99E−05	**2.64E−04**	**1.79E−03**
	Post-monsoon (POM)					
Cr	6.09E−05	3.88E−04	2.99E−05	2.99E−04	9.08E−05	**6.87E−04**
Co	10.22E−06	6.50E−05	6.54E−08	6.54E−07	1.03E−05	6.57E−05
Ni	8.65E−06	5.68E−05	1.65E−05	1.65E−04	2.52E−05	**2.22E−04**
Cd	2.65E−04	1.86E−03	2.12E−06	1.12E−05	**2.67E−04**	**1.87E−03**

Note: Bold values cross the risk mark of cancer

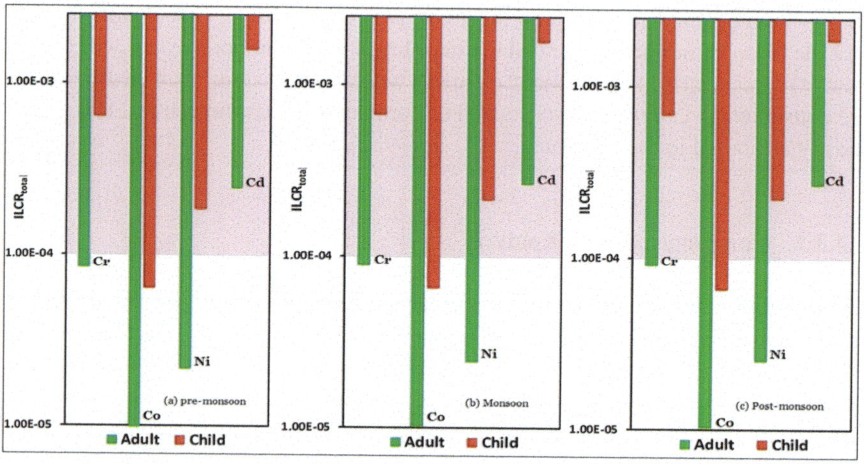

Fig. 6.5 $ILCR_{total}$ values for Cr, Ni, Cd, and Pb metals in the three sampling periods

For a single trace element, an incremental lifetime cancer risk (ILCR) lower than 1×10^{-6} is measured unimportant, and the cancer risk can be abandoned, though an ILCR above 1×10^{-4} is considered injurious, and the risk of cancer is perturbing. For the whole toxic element overall exposure pathways, the satisfactory level is 1×10^{-5} [32, 35]. Mohammadi et al. [36] stated that among toxic metals, Cr has the extreme chance of cancer formation and Ni has the lowest chance of cancer risk. The oral route donated more notably to $ILCR_{total}$ than dermal absorption. Considering both exposure paths, the $ILCR_{total}$ was calculated in the varies of 1.51E−06 to 1.07E−04 with a mean value of 2.51E−05 and dependent on the sampling station. It was observed that the value of $ILCR_{total}$ of metal exposure (between 10^{-6} and 10^{-4}) in the investigated zone was the satisfactory lifetime risk for carcinogens in consumption water. The variation trends of the sample-by-sample results of both risk values are almost the same. The results revealed that Cd crossed the carcinogenic level for the human body of two age groups in all three sampling seasons (Table 6.12). Cr and Ni for the child group crossed the cancer risk level but not for the adult group.

A judgement of carcinogenic and noncarcinogenic risk values of the present investigation with earlier investigations of some countries is presented in Table 6.13. The table indicated that the measured values of the human health risks were not unvarying among these countries. The values of the HQ_{total} for both adult and child groups for the groundwater intake of North China are higher compared to the results of India and this present study. Like this present study, both the HQ_{total} and HI_{total} values for the child group are much higher than those for the adult group in all the studies in Table 6.13. The carcinogenic risk of the water samples of Tamil Nadu (India) and Lagos (Nigeria) is much higher than that in the other studies.

The study followed the extensively used HRA methods emphasized by the IRAC, WHO, US-EPA, and other recognized documents, but these methods have some

Table 6.13 Judgement of carcinogenic and noncarcinogenic risk values of the present study with earlier studies in some different countries

Study area	HQ_{total}		HI_{total} (mean)		Cancer risk (mean)	References
	Adult	Child	Adult	Child	$ILCR_{total}$	
North Rajasthan, India	$HQ_{total} < 1$, for all used metals	$HQ_{total} < 1$, in all samples	7.32E−01	1.72E+00	5.15E−05	Duggal et al. [19]
Tamil Nadu, India	$HQ_{total} < 1$, for all metals, except Pb	–	–	–	2.41E−03	Raja et al. [29]
Lagos, Nigeria	–	–	3.62E+00	12.38E+00	1.31E−03	Ukah et al. [31]
Lianhuashan District, China	$HQ_{total} < 1$, for all metals, except Pb, As, Mn, and Mo	$HQ_{total} < 1$, for all metals, except Fe, Mn, Mo. Co, Ni, Pb, and As	3.50E−01	1.53E+00	3.25E−05	Tian et al. [32]
Khorramabad, Iran	$HQ_{total} < 1$, for all used metals	–	1.10E−04	–	5.05E−04	Mohammadi et al. [36]
Northwest China	$HQ_{total} = 2.6666$ (average)	$HQ_{total} = 5.8013$ (average)	–	–	–	Zhang et al. [37]
North China Plain	$HQ_{total} > 1$, except Cd	–	–	–	4.23E−06	Liu and Ma [38]
This study[a]	***$HQ_{total} < 1$, for all metals, except Mn, Co, and Cu***	***$HQ_{total} > 1$, for Mn, Co, Cd, As, and Cu $HQ_{total} < 1$, for B, Fe, Cr, Pb, Ni, and Zn***	**1.01E+00**	**2.39E+00**	**9.50E−05**[b] **6.71E−04**[c]	–

[a]All values are the average of the three sampling periods
[b]For adults
[c]For the child

doubts for risk calculation. Doubts have been placed on the values of some procedural factors, such as the permeability constant (K_p), dissimilarities in intake conditions due to different ages and consumers, and spatiotemporal variation in heavy metal concentrations in samples. In the recognized noncarcinogenic HRA method, the same emphasis is given to all metals, which is another drawback for risk calculation. Moreover, the exposure parameters used in the investigation were from the IRAC, RAIS, US-EPA, or WHO, which might not be exact for all places in the world. Supplementary precise risk classification should be well defined, and health risk assessment approaches may be improved in view of the above-mentioned uncertainties.

6.4 Summary

Water quality indexing models, including the Canadian Water Quality Index (CWQI), classical WQI, and weighted average water quality index (WWQI), were considered in this study for the evaluation of drinking water suitability. The results of the CWQI, classical WQI, and WWQI models presented that the groundwater was categorized into 'fair' to 'very poor' or 'rejection' (C to D or E) water quality for drinking purposes. Statistically, the findings of the results of both CWQI and WWQI showed a large variance with a highly negative correlation factor ($r = -0.91$), and all values of pair difference were found very irregular. The CWQI used all water parameters in the same index calculation and can be applied effortlessly to water quality data without conveying any weighting factors. However, in the case of the WWQI, limited and fixed water parameters are used, and the equal importance is given to all variables. The study expected that the CWQI was better than the WWQI method for measuring the water suitability.

This study observed the extent of trace metal contamination in water, chose metal pollution indices, and assessed the human health risks accompanying with the ingestion of groundwater in the study zone. According to the outputs of this study, among the analysed trace elements, Fe, Mn, Ni, and Pb are the most predominant, and 50–100% of samples contained those metals over global and national guideline values. Based on the water quality indices (HMPI, HMEP, and C_d), approximately 75% of the samples had a high degree of heavy metal contamination. The results revealed that $HQ_{total} > 1$ for Mn and Co was the potential noncarcinogenic risk compared to other heavy metals from the adult group and $HQ_{total} > 1$ for Pb, Mn, Co, Cd, and Cu in the case of the child group. All HQ_{total} values of the child group are greater than doubled from the adult group for all metals. So, children are more vulnerable to noncarcinogenic chronic health risks than adults. The values of HI_{total} of the adult group in 36 of the 40 sampling stations are less than 1. It also showed that the HI_{total} values of the adult group exceed the unit value in 90% of samples, whereas the child group crossed the unit value of all samples with an average value of 4.25. The results showed that the HI_{total} value of the child group was almost fivefold higher than that of the adult group. So, the child group was at serious health risk in the study area. Besides, $ILCR_{total}$ (carcinogenic) values for Cr, Ni, Ca, and Pb in all samples were observed within the safe ranges. The noncarcinogenic risk values showed that the opposing effect of the toxic metals on the organs and systems of children is more than that of adults. The study suggests that groundwater should be drunk after the necessary treatment for removing trace heavy metals from potable water.

References and Further Study

1. Islam, M. S., & Mostafa, M. G. (2021). Trends of chemical pesticide consumption and its contamination feature of natural waters in especial reference to Bangladesh: A review. *American-Eurasian Journal of Agricultural & Environmental Sciences, 21*(3), 151–167. https://doi.org/10.5829/idosi.aejaes.2021.151.167
2. Najafzadeh, M., Homaei, F., & Farhadi, H. (2021). Reliability assessment of water quality index based on guidelines of national sanitation foundation in natural streams: Integration of remote sensing and data-driven models. *Artificial Intelligence Review, 54*, 4619–4651. https://doi.org/10.1007/S10462-021-10007-1
3. Najafzadeh, M., Homaei, F., & Mohamadi, S. (2022). Reliability evaluation of groundwater quality index using data-driven models. *Environmental Science & Pollution Research, 29*, 8174–8190. https://doi.org/10.1007/s11356-021-16158-6
4. Horton, R. K. (1995). An index number system for rating water quality. *Journal of the Water Pollution Control Federation, 37*(3), 300–306.
5. Brown, R. M., McClelland, N. I., Deininger, R. A., & Tozer, R. G. (1970). A water quality index – Do we dare? *Water and Sewage Works, 117*(10), 339–343.
6. Sarkar, A. M., Lutfor Rahman, A. K. M., Samad, A., Bhowmick, A. C., & Islam, J. B. (2019). Surface and ground water pollution in Bangladesh: A review. *Asian Review of Environmental and Earth Sciences, 6*(1), 47–69. https://doi.org/10.20448/journal.506.2019.61.47.69
7. Saha, N., & Zaman, M. R. (2013). Evaluation of possible health risks of heavy metals by consumption of foodstuffs available in the central market of Rajshahi City, Bangladesh. *Environmental Monitoring and Assessment, 185*(5), 3867–3878. https://doi.org/10.1007/s10661-012-2835-2
8. Islam, M. S., & Mostafa, M. G. (2021). Groundwater quality and risk assessment of heavy metal pollution in middle-west part of Bangladesh. *Journal of Earth and Environmental Science Research, 3*(2), 1–9. https://doi.org/10.47363/JEESR/2021(3)143
9. CCME. (2001). Canadian water quality guidelines for the protection of aquatic life. Canadian environmental quality guidelines CCME water quality index 1.0 technical report, Ottawa.
10. UNEP-GEMSWater. (2007). Global drinking water quality index. Development and sensitivity analysis report. United Nations Environment Programme Global Environment Monitoring System (GEMS)/Water Programme, Burlington.
11. Sahu, P., & Sikdar, P. K. (2008). Hydrochemical framework of the aquifer in and around East Kolkata wetland, West Bengal, India. *Environmental Geology, 55*, 823–835. https://doi.org/10.1007/s00254-007-1034-x
12. Cude, C. G. (2007). Oregon water quality index: A tool for evaluating water quality management effectiveness. *Journal of the American Water Resources Association, 37*(1). https://doi.org/10.1111/j.1752-1688.2001.tb05480.x
13. Backman, B. K., Bodis, D., Lahermo, P., Rapant, S., & Tarvainen, T. (1997). Application of a groundwater contamination index in Finland and Slovakia. *Environmental Geology, 36*, 55–64.
14. US-DOE. (2011). The Risk Assessment Information System (RAIS). U.S. Department of Energy's Oak Ridge Operations Office (ORO), Washington, DC.

15. US-EPA. (1997). *Exposure factors handbook*. Office of Research and Development. EPA/600/P-95/002Fa. U.S. Environmental Protection Agency.
16. US-EPA. (2001). Risk assessment guidance for superfund: Volume III: Part A, process for conducting probabilistic risk assessment. EPA 540-R-02-002. Office of Emergency and Remedial Response. U.S. Environmental Protection Agency, Washington, DC.
17. Rani, A., Mehra, R., Duggal, V., et al. (2013). Analysis of uranium concentration in drinking water samples using ICPMS. *Health Physics, 104*, 251–255. https://doi.org/10.1097/HP.0b013e318279ba05
18. Harries, S., & Harper, B. (2004). Exposure scenario for CTUIR traditional subsistence lifeways, confederated tribes of the Umatilla Indian Reservation. Department of Science and Engineering, Pendleton, Oregon.
19. Duggal, V., Mehra, R., & Rani, A. (2013). Determination of 222RN level in groundwater using a RAD7 detector in the Bathinda district of Punjab, India. *Radiation Protection Dosimetry, 156*, 239–245.
20. US-EPA. (2004). Risk assessment guidance for superfund volume 1: Human health evaluation manual (Part E, Supplemental Guidance for Dermal Risk Assessment). EPA/540/R/99/005 OSWER 9285.7-02EP PB99-963312 July 2004, Office of Superfund Remediation and Technology Innovation U.S. Environmental Protection Agency, Washington, DC.
21. US-EPA. (1989). Risk assessment guidance for superfund volume I: Human health evaluation manual (part A) interim final. EPA/540/1-89/002. U.S. Environmental Protection Agency, Office of Emergency and Remedial Response, Washington, DC.
22. Didar-ul, S. M. I., Bhuiyan, M. A. H., Rume, T., et al. (2017). Hydrogeochemical investigation of groundwater in shallow coastal aquifer of Khulna District, Bangladesh. *Applied Water Science, 7*, 4219–4236.
23. Islam, S. M.-U., Majumder, R. K., Uddin, M. J., Khalil, M. I., & Alam, M. F. (2017). Hydrochemical characteristics and quality assessment of groundwater in Patuakhali District, southern coastal region of Bangladesh. *Water Quality, Exposure and Health, 9*(1), 43–60. https://doi.org/10.1007/s12403-016-0221-y
24. Howladar, M. F., Al-Numan, M. A., & Faruque, M. O. (2017). An application of Water Quality Index (WQI) and multivariate statistics to evaluate the water quality around Maddhapara Granite Mining Industrial Area Dinajpur, Bangladesh. *Environmental System Research, 6*, 13.
25. Islam, M. S., & Mostafa M. G. (2022). Evaluation of hydrogeochemical processes in groundwater using geochemical approaches and geostatistical models in the upper Bengal basin. *Geofluid, 2022*, Article ID 9591717, 1–21. https://doi.org/10.1155/2022/9591717
26. Edet, A., & Offiong, O. (2002). Evaluation of water quality pollution indices for heavy metal contamination monitoring. A study case from Akpabuyo-Odukpani area, Lower Cross River Basin (southeastern Nigeria). *Geo Journal, 57*, 295–304. https://doi.org/10.1023/B:GEJO.0000007250.92458.de
27. Bodrud-Doza, M., Islam, A. R. M. T., Ahmed, F., et al. (2016). Characterization of groundwater quality using water evaluation indices, multivariate statistics and geostatistics in central Bangladesh. *Water Science, 30*, 19–40. https://doi.org/10.1016/j.wsj.2016.05.001
28. Bhuiyan, M. A. H., Parvez, L., Islam, M. A., Dampare, S. B., & Suzuki, S. (2010). Evaluation of hazardous metal pollution in irrigation and drinking water systems in the vicinity of a coal mine area of northwestern Bangladesh. *Journal of Hazardous Materials, 179*, 1065–1077. https://doi.org/10.1016/j.jhazmat.2010.03.114
29. Raja, V., Lakshmi, R. V., Sekar, C. P., et al. (2021). Health risk assessment of heavy metals in groundwater of industrial township Virudhunagar, Tamil Nadu, India. *Archives of Environmental Contamination and Toxicology, 80*, 144–163. https://doi.org/10.1007/s00244-020-00795-y
30. Wagh, V., Muley, A., & Mukate, S. (2018). Health risk assessment of heavy metal contamination in groundwater of Kadava River Basin, Nashik, India. *Modeling Earth Systems and Environment, 4*, 969–980. https://doi.org/10.1007/s40808-018-0496-z

31. Ukah, B. U., Egbueri, J. C., Unigwe, C. O., & Ubido, O. E. (2019). Extent of heavy metals pollution and health risk assessment of groundwater in a densely populated industrial area, Lagos, Nigeria. *International Journal of Energy and Water Resources, 3*, 291–303. https://doi.org/10.1007/s42108-019-00039-3
32. Tian, H., Xiujuan, L., Yan, G., et al. (2020). Risk assessment of metals from shallow groundwater in Lianhuashan District, China. *La Houille Blanche, 1*, 5–15. https://doi.org/10.1051/lhb/2019063
33. IARC. (2011). *Working Group on the evaluation of carcinogenic risks to humans*. WHO Press, World Health Organization.
34. IARC. (2013). *International Agency for Research on Cancer. IARC monographs on the evaluation of carcinogenic risks to humans. Non-ionizing radiation, part 2: Radiofrequency electromagnetic fields* (Vol. 102). International Agency for Research on Cancer.
35. Wu, Y., Zhou, Y., Qiu, Y., et al. (2017). Occurrence and risk assessment of trace metals and metalloids in sediments and benthic invertebrates from Dianshan Lake, China. *Environmental Science and Pollution Research, 24*, 14847–14856. https://doi.org/10.1007/s11356-017-9069-3
36. Mohammadi, A. A., Zarei, A., & Majidi, S. (2019). Carcinogenic and non-carcinogenic health risk assessment of heavy metals in drinking water of Khorramabad, Iran. *MethodsX, 6*, 1642–1651. https://doi.org/10.1016/j.mex.2019.07.017
37. Zhang, Q., Xu, P., & Qian, H. (2020). Groundwater quality assessment using improved water quality index (WQI) and human health risk (HHR) evaluation in a semi-arid region of northwest China. *Exposure and Health, 12*, 487–500. https://doi.org/10.1007/s12403-020-00345-w
38. Liu, Y., & Ma, R. (2020). Human health risk assessment of heavy metals in groundwater in the Luan River catchment within the North China plain. *Geofluids*, ID 8391793, 1–7. https://doi.org/10.1155/2020/8391793

Chapter 7
Irrigation Water Quality

Abbreviations

CCME	Canadian Council of Ministers of the Environment
CWQI	Canadian Water Quality Index
GIS	Geographic information system
$IIWQ_{Index}$	Integrated irrigation water quality index
IWQI	Irrigation water quality index
KR	Kelly's ratio
MAR	Magnesium adsorption ratio
MWQI	Meireles irrigation water quality index
Na%	Per cent sodium
PI	Permeability index
PSS	Potential soil salinity

M. S. Islam, *Hydrogeochemical Evaluation and Groundwater Quality*,
https://doi.org/10.1007/978-3-031-44304-6_7

RSBC	Residual sodium bicarbonate
RSC	Residual sodium carbonate
SAR	Sodium adsorption ratio
$SIWQ_{Index}$	Simsek irrigation water quality index
SSP	Saturated sodium percentage
UCCC	University of California Committee of Consultants
WQI	Water quality index

7.1 Irrigation Water Quality: *Background*

Worldwide, the area equipped for irrigation is presently around 301 million hectares. Of these, 38% are equipped for irrigation with aquifer water. The total groundwater use for irrigation is projected at 545 km^3/year, which is 43% of the total irrigation water use of 1277 km^3/year. Before the 1970s, agriculture in Bangladesh was totally reliant on surface water and seasonal natural rainfall. After that period, millions of drinking and irrigation shallow and semi-deep wells were installed in Bangladesh to harvest high-production rice diversities and to achieve success in the global campaign of safe and clean drinking water decade (1980–1990). Today, groundwater is no longer surface water but the country's main source of irrigation. In 2006, nearly 78% of irrigated paddy fields were supplied by groundwater, with almost 80% of irrigation water coming from shallow (25–40 m) motorized tube wells and the remainder from deep (100–150 m) tube wells. This irrigation water must be of suitable quality. Irrigation water chemistry can affect plant growth and crop quality through nutrient deficiencies and toxicity or indirectly through soil degradation. Poor irrigation water quality reduces soil productivity and alters the soil's physical and chemical properties. It also causes phytotoxicity and ultimately reduces crop yields. Arsenic contamination is now threatening irrigation water, and groundwater is presently believed to be the only source of irrigation water in the country. Therefore, rational use must be ensured both qualitatively and quantitatively. So, information on irrigation water quality is essential for understanding agricultural management needs for long-term productivity.

Natural processes such as varieties of aquifer rocks, groundwater flow direction, water–rock interactions, residence times of water in saturated and unsaturated zones, and anthropogenic activities such as industrialization, urbanization, agriculture, and overwater mining of groundwater resources are very important factors which regulate groundwater geochemistry. Groundwater is the most important and comparatively safe water source for sustainable irrigation practices, especially when surface water is unavailable and polluted by surface contamination. The suitability of irrigation water is highly dependent on the quantity and type of various dissolved substances. Based on these substances, numerous irrigation water quality parameters and index models were developed to assess the quality of irrigation water for soil environment conservation and crop yield enhancement. Groundwater is the only source of irrigation in the present study zone, south-eastern Bangladesh, which is largely irrigated, but the quality of this water is not verified. Several investigations have shown that the groundwater in this region is deteriorating gradually, characteristically due to cyclical geochemical fluctuations, overuse of water, gigantic arsenic pollution, the interaction of the groundwater with saline water, and the deposition of pesticides and fertilizers on the soil. In coastal areas where water sodicity is a serious problem, the quality of irrigation water has been extensively measured. However, in the upper Ganges basin, calcium hardness and excessive trace metal loading in the groundwater are the main quality problems. Therefore, it is vital to appropriately evaluate water quality when implementing surface irrigation in the study area. For this purpose, the newly developed overall irrigation water quality index (overall IWQ_{Index}), integrated irrigation water quality index ($IIWQ_{Index}$), irrigation water quality parameters, and several conventional diagram-based statistical approaches are used in this chapter.

This is a very complex process to justify the irrigation water quality because the negative influence of water on soil and plants depends on the physical and chemical properties of soil, the nature of irrigation practice, local climate, crop diversity, and land management. It is impossible to entirely evaluate the quality of that water through a single indexing method. Several irrigation water quality indices and guidelines for the appropriate justification of irrigation water quality have been documented worldwide by some investigators. For instance, Simsek and Gunduz [1] recognized a GIS-based irrigation water quality index (IWQ) by the incorporation of five most possible geochemical hazards such as salinity hazards, ion toxicity, permeability hazards, trace metal toxicity, and miscellaneous hazards to evaluate the irrigation water suitability of the Simav Plain, Turkey. Another researcher, Meireles et al. [2], also projected a new indexing model for the assessment of irrigation water quality utilizing some water quality parameters. Similarly, Ashraf et al. [3] developed an IWQ index using a geographical information system (GIS) for determining the sodium adsorption ratio (SAR), residual sodium carbonate (RSC), and saturated sodium percentage (SSP) with other parameters. To evaluate groundwater quality for irrigation in the Wet Pampa Plain in Argentina, Romanelli et al. [4] measured the IWQ index by a combination of water geochemical characteristics, topographical nature, and other miscellaneous physical parameters such as hydraulic conductivity of soil and aquifer size. Besides, they included some water parameters: electrical conductivity (EC), total hardness (TH), total dissolved solids (TDS), per cent sodium (Na%), SSP, SAR, and RSC in their index model. Later, together with GIS, Bozdag [5] used an analytic hierarchy process (AHP) to assess irrigation water quality in the central part of Anatolia in Turkey. Those miscellaneous strategies of water suitability measurement associated with irrigated agriculture have some kind of success, but none is completely suitable due to large inconsistencies in cropland situations and crop diversity. In the meantime, those methods comprise of a limited number of water parameters and little of environmental events and applied outdated and traditional rating, scoring, and weighting factors of utilized parameters and hazard classes. Likewise, a large inconsistency was observed, and an inhomogeneous procedure was introduced in the construction of that indexing model. The present study projected an indexing method that considered the maximum number and types of water quality parameters with a total of six hazard classes. In addition, modelling values of water parameters and the weight factor of each hazard class are utilized. The proposed indexing method explores sufficient information on the aptness of various kinds of water for each irrigation practice. The measures presented in this paper have been improved to give more practical actions for evaluating and managing water quality-related hazards of irrigated agriculture. With this probable method, some recognized measurements were used to complete the assessment of irrigation water quality and rationalize the comparative acceptability between each of the groundwater samples in the study zone.

The objective of the study is to evaluate the suitability of groundwater in the Ganges River basin areas of Bangladesh for irrigation purposes. The findings of this study help to evaluate the consequence of irrigation water on soil environment and crops and choose the suitable alternatives that may be helpful to increase crop production in these study areas.

7.2 Irrigation Water Quality and Crop Yield

The cultivable land of the study area (Chap. 3) is very productive, and the main occupation of populations is cropping yield through farming on that land. This study area comprises of total 160,800 hectares of land, of which 104,520 hectares is highly arable. The total irrigated land area is 75,714 hectares in the study area. About 80% of cultivable land grew three crops in a year. Almost all used water in irrigation events is groundwater in this area. Table 7.1 shows the statistics of main crop production in the study area.

About 60–70% of the cultivated land in the study area is irrigated by groundwater, and the quality of the water greatly affects crop yields due to long-term irrigation.

Table 7.1 Crop variety and rate of production in the study area

Crops variety	Cultivation (hectares)	Production (tonnes)/hectare	Total production (tonnes)
Rice	100,192	3.365	337,150
Jute	33,535	13.171 (bales)	563,754 (bales)
Vegetables	12,378	2.453	30,363
Onion	10,119	4.074	41,230
White	10,440	3.541	36,968
Tobacco	14,467	3.285	47,525
Maize	7466	4.280	31,967
Oil seed	5124	1.433	7335
Sugarcane	3506	14.972	52,470

High salinity (mean EC: 813.4 μS/cm) and elevated levels of some toxic metals are the main features of groundwater samples. Higher total salinity in irrigation water creates a risk of accumulation of metal salts in the topsoil and salinity in crops, given the same soil and climatic factors. Soil, crop, climate, and cultural factors that promote the accumulation of soluble salts in the root zone prevent the use of high-salinity water for irrigation. Similarly, factors that promote salt leaching from the root zone by regular leaching drive the use of high-salinity water for irrigation. Under favourable conditions, groundwater with salinity above 800 μS/cm was used to produce semi-tolerant crops such as wheat on coarsely structured soils with only minor yield losses. Another problem arises from the fact that toxic substances such as boron and heavy metals such as Fe, Mn, Pb, As, Cd, and Co are present in some water samples. High concentrations of boron in irrigation water can have toxic effects on the growth of many plants. Similarly, certain other ions such as Cl^-, Na^+, and Ca^{2+} can be toxic to certain crops when present in excess in irrigation water. Therefore, highly mineralized water can affect both the soil environment and the total crop yield.

7.3 Evaluation of Irrigation Water Quality

The assessment of groundwater suitability for irrigation is not possible through a single process. In this present study, various irrigation water quality parameters are calculated and reported methodically. In addition, three irrigation WQIs, Simsek irrigation WQI, Meireles irrigation WQI, and Canadian WQI, are used simultaneously to assess the water quality for irrigation. Also, some diagrams are utilized for the same. Finally, this study developed a new and rationalized index method for evaluating irrigation water suitability.

7.3.1 *Irrigation Water Quality Parameters*

The four most common measures are EC or TDS, sodium adsorption ratio (SAR), residual sodium carbonate (RSC), and concentration of some ions, such as Na^+, HCO_3^-/CO_3^{2-}, and Cl^- for the water quality evaluation of irrigation. Except for the other physicochemical parameters, the following methods were considered for the present irrigation water quality evaluation:

(a) The total hardness (TH) [6] of water in mg/L was calculated by Eq. 7.1:

$$TH = 2.497\left[Ca^{2+}\right] + 4.115\left[Mg^{2+}\right] \tag{7.1}$$

(b) Na% is calculated using Eq. 7.2 [7]:

$$Na\% = \frac{Na \times 100}{(Ca^{2+} + Mg^{2+} + Na^{+})} \tag{7.2}$$

(c) The sodium adsorption ratio (SAR) value of irrigation water estimates the comparative proportion of Na^{+} to Ca^{2+} and Mg^{2+}, and as stated by Richards [8], SAR is expressed as:

$$SAR = \frac{Na^{+}}{\sqrt{\frac{(Ca^{2+}+Mg^{2+})}{2}}} \tag{7.3}$$

(d) The soluble sodium percentage (SSP) was utilized to assess the sodium hazard in irrigation water. Todd [7] recognized SSP as follows:

$$SSP = \frac{(Na^{+} + K^{+}) \times 100}{(Ca^{2+} + Mg^{2+} + Na^{+} + K^{+})} \tag{7.4}$$

(e) Gupta [9] developed a relation to determine the residual sodium bicarbonate (RSBC) as following Eq. 7.5:

$$RSBC = HCO_3{}^{-} - Ca^{2+} \tag{7.5}$$

(f) Doneen [10] established a method to calculate the permeability index (PI) as follows:

$$PI = \frac{Na + \sqrt{HCO_3{}^{-}}}{(Na^{+} + Ca^{2+} + Mg^{2+})} \times 100 \tag{7.6}$$

(g) The magnesium adsorption ratio (MAR) [6] is calculated by the following Eq. 7.7:

$$MAR = \frac{(Mg^{2+} \times 100)}{(Ca^{2+} + Mg^{2+})} \tag{7.7}$$

(h) Finally, Kelley's ratio (KR) [11] is designated as:

$$\mathrm{KR} = \frac{\mathrm{Na}^{+}}{(\mathrm{Ca}^{2+} + \mathrm{Mg}^{2+})} \tag{7.8}$$

In all the above equations, the concentrations are quantified as milli-equivalents per litre (mEq/L) and are considered by dividing the aqueous concentration of the corresponding ion component stated in mg/L by its equivalent weight. The calculated values of these parameters in the three sampling seasons are presented in Appendices XI–XIII.

7.3.2 *Irrigation Water Quality Indices*

Based on the degree of pollution or hygiene, information describing water suitability levels started first in 1848 in Germany. Since then, over a hundred local-, regional-, and global-based water quality indexing methods have been acknowledged. At this time, for easy understanding of the datasets, three (3) different WQI models, the Canadian WQI, Simsek water quality index (SWQI), and Meireles WQI, were used for the designated water quality parameters. These indexing models are discussed below:

7.3.2.1 Simsek Water Quality Index (SWQI)

To launch a quick view of the overall irrigation water suitability, two models, viz. Simsek and Gunduz [1] and Meireles et al. [2], were used to compute the $\mathrm{IWQ}_{\mathrm{index}}$. Here, the first one included trace element toxicity to the crop, but the second one omitted trace metal toxicity. The $\mathrm{IWQ}_{\mathrm{index}}$ was organized based on the linear connection of five sets of irrigation water suitability parameters, which are related to form a single index value to assess the irrigation water quality in the study areas. As stated by the guidance by Ayers and Westcot [12], five groups of irrigation water quality parameters, such as salinity hazard ($w = 5$), penetrability hazard ($w = 4$), fixed ion toxicity ($w = 3$), trace element toxicity ($w = 2$), and miscellaneous belongings to sensitive crops ($w = 1$), are selected (Table 7.2). Simsek and Gunduz [1] selected the detailed standard measures for irrigation water suitability indices by Eq. 7.9:

$$\mathrm{IWQ}_{\mathrm{index}} = \sum_{i=1}^{5} Gi \tag{7.9}$$

where i = incremental index and G = involvement of each one of the five hazard groups that are essential to measure the quality of definite irrigation water resources; G can be intended by Eq. 7.10:

Table 7.2 Categorization for IWQ index parameters

Hazards with weight (w) and rating (r) values				
1. Salinity hazard (N = 2; w = 5)	Equation	***4. Trace metal toxicity (N = 10; w = 2)***		Equation
(a) EC < 700 ($r = 3$)	$W_1 = \frac{w_1}{2} \sum_{j=1}^{1} r_a$	(a) Fe < 5.0 ($r = 3$)	(h) Ni < 0.2 ($r = 2$)	$W_4 = \frac{w_4}{4} \sum_{j=1}^{4} r_c$
700 ≤ EC ≤ 3000 ($r = 2$)		5.0 ≤ Fe ≤ 20.0 ($r = 2$)	0.2 ≤ Ni ≤ 2.0 ($r = 1$)	
EC > 3000 ($r = 1$)		Fe > 20.0 ($r = 1$)	Ni > 2.0 ($r = 1$)	
(b) TDS < 450 ($r = 3$)		(b) Mn < 0.2 ($r = 3$)	(i) Cd < 0.01 ($r = 3$)	
450 ≤ TDS ≤ 2000 ($r = 2$)		0.2 ≤ Mn ≤ 10.0 ($r = 2$)	0.01 ≤ Cd ≤ 0.05 ($r = 2$)	
TDS > 2000 ($r = 1$)		Mn > 10.0 ($r = 1$)	Cd > 0.05 ($r = 1$)	
2. Permeability hazard (N = 1; w = 4)		(c) Cr < 0.1 ($r = 3$)	(j) As < 0.1 ($r = 3$)	
SAR < 3 (highest score)	$W_2 = w_2 r_2$	0.1 ≤ Cr ≤ 1.0 ($r = 2$)	0.1 ≤ As ≤2.0 ($r = 2$)	
EC > 700 ($r = 3$)		Cr > 1.0 ($r = 1$)	As > 2.0 ($r = 1$)	
EC = 700–200 ($r = 2$)		(d) Pb < 5.0 ($r = 3$)		
EC < 200 ($r = 1$)		5.0 ≤ Pb ≤ 10.0 ($r = 2$)		
3. Specific ion toxicity (N = 3; w = 3)		Pb > 10.0 ($r = 1$)	***5. Miscellaneous effects (N = 3; w = 1)***	
(a) SAR < 3.0 ($r = 3$)	$W_3 = \frac{w_3}{3} \sum_{j=1}^{3} r_b$	(e) Co < 0.05 ($r = 3$)	(a) HCO_3^- < 90 ($r = 3$)	$W_5 = \frac{w_5}{5} \sum_{j=1}^{5} r_d$
3.0 ≤ SAR ≤ 9.0 ($r = 2$)		0.05 ≤ Co ≤ 5.0 ($r = 2$)	90 ≤ HCO_3^- ≤ 500 ($r = 2$)	
SAR > 9.0 ($r = 1$)		Co > 5.0 ($r = 1$)	HCO_3^- > 500 ($r = 1$)	
(b) B < 0.7 ($r = 3$)		(f) Cu < 0.2 ($r = 3$)	(b) NO_3^- < 5 ($r = 3$)	
0.7 ≤ B ≤ 3.0 ($r = 2$)		0.2 ≤ Cu ≤ 5.0 ($r = 2$)	NO_3^- = 5–30 ($r = 2$)	
B > 3.0 ($r = 1$)		Cu > 5.0 ($r = 1$)	NO_3^- > 30 ($r = 1$)	
(c) Cl < 140 ($r = 3$)		(g) Zn < 2 ($r = 3$)	(c) 7.0 ≤ pH ≤ 8.0 ($r = 3$)	
140 ≤ Cl^- ≤ 350 ($r = 2$)		2 ≤ Zn ≤ 10 ($r = 2$)	6.5 ≤ pH < 7.0; 8.0 < pH ≤ 8.5 ($r = 2$)	
Cl^- > 350 ($r = 1$)		Zn > 10.0 ($r = 1$)	pH < 6.5 or pH > 8.5 ($r = 1$)	

Based on Ayers and Westcot [12]

$$G = \frac{w}{N} \sum_{k=1}^{N} (r_k) \tag{7.10}$$

Here, $N =$ total number of water parameters for the analysis, $w =$ weight factor of the grouping, $k =$ incremental index, and $r =$ rating values of each parameter.

That is,

$$\mathrm{IWQ}_{\mathrm{index}} = w_1 r_1 + w_2 r_2 + \frac{w_3}{3} \sum_{j=1}^{3} r_a + \frac{w_4}{4} \sum_{j=1}^{4} r_b + \frac{w_5}{5} \sum_{j=1}^{5} r_c \tag{7.11}$$

The sodium adsorption ratio is calculated by Eq. 7.12:

$$\mathrm{SAR} = \frac{\mathrm{Na}^+}{\sqrt{\frac{(\mathrm{Ca}^{2+} + \mathrm{Mg}^{2+})}{2}}} \tag{7.12}$$

To compute the Simsek water quality index (SWQI), the values of parameters such as EC, TDS, Na, Ca, Mg, B, Cl^-, trace metals, HCO_3^-, NO_3^-, and pH were used. These parameter values are described in Chaps. 4 and 5 of this book.

7.3.2.2 Meireles Water Quality Index (MWQI)

Meireles projected a new categorization for irrigation water and measured the water quality index for irrigation purposes [2]. The water parameters that cause more inconsistency in irrigation water quality were designated. In this model, five parameters such as EC, Na^+, sodium adsorption ratio (SAR), Cl^-, and HCO_3^- were quantified. These are the main factorial weights defining the best water quality.

Cataloguing of water quality measurement limits (q_i) and accumulated weights (w_i) was documented. The values of q_i were derived based on each parameter value, in view of the criteria recognized by Ayers and Westcot [12] and irrigation water quality parameters projected by the UCCC [13], and are quantified in Table 7.3. The SAR value of the water sample calculates the comparative proportion of Na^+ to Ca^{2+} and Mg^{2+}, and as stated by Richards [8], SAR was computed by Eq. 7.3.

The values of q_i were calculated by using Eq. 7.13:

$$Q_i = Q_{i\max} - \frac{\left\{ \left(x_{ij} - x_{\mathrm{inf}} \right) Q_{i\mathrm{amp}} \right\}}{x_{\mathrm{amp}}} \tag{7.13}$$

Table 7.3 The limiting values of water parameters for quality measurement (q_i) calculation

Q_i	SAR $(mEq/L)^{1/2}$	EC (μS/cm)	Na^+ (mEq/L)	Cl^- (mEq/L)	HCO_3^- (mEq/L)
85 ≤ 100	2 ≤ SAR < 3	200 ≤ EC < 750	2 ≤ Na < 3	1 ≤ Cl < 4	1 ≤ HCO_3 < 1.5
608 ≤ 5	3 ≤ SAR < 6	750 ≤ EC < 1500	3 ≤ Na < 6	4 ≤ Cl < 7	1.5 ≤ HCO_3 < 4.5
35 ≤ 60	6 ≤ SAR < 12	1500 ≤ EC < 3000	6 ≤ Na < 9	7 ≤ Cl < 10	4.5 ≤ HCO_3 < 8.5
0 ≤ 35	SAR ≥ 12 or SAR < 2	EC < 200 or EC ≥ 3000	Na < 2 or Na ≥ 9	Cl ≥ 10 or Cl < 1	HCO_3 < 1 or HCO_3 ≥ 8.5
Weight value (w_i)	*0.211*	*0.189*	*0.204*	*0.194*	*0.202*

where $Q_{i\max}$ = the highest value of Q_i for the equivalent class

x_{ij} = measured value of the parameter
$x_{\inf}$ = lower value of the parameter to which the class belongs
Q_{iamp} = class capacity
x_{amp} = class capacity to which the parameter belongs

To determine x_{amp} in the case of the last class of each parameter, the maximum value was obtained from the chemical analysis of the water samples measured to be the higher limit. The weight of each parameter used in computing MWQI was normalized so that their sum was equal to one. Table 7.3 demonstrates the weights of the water quality parameters. Lastly, the MWQI was measured utilizing Eq. 7.14 as follows:

$$\text{MWQI} = \sum_{i=1}^{n} q_i w_i \tag{7.14}$$

where q_i signifies the quality of the ith parameter, which is a function of its measurement or concentration and ranges among 0 to 100, and w_i is the normalized weight of the ith parameter, which is significant in the inconsistency of the water quality.

7.3.2.3 Canadian Water Quality Index (CWQI)

The Canadian water quality index (CCME) is used for the evaluation of drinking, irrigation, and aquatic life water quality. Here, the CWQ index model is used for the assessment of irrigation water suitability. This method is discussed in Chap. 6. The calculated CWQI and MWQI values for all samples (n = 40) in three different sampling seasons are presented in Appendix XIV.

7.4 Evaluation of Irrigation Water Suitability

Irrigation water quality depends on both the nature and amount of nutrients or solutes dissolved. These solutes are mostly conceived for the water phase from the natural dissolution of minerals and soil, and some portion comes from industrial and domestic discharge leakage. It is typically acknowledged that the problems created by irrigation water suitability fluctuate in type and severity as a function of frequent factors, together with the types of soil and crop diversity, the weather conditions of the region, and the water used. However, there is a general consideration that these problems can be categorized as salinity hazards, penetrability problems, ion toxicity hazards, and miscellaneous problems [11]. These problems were assessed by numerous irrigation water suitability parameters, including the IWQ indices.

7.4.1 Using Irrigation Water Quality Parameters

The pH, salinity hazard, Na hazard, free Cl, some trace metals, HCO_3^-, and CO_3^{2-} in combination with the Ca and Mg content, toxic anions, and several types of nutrients are the vigorous factors controlling the appropriateness of water use in irrigation. Instead, as stated by Raghunath [6], Michael [14], Matthess [15], and Hem [16], irrigation water quality is determined by the four most acceptable criteria: total dissolved solids (TDS) or electrical conductivity (EC); the concentration of certain definite ions like Na^+, K^+, Ca^{2+}, Mg^{2+}, Cl^-, and B contents; comparative quantity of Na^+ to other cations, denoted by the sodium adsorption ratio (SAR); residual sodium carbonate (RSC); and residual sodium bicarbonate (RSBC). Except from these, other measures, such as the presence of soil caliche, depth of the water table, $CaCO_3$ concentration in the soil, and K^+ and NO_3^- ions, also indirectly influence the suitability of irrigation water. In addition, Matthess [14] and Ayers and Westcot [11] designated that poor-quality irrigation water generates four types of problems: definite ion toxicity (affects delicate crops), salinity (affects crop water availability), water penetrability (affects infiltration rate of water into the soil), and miscellaneous.

From 1950 to 2000, numerous equations were developed for the calculation of irrigation water quality parameters. Such parameters are the sodium adsorption ratio, SAR [8]; potential soil salinity, PSS [10]; Kelly's ratio, KR [11]; permeability index, PI [10]; per cent sodium, Na%; saturated sodium percentage, SSP; magnesium adsorption ratio, MAR [7]; residual sodium bicarbonate, RSBC [9]; residual sodium carbonate, RSC [6]; and total hardness, TH [16]. Some geochemical components such as Na^+, K^+, Ca^{2+}, Mg^{2+}, CO_3^{2-}, and HCO_3^- in the mEq/L unit (except total hardness, TH in mg/L) are used to measure those parameters. The values of previously discussed parameters and other parameters, such as SSP, RSC, KR, PI, MAR, Mg:Ca, and Na:Ca, of irrigation groundwater samples were measured (Appendices XI–XIII). Also, the limits of some imperative indices for rating water quality and its suitability for irrigational use are shown in Table 7.4.

Table 7.4 Limits of important parameter indices for rating groundwater quality and its suitability in irrigation use

Category (water class)	Irrigation water quality parameters								Degree on restriction on use
	EC (µS/cm)	SAR (%)	SSP (me/L)	RSC (me/L)	PI (%)	MAR (%)	TDS (mg/L)	Salinity (mg/L)	
A	<700	<10	<20	–	>75	–	–	–	Excellent
B	700–3000	10–18	20–40	<1.25	25–75	<50	>2000	<450	Good
C	>3000	18–26	40–80	1.25–2.5	<25	–	450–2000	450–2000	Fair
D	–	>26	>80	>2.5	–	–	<450	>2000	Poor/rejection
References	UCCC [13]	Fipps [17]	Wilcox [18]	Gupta and Gupta [19]	Hem [16]	Gupta and Gupta [19]	Bauder et al. [20]	UCCC [13]	

The important feature of the present water chemistry for irrigation was found as the elevated level of Ca^{2+}, Mg^{2+}, and HCO_3^-; no CO_3^{2-} concentration; lower pH; higher EC/TDS value; less concentration of Na^+, Cl^-, and SO_4^{2-}; and Ca–HCO_3 type water facies. Those characteristics of water would be controlled by all detected irrigation water quality parameters. According to Ayers and Westcot [12] and FAO [21], based on the above-mentioned quality parameters, a total of six hazard classes for irrigation water were identified. The characteristics and impacts on soil environments and crop productivity are discussed below:

7.4.1.1 Salinity Hazard

Salinity hazard measured by EC and/or TDS (both are used if the correlation matrix, r, is less than 0.95) of irrigation water considering the most priority (scoring rate: $s = 6$) to calculate the index value that affects plants and leads to saline soil conditions. Soils with high salinity generally have a pH value under 8.5, are relatively low in Na, and contain typically Ca, Mg, and Na salts of Cl^-, SO_4^{2-}, CO_3^{2-}, HCO_3^-, NO_3^-, and PO_4^{3-}. Heavy calcite and sodic water presented higher values of EC as well as TDS. Typically, waters with EC values under 700 μS/cm are considered as good quality irrigation water. A salinity hazard exists if salt gathers in the crop root zone to a concentration that causes a loss in yield. Yield reductions occur when the salts accumulate in the plant root zone to such an extent that the plant is no longer able to extract adequate water from the salty soil environment, resulting in water shortage due to drought/osmotic stress for a substantial period. If water uptake is noticeably reduced, the crop slows down its rate of growth. Overall, the potential for salinity hazards depends on the depth of the water table, local climate, soil structure, type of irrigation (surface or sprinkler), rooting depth, and permeable capacity of the soil.

7.4.1.2 Sodicity Hazard

Though EC is a valuation of all soluble salts in natural water, sodicity problems are generally defined distinctly owing to the specific detrimental effects of Na^+ on the physical properties of soil and plant beings. Sodicity problems ($s = 5$) are typically measured by the values of Na%, SSP, and SAR (or SAR_{Adj}) that affect soil and lead to sodic soil conditions. Na% and SAR (or SAR_{Adj}) values potentially impact the Na hazard, soil texture, and infiltration rate and create saline (with Cl^-) and alkaline environments (with HCO_3^-/CO_3^{2-}). CO_3^{2-} and HCO_3^- ions can bind some Ca and Mg, thereby dropping their obtainability to counter the hazardous effects of Na^+. For irrigation waters containing significant amounts of CO_3^{2-} and HCO_3^-, instead of SAR, the adjusted SAR (SAR_{Adj}) should be used to evaluate the potential sodicity hazard. Higher contents of Ca and Mg in irrigation water tend to counter the negative effects of Na^+ and endorse good soil collections. This hazard can lead to adverse impacts on both soil and crop yields. High salinity affects plants, with a tendency toward saline soil, but high Na and low salinity affect soil, with a tendency toward sodic soil.

7.4.1.3 Water Infiltration or Permeability Rate

The rate of permeability ($s = 4$) of irrigation water into the soil is significantly influenced by the water quality; soil consistency or the relative ratio of silt, sand, and clay in the soil; organic and humic material content; grade of compaction; and chemical makeup. It is just a mechanical measurement of soil impacted by dissolved minerals of external water. If mutual salinity and sodicity hazards function during the same period, the level of Na^+ in water is the influential parameter for potential permeability problems. It is significant to note that this hazard characteristically occurs in the first few cm of the topsoil and is strongly associated with the texture and stability of the topsoil and with low Ca and Mg contents compared to that of Na [12]. Accordingly, combined SAR(EC), PI, SSP, and Na% parameters are used to evaluate the probable infiltration hazard that might occur in soil. Though low saline water with high Na concentrations has a severe infiltration hazard, the high salinity of water with low concentration of Na does not experience any penetration problem.

7.4.1.4 Toxicity to Crops

Some water parameters are very toxic to crop, and this hazard class ($s = 3$) remains very flexible, and the researcher can exclude or include any other components of irrigation water as local demands. In this proposed model, certain substances, such as Na^+, K^+, Fe, Mn, B, As, Cu, Zn, and Cl^-, are considered to cause toxicity problems for plants when they are observed in higher levels in water as well as in soil. Perennial-type and permanent crops are greatly delicate to this kind of toxicity related to seasonal crops. Generally, ion toxicity is complicated by other hazards, such as salinity hazards and infiltration problem. Also, toxicity can occur from the direct absorption of toxic substances through leaves by overhead sprinkler irrigation systems. Fe, Mn, Cu, and Zn are very common trace metals in groundwater, while the toxicity of other trace elements, such as Pb, As, Hg, Cd, Se, Co, and Cr, is very high, but they are not found in enough amounts in all places. These trace elements directly impact the soil environment, plant growth, and the quality of yields, and through bioaccumulation, they accumulate in the human body cell by food waves.

7.4.1.5 Changing Soil Structure

Soil properties mostly depend on the amount of Ca, Mg, Na, organic materials, and microorganisms in the soil mixture. It has been found that once the soil is irrigated with high concentration Na in waters, by cation exchange, Na possibly removes the Ca and Mg of soil, which in turn decreases the soil texture and structure. The high Na and low Ca and Mg harm the soil structure due to the dispersal of clay particles. Excess Na is adsorbed on the clay particle surface, and the soil becomes harder. The comparative quantity of Na to Ca and Mg in irrigation water is measured by SAR, SSP, and Na%, and those parameters' value is used to determine the hazard class of changing soil properties ($s = 2$). Continued use of irrigation water with Na more than Ca and Mg leads to a very tight soil structure and total breakdown of the soil texture with poor water infiltration, poor aeration, and increased surface crusting; these conditions make ploughing difficult and restrict seedling emergence and root growth.

7.4.1.6 Miscellaneous

The hazards and effects deliberated in the previous subsections, one more flexible hazard class, that is, miscellaneous effects ($s = 1$) on sensitive crops and soil, comprised the rest of the water quality parameters, such as pH, TH, Ca, Mg, NO_3^-, SO_4^{2-}, PO_4^{3-}, CO_3^{2-}, and HCO_3^-. In this group, irrigation water suitability parameters such as TH, RSC, RSBC, and MAR are also included. Higher values of RSC and RSBC indicate the increased Ca- and Mg-carbonate/bicarbonate precipitation and increased $NaHCO_3$ alkalinity. High nutrient levels in the water that supplies N and P to the crop may cause unnecessary vegetative growth and delayed crop maturity. An unusual problem faced by some growers practising irrigation (overhead) is the corrosion of equipment by crusting and scaling due to water hardness. Besides, TH increases the soil's basicity and enhances micronutrient toxicity. Ca, Mg, CO_3^{2-}, and HCO_3^- were also counted in the calculation of combined irrigation water quality parameters (such as SAR, SSP, RSC, etc.), but those are slightly affecting the plant and/or soil at certain higher concentrations. The definite standard values of those parameters are recognized by FAO and other monitoring bodies. According to this standard value, the calculated values of those parameters with statistical results of the groundwater samples for different periods are revealed in Table 7.5 and Appendices XI–XIII.

The destructive result of the computed irrigation water quality parameters is shown in Table 7.5. For more visualization of the result presentation, the values of the parameters are displayed by box plot. The values of physicochemical water parameters such as EC, TDS, and TH in the three sampling periods are shown in Fig. 7.1.

Salinity hazard, with the maximum score ($s = 6$), is the top donor in any index calculation for irrigation water suitability. Usually, it is measured by the electrical conductivity (EC) and total dissolved solid (TDS) values of irrigation water as well as soil solution. The water chemistry of the analysed samples exhibited heavy mineralization, and elevated values of EC and TDS (particularly in the MON and POM seasons) were recorded to determine the salinity hazard in groundwater (Table 7.5 and Fig. 7.1). EC is the most vital parameter that is needed for assessing the suitability of water for irrigation purposes. According to EC values, water is classified into five categories from good to inappropriate for irrigation based on the FAO. The five classes comprise excellent category (<250 μS/cm), good category (250–750 μS/cm), acceptable (750–2000 μS/cm), undecided (2000–3000 μS/cm),

Table 7.5 Destructive statistics of irrigation water suitability parameters in groundwater throughout PRM, MON, and POM and per cent of suitability for irrigation use

Indices	Average	Min.	Max.	±SD	% Sample	Remark
Pre-monsoon, PRM (n = 40)						
EC	670	367.5	1043.5	172.3	Good – 55 Fair – 45	Water availability problems for some crops
TDS	413.4	220.3	676.5	113.3	Excellent – 65 Good – 35	Water availability problems for some crops
TH	374.66	198.72	571.80	92.02	Hard – 5 Very hard – 95	Form a coating on the roots and the leaves; hindering photosynthesis
PSS	1.222	0.675	2.163	0.352	Excellent to good – 100	No water permeability problem
Na%	7.44	3.36	30.48	4.824	Excellent – 100	No water permeability problem
SAR	0.11	0.04	0.58	0.092	Excellent – 100	No water permeability problem
SSP	7.80	3.51	31.00	4.883	Excellent – 97.5 Permissible – 2.5	No effects on sensitive crops
RSBC	2.07	−0.52	5.35	1.316	Safe – 97.5 Marginal – 2.5	No effects on sensitive crops
PI	42.47	29.61	77.54	10.2	Excellent – 2.5 Fair – 97.5	No water permeability problem
MAR	35.67	20.76	51.07	7.562	Suitable – 97.5 Unsuitable – 2.5	No effects on sensitive crops
KR	0.08	0.03	0.44	0.07	Suitable – 100	No sodium toxicity
Mg:Ca	0.35	0.16	0.63	0.114	No permeability problem	No water permeability problem
Na:Ca	0.15	0.05	0.73	0.112	No permeability problem	No water permeability problem
Monsoon, MON (n = 40)						
EC	867.5	560.5	1237.0	179.7	Good – 7.5 Fair – 92.5	Water availability problems for some crops
TDS	558.2	369.0	815.9	114.7	Good – 100	Water availability problems for some crops
TH	403.57	200.48	615.10	92.462	Very hard – 100	Form a coating on the roots and the leaves; hindering photosynthesis
PSS	1.079	0.499	1.976	0.309	Excellent to good – 100	No water permeability problem

Na%	6.20	2.14	28.47	4.539	Excellent – 97.5 Good – 2.5	No water permeability problem
SAR	0.10	0.03	0.55	0.087	Excellent – 100	No water permeability problem
SSP	6.49	2.46	28.92	4.596	Excellent – 97.5 Permissible – 2.5	No effects on sensitive crops
RSBC	1.69	−0.56	4.88	1.288	Safe – 100	No effects on sensitive crops
PI	38.24	28.27	62.67	7.501	Fair – 100	No water permeability problem
MAR	31.45	17.73	44.30	6.923	Suitable – 100	No effects on sensitive crops
KR	0.07	0.02	0.40	0.062	Suitable – 100	No sodium toxicity
Mg:Ca	0.28	0.13	0.48	0.090	No permeability problem	No water permeability problem
Na:Ca	0.11	0.04	0.62	0.096	No permeability problem	No water permeability problem
Post-monsoon, POM (*n* = 40)						
EC	947.1	663.0	1710.0	220.0	Good – 7.5 Fair – 92.5	Water availability problems for some crops
TDS	600.6	449.0	1108.1	156.0	Good – 100	Water availability problems for some crops
TH	404.71	225.66	607.73	92.44	Very hard – 100	Form a coating on the roots and the leaves; hindering photosynthesis
PSS	1.079	0.499	1.976	0.309	Excellent to good – 100	No water permeability problem
Na%	5.66	2.51	21.56	3.38	Excellent – 97.5 Good – 2.5	No water permeability problem
SAR	0.09	0.03	0.39	0.06	Excellent – 100	No water permeability problem
SSP	5.94	2.65	22.07	3.44	Excellent – 97.5 Permissible – 2.5	No effects on sensitive crops
RSBC	1.65	−0.86	4.16	1.30	Safe – 100	No effects on sensitive crops
PI	39.54	28.17	65.69	7.99	Fair – 100	No water permeability problem
MAR	29.54	20.23	41.22	5.72	Suitable – 100	No effects on sensitive crops
KR	0.06	0.03	0.27	0.04	Suitable – 100	No sodium toxicity
Mg:Ca	0.26	0.15	0.42	0.07	No permeability problem	No water permeability problem
Na:Ca	0.10	0.04	0.40	0.06	No permeability problem	No water permeability problem

Note: *EC* electrical conductivity (μS/cm), *TDS* total dissolved solids (mg/L), *TH* total hardness (mg/L)

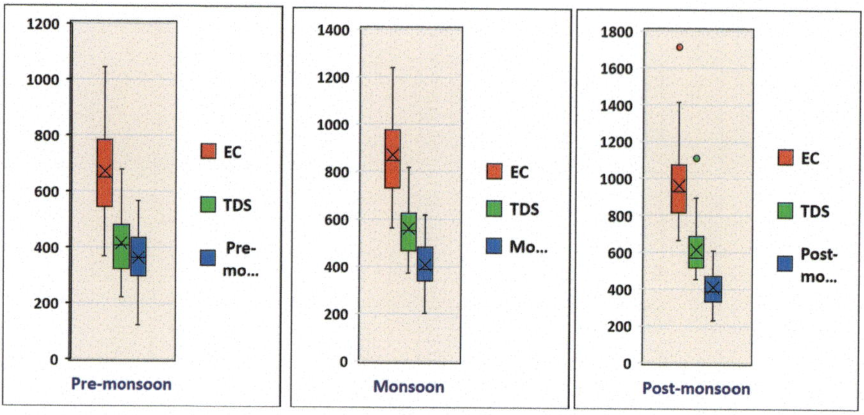

Fig. 7.1 Box plots of irrigation water suitability parameters as %Na, SSP, RSBC, PI, and MAR (*Y*-axis in the above figure denotes the value of the corresponding parameter)

and rejected (>3000 μS/cm). In maximum samples of the study zone, water suitability differs from fair to good for irrigation (Table 7.5). So, samples from the pre-monsoon period are better than those from the monsoon and post-monsoon seasons in terms of EC.

Soils with higher load of salinity typically have a pH below 8.5, are relatively low in Na, and heavily contain Ca, Mg, and Na salts of Cl^-, SO_4^{2-}, CO_3^{2-}, HCO_3^-, NO_3^-, and PO_4^{3-}. Absolute calcite water ($Ca–HCO_3$) with low Na concentration and increased total hardness (TH) in the study area may increase the risk of topsoil salinity. This hazard occurs when salt builds up in the root zone of crops to levels that cause yield loss because in saline soils, the plant roots cannot absorb enough water to keep the plants moist. Decreased water intake in crops affects the crop growth rate. Overall, the potential hazard from salinity depends on the water table depth, kind of irrigation (surface or sprinkler types), local climate, soil structure, root depth, and soil leakage capacity. The soil in the study area consists of gravel and coarse alluvial sand, which increases the risk of salinization in the study area due to the high rate of water infiltration into the soil.

Of the six hazard classes, sodicity hazard, permeability hazard, and soil structure changes may occur generally, and they depend on the value of Na%, saturated sodium per cent (SSP), sodium adsorption ratio (SAR), and permeability index (PI). These parameter values indicate the general Na hazard in soil that makes the soil rigid, and a lower penetration rate of the soil decreases the competence of water adsorption by plant in the root zone and creates alkaline (with HCO_3^-/CO_3^{2-} and high pH) and saline (with Cl^-) environments in the soil. Compared to Ca^{2+} and Mg^{2+} in the samples, the levels of Na^+ and Cl^- are very low, which lowers the values of the above-stated parameters. The elevated load of HCO_3^- ions in samples can bind some Ca^{2+} and Mg^{2+}, in that way dropping their availability to counter the hazardous properties of Na^+ ions in soil systems. Higher contents of Ca^{2+} and Mg^{2+} in the irrigation water of the study area tend to counter the opposing impacts of Na^+

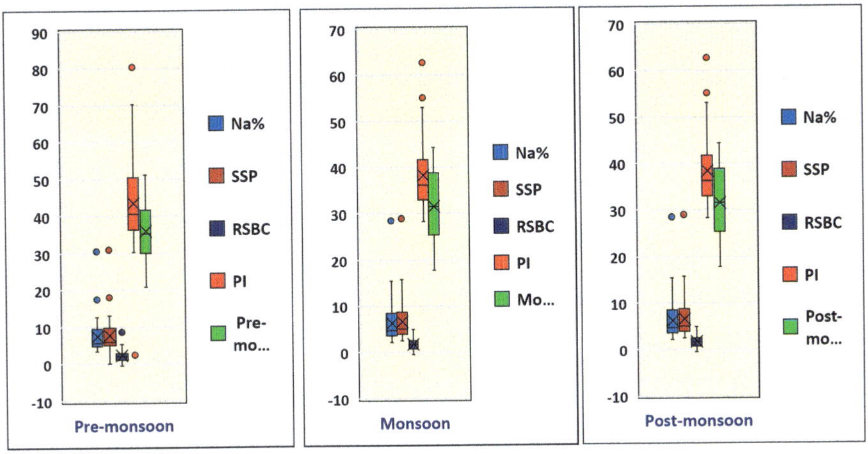

Fig. 7.2 Box plots of irrigation water quality parameters as %Na, SSP, RSBC, PI, and MAR (*Y*-axis in the above figure denotes the value of the corresponding parameter)

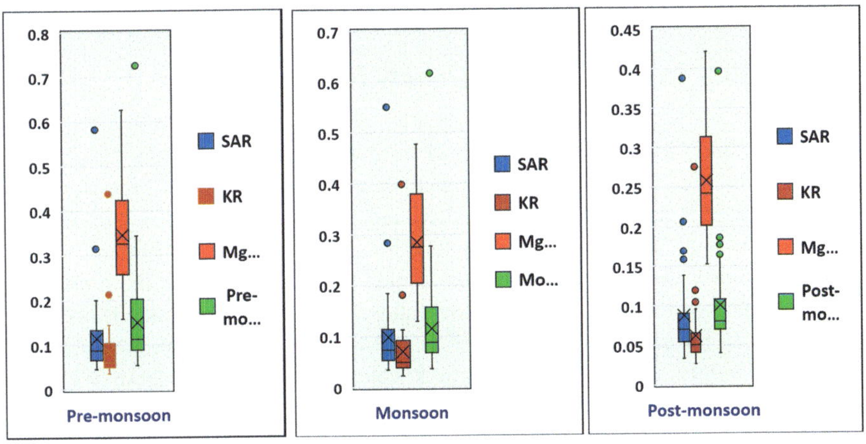

Fig. 7.3 Box plots of irrigation water suitability parameters such as SAR, KR, Mg:Ca, and Na:Ca (the *Y*-axis in the above figure denotes the value of the corresponding parameter)

and endorse good soil aggregates. Table 7.5 and Figs. 7.1 and 7.2 represent the well-favourable values of Na%, SAR, SSP, and PI in all three sampling seasons. Thus, the groundwater of the study zone is very good in position regarding the Na hazard. Instead, the ratios among Mg with Ca and Na with Ca are important measures of irrigation water quality. If the values of Na:Ca and Mg:Ca are surpassed by 3 and 4, respectively, that water is very unsuitable for irrigation practice. The groundwater in the study zone has very low ratio of Mg:Ca and Na:Ca compared to appropriate limits, which demonstrates no threat of permeability problems for soil (Fig. 7.3). When the ratio of Na:Ca crossed the value of 3, excess amounts of Na^{+} in water encourage soil scattering and structural interruption. Due to inadequate Ca^{2+} or Mg^{2+}

to counter the diffusive effect of Na^+, penetration problems would occur initially from such a high ratio. These problems are very common in coastal zones where Na–Cl salinity is a thoughtful threat to agriculture.

Residual sodium bicarbonate (RSBC) designates the higher level of HCO_3^- over Ca^{2+}. The water holding RSBC<5, 5–15, and <15 mEq/L are measured as safe, borderline, and hazardous categories, respectively [19]. Table 7.5 and Fig. 7.2 show that the RSBC values fluctuated from −0.483 to 5.32 with a mean value of 2.09 in the PRM and 0.751 to 4.22 with a mean value of 1.63 in the POM sampling period. The results presented that the water samples were safe and suitable for irrigation practices. The negative values of RSBC indicate additional HCO_3^- in the groundwater in the study zone. As an alternative, Kelley's ratio (KR) displayed a balance among Ca^{2+}, Mg^{2+}, and Na^+ ions in the samples. Once KR is greater than 1, it determines the extra concentration of Na^+ in the water. As expected, the KR value of the samples was less than 1, which showed a high balance between the levels of Na^+, Ca^{2+}, and Mg^{2+} in the water samples. Another parameter, the magnesium adsorption ratio (MAR), showed a lower value that designated a lower value of Mg^{2+} over Ca^{2+}, which was a very good sign of irrigation water quality (Table 7.5).

7.4.2 *Using Irrigation Water Quality Indices (IWQI)*

The present study followed two commonly used water quality index models, which were then modified by numerous researchers or water scientists. First, the weighted average method is used, whereby subindices are generated and further joint into an overall WQI value. Subindices are value functions used to adapt the different units of water suitability parameters to a mutual scale. The second index, the amplitude method (objective-based), is where the overall water quality index value is invented by quantifying the extent to which water quality variables deviate from the standard or objective values. In this study, the Simsek water quality index model ($SIWQ_{index}$) and Meireles water quality index model (MWQI) as the first category and the Canadian water quality index (CWQI) as the second type of method were applied to investigate the groundwater suitability of the study area. Using numerous statistical procedures, the results of those methods were compared with each other, and one was selected to appropriately determine the groundwater quality for both purposes.

7.4.2.1 Simsek Method ($SIWQ_{index}$)

Using Eqs. 7.9 to 7.11, the Simsek irrigation water quality index ($SIWQ_{index}$) was calculated for the 40 collected groundwater samples from the pre-monsoon (PRM), monsoon (MON), and post-monsoon (POM) seasons. The calculated $SIWQ_{index}$ values varied from 36.04 to 41.94 with a mean value of 38.89(±2.44) throughout the PRM, 35.74 to 39.21 with a mean value of 35.46(±2.04) in the PRM, and 35.83 to 38.76 with a mean value of 34.12(±2.98) during the POM (Table 7.5). According

Table 7.6 Statistical results of the Simsek irrigation water quality index ($SIWQ_{index}$)

IWQ index	Category of irrigation water	Pre-monsoon (PRM)	Monsoon (MON)	Post-monsoon (POM)
<22	Low	Range: 36.04–41.94 **Mean: 38.89** 7.5% of samples: medium 92.5% of samples: high	Range: 35.74–39.21 **Mean: 35.46** 67.5% of samples: medium 32.5% of samples: high	Range: 35.83–38.76 **Mean: 34.12** 72.5% of samples: medium 27.5% of samples: high
22–37	Medium			
>37	High			

to the suitability range of the $SIWQ_{index}$, the irrigation water quality decreases as PRM > MON > POM. Maximum water samples for the PRM season (92.5%) were found to be of excellent quality, and 67.5% and 72.5% of samples in the MON and POM periods were of medium quality for irrigation uses (Table 7.6). So, the waters of MON and POM are not suitable for the same purpose. Several investigators in Bangladesh have evaluated irrigation water suitability using other index methods and suitability parameters, such as the $SIWQ_{index}$, soluble sodium percentage (SSP), sodium adsorption ratio (SAR), residual sodium carbonate (RSC), residual sodium bicarbonate (RSBC), magnesium adsorption ratio (MAR), Kelly's ratio (KR), salinity hazard (SH), and permeability index (PI), in neighbouring areas of the present study [22]. The complete results of their studies were almost analogous to this present study.

7.4.2.2 Meireles Water Quality Index (MWQI)

The Meireles water quality index model is used only to determine irrigation water suitability for better crop yield. The mean value of the MWQI of the PRM, MON, and POM seasons in the groundwater of the study zone was 85.01(±4.02), 81.59 (±3.85), and 80.65(±3.97), representing a 'no' and 'low' restriction in use for irrigation (Tables 7.7 and 7.8). But in the case of the PRM period, out of 40 samples, 80% of samples fall under the 'no' restriction, while 37.5% and 22.5% of samples of the MON and POM seasons were in the same category. Averagely, the MWQI value was 82.42, of which 45% and 55% of the samples fell to 'no restriction' and 'low restriction', respectively (Table 7.8). The results appeared to be accompanying with seasonal effects, as waters exhibited a higher load of HCO_3^- throughout the MON and POM seasons (after heavy rain) and Na^+ and Cl^+ enrichment in the PRM period (dry season) owing to strong evaporation. The water with 'no restriction' (MWQI: $85 \leq 100$) category may be used for maximum soils with a less opportunity of causing salinity and sodicity problems and is recommended for leaching within irrigation applies, apart from soils with tremendously low penetrability. This water has no chemical toxicity risk for maximum plants. In addition, the water with a 'low restriction' (MWQI: $70 \leq 85$) category is suggested for use in irrigated soils with very weak texture or moderate penetrability and is recommended for salt leakage.

Table 7.7 Statistical summary of WQI (CWQI and MWQI) values for irrigation of collected water samples

	CWQI (*irrigation*)				MWQI (*irrigation*)			
Value	PRM	MON	POM	Average	PRM	MON	POM	Average
Min.	65.4	61.7	60.0	62.4	72.50	69.40	68.20	70.03
Max.	82.10	79.00	78.00	79.40	91.80	88.00	88.80	88.00
Mean	74.47	71.50	70.21	**72.06**	85.01	81.59	80.65	**82.42**
Variance	19.99	20.30	20.48	19.93	17.66	14.85	15.74	14.59
SD(±)	4.47	4.51	4.53	4.46	4.20	3.85	3.97	3.82
Skewness	−0.30	−0.29	−0.28	−0.30	−0.67	−0.70	−0.82	−0.93

Table 7.8 WQI descriptions and summarized results of water samples

WQI	Parameter uses	Scale	Index value	Category	Description	Average result (% sample)
CWQI (irrigation)	EC, Na^+, HCO_3^-, Cl^-, and SAR	0–100	85–100	A	Excellent	–
			70–84	B	Good	**75**
			55–69	C	Fair	**25**
			40–54	D	Poor	–
			0–39	E	Restrict	–
MWQI (irrigation)	EC, Na^+, HCO_3^-, Cl^-, and SAR	0–100	$85 \leq 100$	A	No restriction	**45**
			$70 \leq 85$	B	Low restriction	**55**
			$55 \leq 70$	C	Moderate restriction	–
			$40 \leq 55$	D	High restriction	–
			$0 \leq 40$	E	Severe restriction	–

Soil sodicity in strong-texture soils may occur, and it is advised to avoid its use in soils with high clay levels of 2:1. In this case, salt-sensitive plants should be avoided.

7.4.2.3 Canadian Water Quality Index (CWQI)

The CWQI [23, 24] is applied for the assessment of domestic water, but some researchers and water programmers have comfortably utilized this method for the estimation of irrigation water suitability. Similar to the MWQI, there are five water parameters, EC, SAR, Na^+, Cl^-, and HCO_3^-, considered in this method for the evaluation of irrigation water quality, and the results for the three sampling periods are shown in Table 7.7. The mean value of the CWQI in the PRM, MON, and POM periods in the groundwater samples of the study zone was 64.47(±4.47), 71.50 (±4.51), and 70.21(±4.53), indicating a 'good' and 'fair' category in use for irrigation purposes (Tables 7.7 and 7.8). It was found that in the case of the PRM season, out of the total 40 samples, 87.5% of fall under the 'good' category, while 25% and 22.5% of the MON and POM seasons were in the same category. Averagely, the CWQI value was 72.06, of which 75% and 25% of the samples fell to 'good' and 'fair', respectively (Table 7.8).

7.4.3 *Judgement of Irrigation WQIs and Their Use in Practice*

Concerning water chemistry and pollution levels, in practice, there are some differences observed between surface and underground water. Most of the indexing models were computed using parameters of surface water quality such as microbial loads and biochemical contents, which are unimportant for the measurement of groundwater quality. Due to the insufficient index method, it is very problematic to properly evaluate the suitability of groundwater for drinking, industrial, and irrigation purposes. This present investigation used two indexing methods for the assessment of irrigation water quality. It made a comparison between the CWQI (irrigation) and MWQI methods and among them, find a better irrigation water quality index.

7.4.3.1 Comparison Between CWQI (Irrigation) and MWQI

The Canadian water quality index model was not only used to evaluate the water suitability for drinking purposes but also to deliver appropriate information about the irrigation water quality. Here, this study compares the CWQI with another irrigation water quality subindex model such as the Meireles WQI. The dataset of the WQIs for irrigation of the two models was verified by the SPSS statistical technique. There is a substantial variance among the two models at a specific significant limit of the paired *t*-test. It is a statistical method to determine the zero of the mean difference among the two sets. The results obtained of the running paired *t*-test are shown in Table 7.9.

In Table 7.9, the CWQI has a somewhat lesser mean score (72.06) than the Meireles WQI (82.42). This table shows that the standard error, standard deviation, and variance of both models were statistically very analogous to each other. It demonstrates that there is a strong positive association ($r = 0.713$) wherever the population correlation coefficient is zero ($p = 0$). Therefore, a zero correlation could occur (significant = 0.000). This substantial connection makes the *t*-test important. In addition, the mean difference was −12.665, and the standard error mean and

Table 7.9 Paired sample *t*-test (CWQI vs. MWQI) for comparative study

Pair CWQI-MWQI (irrigation)	Paired differences							*T*	*df*	Significance (two-tailed)
	Mean diff.	Std. deviation	Std. error mean	Paired sample correlations		95% confidence interval (*CI*) of the difference				
				Correlation	Sig.	Lower	Upper			
	−12.67	3.384	0.535	**+0.713**	0.000	−13.747	−11.582	−23.67	39	0.000

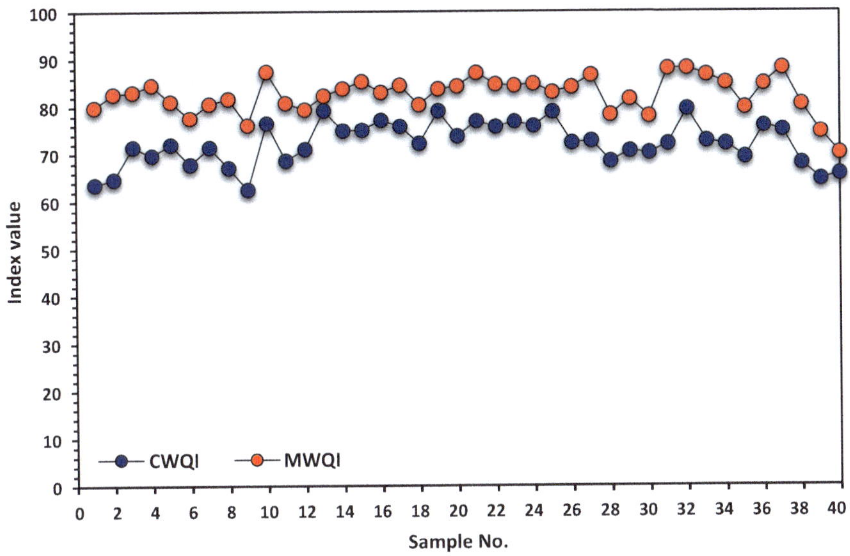

Fig. 7.4 The pair difference statistics between CWQI (irrigation) and MWQI values for all water samples

standard deviation of the difference were 0.535 and 3.384, respectively. Besides, about 95% of the confidence interval (CI) for the mean difference was −13.747 (lower) to −11.582 (upper). The computed *t*-value (*T*) and degrees of freedom (*df*) were −23.671 and 39, respectively. The *p*-value signified by 'significant (two-tailed)' was 0.000. Thus, the findings of the relative study results designated that the irrigation water suitability of both indexing models was statistically similar and significant.

Similar to the MQWI model, only five water parameters were utilized for the CWQI calculation, but the procedures of index calculation were totally different. Figure 7.4 shows that the similarities in the sample-by-sample index value and trend line of both models were nearly the same, and the CWQI values were repeatedly 10–15 points less than the MWQI values. Likewise, the standard deviation and variance of both methods were almost the identical (Table 7.9). There were two ways to adjust the results, for example, increasing the test volume and/or increasing the input parameters. It was tested that if the study considers boron (B), pH, and TDS in the CWQI method, then the values of this index rise to 10–15 points from the former values, and the result would be analogous to the MWQI. Further study is needed to improve both indexing models.

7.4.4 *Using the Diagram Method*

The irrigation water suitability of the study area was approximately determined by the above-mentioned techniques. Finally, the study used some diagram-based procedures, such as the Wilcox diagram, US salinity hazard diagram, and permeability index (PI) diagram to completely evaluate the water's suitability for irrigation.

7.4.4.1 US Salinity Hazard Diagram

This study placed the EC and SAR values of total 40 groundwater samples from three sampling periods in the US salinity hazard diagram which offered by Richards [8]. It was shown that more than 70% of samples for the PRM season fall in the category of C2S1 (C2 means medium salinity hazard and S1 means lowermost sodicity hazard) for irrigation practice. As well, approximately 90% of samples for the MON and POM sampling seasons fall in the category of C3S1 (C3 denotes higher salinity hazard) (Fig. 7.5). This outcome designates that the groundwaters of the study zone are used for all kinds of soil without the hazard of transferrable Na^+ ions. The excess salinity of the groundwater, particularly in the MON and POM periods, may be the cause of the higher level of divalent alkaline earth cations (Ca^{2+} and Mg^{2+}), i.e. the high salinity was not the cause of Na salt, but it happened by the higher concentration of Ca and Mg salt. The detected values of EC, TDS, and SAR of water samples in three seasons support the result of this salinity hazard diagram.

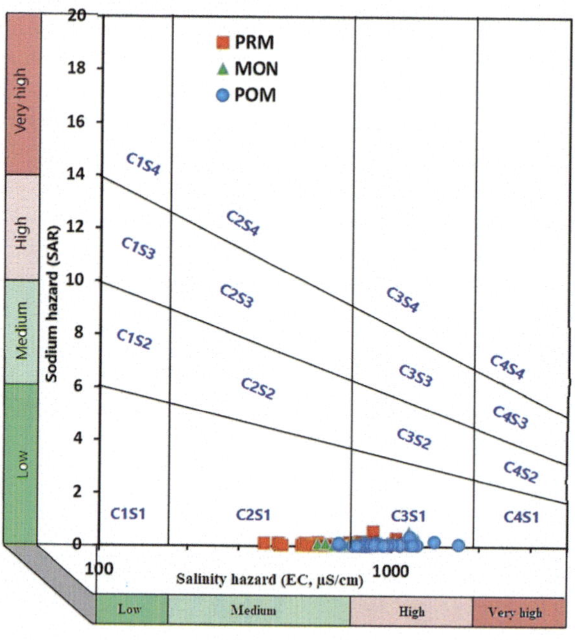

Fig. 7.5 US salinity hazard diagram

Furthermore, owing to the variation in salinity conditions, both the IWQ_{Index} and irrigation water quality parameter values of the MON and POM are less favourable than those of the PRM season, which supports the outcomes of the US salinity hazard diagram.

7.4.4.2 Wilcox Diagram

Based on the above deliberations of US salinity hazard diagram, it was expected that the quality of water for irrigation typically is subject to the comparative values of EC (generate salinity hazard) and concentrations of Na^+ ion (make sodicity hazard). Henceforward, the percentages of Na (Na%) were plotted compared to the EC on the Wilcox diagram [18], which designated that almost all samples from the PRM period fell into the place of excellent to good categories (Fig. 7.6). But about 75% of samples of the MON and POM seasons fell in the excellent to good class, and 25% were in the good to acceptable category. Similar to the findings of the US salinity hazard diagram, the Wilcox diagram also showed that the general EC value of the samples was parallel to the Na% values. This designated that the EC is not reliant on the Na% of the water samples.

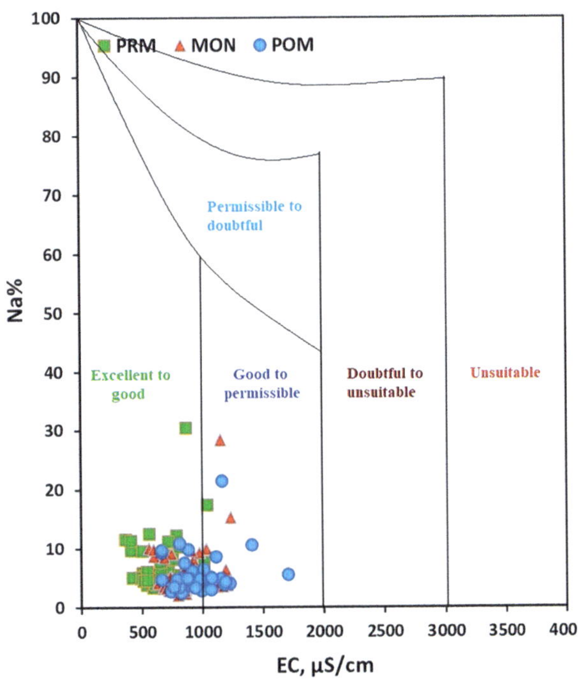

Fig. 7.6 Wilcox diagram for the assessment of irrigation water quality

7.4.4.3 Permeability Index (PI) Diagram

The penetration rate of water or permeability rate of soil directly is subject to the concentrations of Na^{+}, Ca^{2+}, Mg^{2+}, and HCO_3^{-} in irrigation water and soil. The higher amount of Na^{+} ions in water is a major threat to soil penetrability. According to the permeability index (PI), Doneen [10] used a measure for assessing the quality of groundwater for irrigation purposes. The PI value of 40 samples of three sampling seasons was plotted against the entire anionic and cationic concentration in mEq/L unit of the water samples on Doreen's diagram (Fig. 7.7). Three water categories were used to designate the suitability based on the diagram: category I water showed excellent quality for irrigation and low PIs and low to very high ion concentrations, category II water was mostly suitable for irrigation, and category III water was inappropriate for irrigation use. The PI value of the groundwater samples ranges from 42.47% to 77.54% with a mean value of 42.47% in the PRM season (Table 7.6). Also, somewhat fewer values were found in the MON (39.54%) and POM (38.24%) periods. Along with the PI diagram, all the samples are placed in class I and II, which displayed that the groundwater was of 'good' quality (25% to 75%) for irrigation in the study area and had no water penetration problem.

According to the above discussion, it was clear that there was a large difference in irrigation water suitability throughout the pre-monsoon (PRM) and post-monsoon (POM) sampling periods. This is because of the climatic variation, rate of rainwater

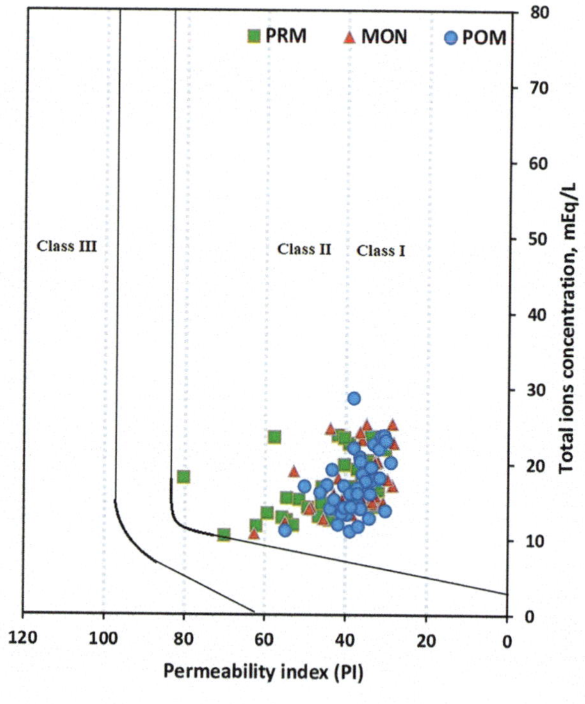

Fig. 7.7 Permeability index (PI) diagram and Wilcox diagram for the assessment of irrigation water quality

filtration, and imbalanced recharge/discharge ratio in the study zone. In the dry season (PRM), only a few mm of rainfall occurs, but in the rainy season, heavy rainwater (~1200 mm/y) is received by this area. Elevated EC, TDS, total hardness, and some ionic component concentrations were found in the POM period compared with the PRM season, which may be due to the heavy mineralization of water during surface runoff and percolation in the rainy monsoon period. The large variations in conductivity values (±SD: 172.5 to 206.1) are typically attributed to geochemical activities such as percolation and infiltration of rainwater, evaporation, exchange of ions, and soil/sediment dissolution. On the other hand, when the water moved through the recharge to the discharge region in the study area, it dissolved additional components along its flow path of journeys. For these reasons, as expected, a large variation was observed in the irrigation water quality for the different seasons.

To date, irrigation water quality has been explored in 14 zones of Bangladesh. The results of those studies revealed that the irrigation water of coastal areas such as the districts of Satkhira [25], Barguna [26], Lakshipur [27], and Khulna [28] is enormously sodicity affected, and those waters are not appropriate for irrigation use. Other irrigation water of the east-northern and northern regions of the country has a slight salinity problem. But the present study exposed that except for EC and temporary Ca-hardness problems, the irrigation water of the study area is very good in position.

7.5 Newly Developed IWQ Index

Several regional and global indexing methods for the calculation of irrigation water suitability are recognized. But it is very difficult to establish a single method that can assess all types of water for every plantation and under any weather condition. In this section, this study proposed two very rational and balanced separate indexing methods for the same assessment.

The Drawback of Existing Models

From 1950 to the 2000s, several researchers, including Eaton [29], Wilcox [18], Doonen [10], Kelley [11], Richard [8], Hem [16], UCCC [13], Todd [7], Matthess [15], Ayers and Westcot [12], Gupta and Gupta [19], Raghunath [6], and Rhoades [30], have been working on agricultural water quality and suitability. After 2000, some irrigation water quality indices were established by CCME [23], Simsek and Gunduz [1], Meireles et al. [2], Maia et al. [31], Hussain et al. [32], etc. Among them, only the Canadian WQI is used for drinking, irrigation, and aquatic culture purposes, but others are applied for irrigation water quality only. The present indexing models considered only some selected and limited numbers of parameters. Yet now, no one model has been recognized that equally considers the water quality parameter value, the allowable range of parameters, and numerous ratings with scoring values of parameters and hazard classes. For instance, Simsek and Gunduz [1] used only scoring of parameters and weight factor of a hazard class; however, Meireles et al.

[2] used a limited number of water parameters, but rating or weight factors were not included in their model. Furthermore, it was seen that these models exposed dissimilar indexing values and statuses for irrigation water for the same samples. For example, one model showed the water quality status of a water sample as 'good', but another model presented it as 'poor' for the same sample. One of the best answers to these anomalies is the different types and numbers of parameters used in these models. Table 7.10 demonstrates the different documented index methods with their characteristics of related equations for evaluating irrigation water suitability. This table displays that the assessment techniques of water cataloguing are different between these considered methods. Every model partly fulfilled the requirements of a well-fitted and complete indexing equation. So, the uncertainty of water quality indexing inspires the development of a new model considering maximum water quality parameters and hazard classes in the proposed model to obtain the best solution for irrigation.

The maximum index models considered the lowest rating value of the parameters to be 1. But the newly proposed method considered that the rejected category value of a parameter can influence the total index value. Hence, the lowest score of the parameter in the rejected class should be zero. Besides, the satisfactory and allowable ranges of water quality parameters for irrigation are very backdated and nonuniform. Therefore, because of present research results, the score of some parameter ranges should be modified and converted into a uniform pattern. Here, two newly developed irrigation water quality index methods, the overall irrigation water quality index (overall IWQ_{Index}) and integrated irrigation water quality index ($IIWQ_{Index}$), are described.

7.5.1 Overall Irrigation Water Quality Index (Overall IWQ_{Index}): A Newly Proposed Method

Generally, there are four steps considered for the establishment of any water quality index equation. Among these steps, the selection of parameters, weightage of variables, subindexing calculation, and aggregation of subindex is vital for all WQIs, while the scoring and weight of the hazard class, selection of hazard class, and rating value of each parameter are utilized in developing an irrigation WQI. The details of these steps are deliberated here.

7.5.1.1 Selection of Hazard Class and Scoring of Water Parameters

Irrigation water hazards are considered approximately into six classes: (i) salinity hazards, (ii) sodicity problems, (iii) infiltration rate of water or permeability of the soil, (iv) ion and heavy metal toxicity to crops, (v) changes in soil structure, and (vi) miscellaneous effects on plants. The rating value of parameters and score of

Table 7.10 Indexing model of irrigation water quality evaluation

Name of index	Final equation	Selected parameters	Subindex used	Uses of scoring/rating/weights[a]	Aggregation method used
CCME WQI [23]	$CWQI = 100 - \left(\frac{\sqrt{F_1^2+F_2^2+F_3^2}}{1.732}\right)$	The maximum number of parameters is not quantified At least 4 parameters	No subindex used	No weights used No scoring factors used No rating factors used	Aggregation method is not used
Simsek and Gunduz index [1]	$IWQ = \sum_{i=1}^{5} G_i$	No. of parameters is not quantified Nine geochemical parameters and 17 trace elements	Mixed method	Scoring factors used Unequal weights used Rating factor used	Additive method
Meireles index [2]	$MWQI = \sum_{i=1}^{n} q_i w_i$	5 parameters: SAR, EC, Na^+, Cl^-, and HCO_3^- No. of parameters is quantified	No subindex used	No scoring used Unequal weights used No rating factors used	Additive method
Maia method [31]	$IWQI = \frac{1}{N} \sum_{i=1}^{N} \sqrt{Z_i^2}$	9 parameters are used: SAR, Ca, Mg, Na, K, Cl, SO_4, and $HCO_3 + CO_3$. No. of parameters is not quantified	Parameters are directly taken as the subindex	No scoring factors used No rating factors used No weights used	Additive method
Hussain method [32]	$IAWQ = \frac{C}{n} \sum_{i=1}^{n} (W_{ri} \times Y_{ri})$	12 parameters are used: pH, EC, TDS, SO_4, Na, K, Ca, Mg, Cl, Na%, SAR, and TH No. of parameters is not quantified	Parameters are directly taken as subindex	No scoring factors used No rating factor used Relative weight used	Additive method

[a]Scoring of hazard class; the rating value of parameters in each hazard class; weight value of hazard class or parameters of water samples

Table 7.11 The scoring and weight factor of hazard classes and involved parameters in each class

Name of hazard class	Scoring value (s_x)	Relative weight value (W_i)	No. parameters (N)	Included parameters
1. Salinity hazard	6	0.286	2	EC and TDS
2. Sodicity hazard	5	0.238	3	SAR, Na%, and SSP
3. Infiltration/permeability rate	4	0. 190	4	SSP, Na%, SAR, and PI
4. Trace metal and ion toxicity to crop	3	0.143	9	K, B, Na^+, Cl^-, Fe, Mn, Cu, Zn, and As
5. Soil structural change	2	0.095	3	Na%, SSP, and SAR
6. Miscellaneous impact on crop	1	0.048	12	pH, Ca, Mg, NO_3^-, SO_4^{2-}, PO_4^{3-}, CO_3^{2-}, HCO_3^-, TH, RSBC, RSC, and MAR
	Total score = 21	$\sum W_i = 1$		

hazard classes are significant factors in the index calculation. This projected model set a score from 1 to 6 for each hazard class in view of their rank in the index calculation (Table 7.11). The salinity hazard is considered the greatest important issue in irrigation water quality assessment, and its score is obviously 6 ($s = 6$). The miscellaneous hazard class impacts on plants are measured as the least vigorous factor, and their score is considered 1 ($s = 1$). The scores for the remaining four hazard classes are set to 5, 4, 3, and 2 for sodicity hazard, the water permeability rate of the soil, ion toxicity to crops, and changing soil structure, respectively. Table 7.12 was created using the improved concept of FAO guidelines for agricultural water suitability.

7.5.1.2 Selection of Parameters

Twenty-seven irrigation water quality parameters are considered in the six hazard classes (Tables 7.11 and 7.12). Salinity hazard measured by EC and/or TDS (both are used if correlation matrix $r < 0.95$) has given the highest score to calculate the index value that affected plants and led to saline soil condition. Some parameters are commonly considered for different categories of hazards because of their different harmful influences. For instance, %Na, SAR, and SSP of water can concurrently change the sodicity nature and water penetrability of the soil, the soil structure, etc. Therefore, these are involved as operating parameters for subindex calculation in the hazard classes (Table 7.11). Instead, the hazard class, toxicity to crops, is very flexible in choosing water quality parameters. If needed, any components related to toxicity to crops for irrigation water can be excluded or included. The proposed

Table 7.12 Cataloguing of irrigation water quality parameters with the rating value (r)

Parameter	Rating value, r	Degree of restriction on use	Parameter	Rating value, r	Degree of restriction on use
EC (μS/cm)			**Mn (mg/L)**		
<700	3	Excellent	<0.5	3	Excellent
700–1500	2	Good	0.5–2	2	Good
>1500–3000	1	Fair	>2–20	1	Fair
>3000	0	Rejection	>20	0	Rejection
TDS (mg/L)			**Cu (mg/L)**		
<450	3	Excellent	<2	3	Excellent
450–900	2	Good	2–5	2	Good
>900–2000	1	Fair	>5–30	1	Fair
>2000	0	Rejection	>30	0	Rejection
%Na and **SSP** (%)			**Zn (mg/L)**		
<20	3	Excellent	<3	3	Excellent
>20–40	2	Permissible	3–7	2	Good
>40–80	1	Doubtful	>7–35	1	Fair
>80	0	Unsuitable/ rejection	>35	0	Rejection
SAR (meq/L)$^{1/2}$			**As (mg/L)**		
<10	3	Excellent	0–0.01	3	Suitable
10–18	2	Good	>0.01–0.025	2	Marginal
>18–30	1	Fair	>0.025–0.05	1	Poor
>30	0	Rejection	>0.05	0	Rejection
RSC (meq/L)			**pH**		
<0.5	3	Excellent	>7–7.5	3	Excellent
0.5–2	2	Suitable	>7.5–8	2	Good
>2–3	1	Marginal	6.5–7; >8–8.5	1	Fair
<3	0	Rejection	6.5 > pH > 8.5	0	Rejection
RSBC (meq/L)			**Ca (mg/L)**		
<1	3	Excellent	>50–75	3	Excellent
1–5	2	Suitable	>75–150	2	Good
>5–15	1	Marginal	50 > Ca > 150	1	Fair
>15	0	Rejection	>400	0	Rejection
PI (%)			**Mg (mg/L)**		
>90	3	Excellent	>10–20	3	Excellent
90–75	2	Good	>20–30	2	Good
<75–30	1	Fair	10 < Mg > 60	1	Fair
<30	0	Rejection	>60	0	Rejection
MAR (%)			**TH (mg/L)**		
<10	3	Excellent	<75	3	Soft

(continued)

Table 7.12 (continued)

Parameter	Rating value, *r*	Degree of restriction on use	Parameter	Rating value, *r*	Degree of restriction on use
>10–30	2	Good	75–150	2	Moderately hard
>30–50	1	Fair	150–300	1	Hard
>50	0	Rejection	>300	0	Very hard/ rejection
Na (mg/L)			**CO_3^{2-} (mg/L)**		
<50	3	Suitable	<1	3	Suitable
50–150	2	Marginal	1–3	2	Marginal
>150–400	1	Poor	>3–15	1	Fair
>400	0	Rejection	>15	0	Rejection
Cl^- (mg/L)			**HCO_3^- (mg/L)**		
<30	3	Suitable	<50	3	Suitable
30–150	2	Marginal	50–150	2	Marginal
150–300	1	Poor	>150–600	1	Fair
>300	0	Rejection	>600	0	Rejection
B (mg/L)			**NO_3^- (mg/L)**		
<0.5	3	Suitable	<2	3	Excellent
0.5–1	2	Marginal	2–5	2	Good
>1–2	1	Poor	>5–30	1	Fair
>2	0	Rejection	>30	0	Rejection
K (mg/L)			**SO_4^{2-} (mg/L)**		
<2	3	Excellent	<10	3	Excellent
2–5	2	Good	10–50	2	Good
>5–35	1	Fair	>50–200	1	Fair
>35	0	Rejection	>200	0	Rejection
Fe (mg/L)			**PO_4^{3-} (mg/L)**		
<2.5	3	Excellent	<0.5	3	Excellent
2.5–5	2	Good	0.5–5.0	2	Good
>5–30	1	Fair	>5–20	1	Fair
>30	0	Rejection	>20	0	Rejection

model considered some common components, such as B, K, Na^+, Cl^-, Fe, Mn, As, Cu, and Zn, that are toxic to crops if they are present in higher levels in water and soil. The toxicity of other trace elements, such as Se, Sb, Pb, Hg, Cd, Co, and Cr, is very high but is not found in enough quantities all over the world. Anyone may include these metals in this hazard class. One more flexible hazard class named miscellaneous included the rest of the water quality parameters, such as pH, Ca, Mg, NO_3^-, SO_4^{2-}, PO_4^{3-}, CO_3^{2-}, and HCO_3^- which are considered less sensitive to crops and soil. In this class, irrigation water quality parameters such as total hardness (TH), magnesium adsorption ratio (MAR), residual sodium carbonate (RSC), and residual sodium bicarbonate (RSBC) were also included (Table 7.11).

7.5.1.3 Weight Factors of Hazard Class and Rating of Parameters

Then, the weight value of each hazard class was calculated by dividing 21 (total score) by each hazard score. The total weight is added up as equal to 1 (Table 7.11). Let us say that if the scoring of the sodicity hazard is 5, then the weight value obtained 0.238. In each class, the rating value *r* of all parameters is measured as 3 to 0 (Table 7.12). The parameter value with rating of 3 represented the maximum value of the excellent range, but rating of 0 indicated the fully rejected class. The values of parameters with ratings of 2 and 3 signified the good and poor/fair ranges of water suitability for irrigation purposes.

7.5.1.4 Subindex and Final Index Calculation

This step objects to convert the water suitability parameters into a common scale. Usually, at the maximum of the WQI, the parameters can only be combined when they have analogous common scales, but irrigation water quality parameters have different scales and units. Therefore, rescaling or normalizing to form a subindex value is needed. To develop the subindex and final index, the subsequent steps are followed as steps 1 and 2:

Step 1. It is very difficult to simultaneously count the values of parameters, acceptable ranges of water parameters, and other associated factors in the same equation. In this step, the subindex value of hazard classes was calculated by the best fitted Eq. 7.15, in which rating scores, weight factors, and three kinds of parameter values are comprised at once:

$$S_i = \frac{s_x}{N} \sum_{\mathrm{i}=1}^{6} \frac{\mathrm{C}_{\max}}{\mathrm{C}_{\min} + C_i} \times r_i \tag{7.15}$$

where $\mathrm{S_i}$ = subindex of *i*th hazard class

s_x = scoring value of hazard class
N = number of parameters in each class (Table 7.11)
r_i = rating value of *i*th parameter (Table 7.12)
$C_{\max}$ = maximum value of the parameter (at $r = 1$)
$C_{\min}$ = minimum value of the parameter (at $r = 3$)
C_i = measured value of the *i*th parameter

Step 2. The overall $\mathrm{IWQ_{Index}}$ is computed by the summation of the subindex values of six hazard classes multiplied by the weight factors (W_i) of each class:

$$\text{Overall } \mathrm{IWQ_{Index}} = \sum_{i=1}^{n} S_i \times W_i \tag{7.16}$$

Table 7.13 Irrigation water classification according to the overall IWQ_{Index}

Overall IWQ_{Index} value	Category
<3.5	Rejection
3.5 to <6	Poor
6 to <12	Fair
12 to <23	Moderate
23 to <35	Good
≥35	Excellent

7.5.1.5 Water Categorization

After the total index value of the water samples was calculated, it was crucial to classify the water quality grade for appropriateness for irrigation purposes. Depending on substantial *r* values (0 to 3), the water samples are categorized into six classes (Table 7.12). The values given in the table are obtained by assigning different rating factors (*r*) (at a minimum and maximum value) to the parameters under each hazard class without changing their weighing and scoring factors. In this presentation, the fraction values of suitability ranges are transformed to the nearest round figures. The obtained water classification is shown in Table 7.13. Areas with overall IWQ_{Index} values of less than 3.5 are decided as 'rejection' quality and are not appropriate for application in irrigation practice. Such waters could destroy soil quality enormously and result in production loss. When the computed index value is between 3.5 and 6, the equivalent waters determine 'poor' suitability for irrigation. In the range, values higher than 6 to 23 are considered to designate waters that could be fair to moderately used on limited types of crops. In contrast, a score value of 23 is measured as 'good' and 'excellent' classes. The computed minimum and maximum values of this index are 0 and 60 at ratings $r = 0$ and $r = 3$, respectively.

7.5.1.6 Utility of Overall IWQ_{Index}

It looked that the limitations of the former existing index models are determined by this newly proposed model. This proposed method is better quality and updated in six characteristics: easy to calculate, considering the more substantiated required and acceptable value of parameters, counting maximum hazard class, using an improved rating system of parameters in each class, water categories into maximum classes, and large number of diversified parameters are considered that affect both soil and crop yield. This method is appropriate not only for groundwater but also for surface water quality in irrigation. Typically, four classes of groundwater, viz. calcite ($Ca–HCO_3$), sodic (Na–Cl), calcite–dolomite ($Ca–Mg–HCO_3$), and mixed ($Ca–Mg–Na–Cl–HCO_3–SO_4$), were used in agricultural divisions. The evaluation of the quality for those types of water in an irrigation system is correspondingly made by this method.

7.5.1.7 Water Quality Assessment Using the Overall IWQ_{Index}

The analysed geochemical water parameters and calculated irrigation water suitability parameters such as Na%, SAR, SSP, RSC, RSBC, MAR, and PI are utilized to compute the proposed index method. Using the overall IWQ_{Index} model, the index value of calcite type ($Ca–HCO_3$) water in the study zone is obtained at 34.57 and 29.65 in the sampling periods of pre-monsoon (PRM) and post-monsoon (POM), respectively. According to water cataloguing, the results revealed that the samples of both seasons fall in the 'good' category (Table 7.13). However, the irrigation water quality of the PRM period is comparatively better, close to the 'excellent' category, than that of the POM period. Throughout the rainy monsoon, heavy percolation of rainwater, increasing water–rock interface, and extra mineral dissolution occurred. For these reasons, after the rainy season (POM season), the groundwater of the study zone becomes highly mineralized and rises the EC as well as TDS value, which leads to the dropping the overall IWQ_{Index} value. In addition, the raised amounts of Ca, Mg, and HCO_3^- during the POM period (Chap. 4) make groundwater less suitable for irrigation.

7.5.2 Integrated Irrigation Water Quality Index ($IIWQ_{Index}$)

This is another newly developed model that comprises all the water parameters and is single-valued, showing a very rational water category. The $IIWQ_{Index}$ method is very updated and may be used for every type of water. The detailed procedure to develop the indexing equation is presented here.

7.5.2.1 Irrigation Water Quality Criteria

Irrigation water quality knowledge is critical to understanding management for long-term crop yield. However, to develop a relatively more effective water quality assessment model, we considered the six hazard classes, in which 'changing in soil structure' is taken into account and others are traditionally utilized in existing models. But the type and number of water quality parameters standardized for irrigation uses by international authorities, such as the WHO, FAO, and US-EPA, were changed in every hazard class of this proposed model. These water quality standards can never be absolute because soils, management, crop diversity, and the type of irrigation practice can impact water suitability. The interrelation of considered criteria for irrigation water quality and system management is crucial to establishing an effective suitability measurement procedure for water. Chemical rather than mechanical hazards for both soil and crops are categorized into six classes: (1) salinity hazards, (2) sodicity problems, (3) infiltration rate of water/penetrability of the soil, (4) ion and trace element toxicity to crops, (5) changes in soil structure, and (6) miscellaneous effects on crops. Under each hazard class, there are some parameters included with specific rating values depending on categorization. The same type of parameter is also included in a different category of hazard class because of their different harmful influences. For instance, %Na, SAR, and SSP of water can simultaneously alter the sodicity nature, water penetrability of the soil can change the soil structure, etc., and for this reason, they are also involved as an operating parameter to subindex calculation in those hazard classes.

The scoring value of hazard classes and the rating value of parameters included in each class are very important factors and are the main operators of the final index value calculation. In the proposed method, each one of these classes is given a scoring number from 1 to 6 such that the most and the least vital classes in water quality for irrigation are given the maximum (6) and lowermost (1) points (Table 7.14). The salinity hazard is the most imperative issue in irrigation water quality valuation and is given the maximum importance ($s = 6$). Instead, the miscellaneous effects ($s = 1$) on plants are considered the least vigorous factor affecting water quality for irrigation. Among these two extremes, sodicity hazard, the penetration rate of the soil, specific ion/metal toxicity, and changes in soil properties are rated in decreasing order of meaning for irrigation water quality. Whereas one can claim that these issues might show radical differences, unlike topographical settings with different soil conditions and dissimilar crop patterns. Importance cataloguing (Table 7.14) could be used for a typical agricultural pattern as a general quality valuation tool. Furthermore, to the scoring values, the procedure also assigns rating factors for each parameter, as revealed in the same table. Previous studies [12, 18, 33] and numerous recent investigations considered the lower rating value to be 1, but we assumed that it should be zero. The worst or most rejected value of categories or parameters can unnecessarily affect the total quality index value. Besides, we modify the score point of some parameters by rearranging the range of parameter values, and all scorings are brought to a uniform pattern. In the case of all parameters, four types of scoring are settled (viz. 3 to 0), with 3 being the highest quality and 0 being the rejection quality of water (Table 7.14). This table is constructed from the modified concept of the FAO guiding principle for agricultural water quality, which was authored by Ayers and Westcot [12].

7.5.2.2 Approach to Developing $IIWQ_{Index}$

There are four steps considered for the development of an integrated WQI for irrigation. Of these steps, parameter selection, weightage, subindexing values, and aggregation of subindex are important for all WQIs; however, the selection of hazard classes, scoring of that class, and rating value of each parameter are especially used in developing an irrigation WQI. Details of these steps, counting the selection method used under each step, are deliberated in this section.

Table 7.14 Scoring of hazard class and rating of parameters in the proposed new indexing model for the assessment of irrigation water suitability

Hazard class (scoring value, *s*)	Parameter (rating value, *r*)		Degree of restraint on use	Parameter (rating value, *r*)	Degree of restraint on use
(1) Salinity ($s = 6$) (2) Sodicity ($s = 5$) (3) Water infiltration rate ($s = 4$) (4) Toxicity to plants ($s = 3$) (5) Changing soil structure ($s = 2$) (6) Miscellaneous effect ($s = 1$)	**EC (μS/cm)**			**Mn (mg/L)**	
	<700 ($r = 3$)		Excellent	<0.5 ($r = 3$)	Excellent
	700–1500 ($r = 2$)		Good	0.5–2 ($r = 2$)	Good
	>1500–3000 ($r = 1$)		Fair	>2–20 ($r = 1$)	Fair
	>3000 ($r = 0$)		Rejection	>20 ($r = 0$)	Rejection
	TDS (mg/L)			**Cu (mg/L)**	
	<450 ($r = 3$)		Excellent	<2 ($r = 3$)	Excellent
	450–900 ($r = 2$)		Good	2–5 ($r = 2$)	Good
	>900–2000 ($r = 1$)		Fair	>5–30 ($r = 1$)	Fair
	>2000 ($r = 0$)		Rejection	>30 ($r = 0$)	Rejection
	%Na and **SSP (%)**			**Zn (mg/L)**	
	<20 ($r = 3$)		Excellent	<3 ($r = 3$)	Excellent
	>20–40 ($r = 2$)		Permissible	3–7 ($r = 2$)	Good
	>40–80 ($r = 1$)		Doubtful	>7–35 ($r = 1$)	Fair
	>80 ($r = 0$)		Unsuitable/ rejection	>35 ($r = 0$)	Rejection
	SAR (meq/L)$^{1/2}$	**SAR_{Adj}**		**As (mg/L)**	
	<10 ($r = 3$)	<8 ($r = 3$)	Excellent	0–0.01 ($r = 3$)	Suitable
	10–18 ($r = 2$)	8–16 ($r = 2$)	Good	>0.01–0.025 ($r = 2$)	Marginal
	>18–30 ($r = 1$)	>16–28 ($r = 2$)	Fair	>0.025–0.05 ($r = 1$)	Poor
	>30 ($r = 0$)	>28 ($r = 0$)	Rejection	>0.05 ($r = 0$)	Rejection
	RSC (meq/L)			**pH**	
	<0.5 ($r = 3$)		Excellent	>7–7.5 ($r = 3$)	Excellent
	0.5–2 ($r = 2$)		Suitable	>7.5–8 ($r = 2$)	Good
	>2–3 ($r = 1$)		Marginal	6.5–7; >8–8.5 ($r = 1$)	Fair
	<3 ($r = 0$)		Rejection	6.5 > pH > 8.5 ($r = 0$)	Rejection
	RSBC (meq/L)			**Ca (mg/L)**	
	<1 ($r = 3$)		Excellent	>50–75 ($r = 3$)	Excellent
	1–5 ($r = 2$)		Suitable	>75–150 ($r = 2$)	Good
	>5–15 ($r = 1$)		Marginal	50 > Ca > 150 ($r = 1$)	Fair
	>15 ($r = 0$)		Rejection	>400 ($r = 0$)	Rejection

(continued)

Table 7.14 (continued)

Hazard class (scoring value, *s*)	Parameter (rating value, *r*)	Degree of restraint on use	Parameter (rating value, *r*)	Degree of restraint on use
	PI (%)		**Mg (mg/L)**	
	>90 ($r = 3$)	Excellent	>10–20 ($r = 3$)	Excellent
	90–75 ($r = 2$)	Good	>20–30 ($r = 2$)	Good
	<75–30 ($r = 1$)	Fair	10 < Mg > 60 ($r = 1$)	Fair
	<30 ($r = 0$)	Rejection	>60 ($r = 0$)	Rejection
	MAR (%)		**TH (mg/L)**	
	<10 ($r = 3$)	Excellent	<75 ($r = 3$)	Soft
	>10–30 ($r = 2$)	Good	75–150 ($r = 2$)	Moderately hard
	>30–50 ($r = 1$)	Fair	150–300 ($r = 1$)	Hard
	>50 ($r = 0$)	Rejection	>300 ($r = 0$)	Very hard, rejection
	B (mg/L)		**CO_3^{2-} (mg/L)**	
	<0.5 ($r = 3$)	Excellent	<1 ($r = 3$)	Suitable
	0.5–1 ($r = 2$)	Marginal	1–3 ($r = 2$)	Marginal
	>1–2 ($r = 1$)	Poor	>3–15 ($r = 1$)	Fair
	>2 ($r = 0$)	Rejection	>15 ($r = 0$)	Rejection
	Cl^- (mg/L)		**HCO_3^- (mg/L)**	
	<30 ($r = 3$)	Suitable	<50 ($r = 3$)	Suitable
	25–150 ($r = 2$)	Marginal	50–150 ($r = 2$)	Marginal
	150–300 ($r = 1$)	Poor	>150–600 ($r = 1$)	Fair
	>300 ($r = 0$)	Rejection	>600 ($r = 0$)	Rejection
	Na (mg/L)		**NO_3^- (mg/L)**	
	<50 ($r = 3$)	Suitable	<2 ($r = 3$)	Excellent
	50–150 ($r = 2$)	Marginal	2–5 ($r = 2$)	Good
	>150–400 ($r = 1$)	Poor	>5–30 ($r = 1$)	Fair
	>400 ($r = 0$)	Rejection	>30 ($r = 0$)	Rejection
	K (mg/L)		**SO_4^{2-} (mg/L)**	
	<2 ($r = 3$)	Excellent	<10 ($r = 3$)	Excellent
	2–5 ($r = 2$)	Good	10–50 ($r = 2$)	Good
	>5–35 ($r = 1$)	Fair	>50–200 ($r = 1$)	Fair
	>35 ($r = 0$)	Rejection	>200 ($r = 0$)	Rejection
	Fe (mg/L)		**PO_4^{3-} (mg/L)**	
	<2.5 ($r = 3$)	Excellent	<0.5 ($r = 3$)	Excellent
	2.5–5 ($r = 2$)	Marginal	0.5–5.0 ($r = 2$)	Good
	>5–30 ($r = 1$)	Fair	>5–20 ($r = 1$)	Fair
	>30 ($r = 0$)	Rejection	>20 ($r = 0$)	Rejection

7.5.2.2.1 Selection of Hazard Class and Parameters

The selection of parameters is an important step in the constituents of a new water quality index. Conventionally, the indices have a different number of designated parameters, changing from 4 (four) to 26 (twenty-six). Regarding the class of the system used for the choice of parameters, generally, it can be separated into three types, viz. fixed, open, and mixed types. The maximum number of WQIs studied has used a fixed number of parameters. Thus, the operator can only apply the selected parameters for the final index computation. For the fixed and mixed types, the selection of parameters objects to select the parameters that have the highest influence on the water quality. But Abbasi and Abbasi [34] highlighted that there is no way by which 100% impartiality or precision can be achieved in the collection of parameters. In this proposed model, we followed the mixed system to choose the parameters; the fixed system was allowed for the first four hazard classes, and the open system was allowed for the last two hazard classes (Table 7.14).

7.5.2.2.2 Establishing Weight Value

Scoring of hazard classes, rating of parameters, and weight values are very important factors that are the main operators of the final index value. Scoring and rating values were previously discussed in Sect. 7.5.2.1. For index calculation, the desirable and permissible limit values or acceptable ranges of parameters are very important factors, but those values are not absolute and vary with counting authorities and places. Here, in the subindex calculation, to avoid these problems, the study considered the desired and permissible values as the maximum values at $r = 3$ and $r = 1$ of the parameter rating, respectively. For example, the desired and permissible values for EC are 700 and 3000 μS/cm, respectively (Table 7.14). On the other hand, the weights of all hazard categories can be equal or unequal. If the parameters of an index are correspondingly important, equal weights are allotted, while if some parameters have greater or minor importance than others, then unequal weights are allotted. In this study, the well-recognized analytic hierarchy process (AHP) was followed to select the weight value of each hazard class. The AHP method is very suitable for regulating the weights of either discrete or combined parameters. Here, the weight value of each class was measured by dividing the total score by 21 of each score value. For instance, if the scoring of the sodicity hazard is 5, then the weight value is 0.238. The total weight is counted as equal to 1. For situations that depart from the overview presented in this study, the applier could easily adapt the technique and use alternative weighting factors for the six hazard classes introduced in this study.

7.5.2.2.3 Obtaining Subindex Values

This step aims to transform the water suitability parameters into a common scale since the real values of the parameters have their different units and the ranges of values to which different parameters can occur differ significantly from parameter to parameter. At the maximum of the WQIs, the parameters can only be combined when they have the same common scales; consequently, rescaling or normalizing them to form subindices is needed. To develop the subindex functions of different parameters, three different methods are usually applied: (1) judgement by a skilled person in the field, (2) use of the water suitability standards, and (3) numerous statistical approaches. At this point, creating rating curves or subindex functions is based on the suitable limits from the legislated standards, such as technical guidelines, national water necessities, and the US-EPA/WHO/FAO standards or any international directions. First, rating factors for each parameter are intended in step 1, and then the subindex value is calculated in step 2.

Step 1. It is very difficult to simultaneously count the values of parameters, tolerable ranges of parameters, and other associated factors in the same equation. In this stage, it was calculated the rating factor using Eq. 7.17 in which rating scores, rating coefficient, and three types of values of parameters are included at once:

$$Q_i = \frac{2V_i}{V_{\max}} \times R_c \times \frac{|100 - V_{\min}|}{(V_i + V_{\max})} \times r \times 100 \tag{7.17}$$

where

Q_i = rating factor of the ith parameter in respectively hazard class
R_c = rating coefficient
r = rating score of parameters
V_i = analysed value of the parameter
$V_{\min}$ = maximum value of the individual parameter at $r = 3$
$V_{\max}$ = maximum value of the individual parameter at $r = 1$

The rating coefficient (R_c) in this equation is the dimensionless and unitless factor. For $r = 1$, 2, and 3, R_c is 0.167, 0.333, and 0.5, respectively, but at $r = 0$, R_c may be excluded from the equation. For example, in the case of water TDS, 488.5, 450, and 2000 mg/L are the values of V_i, $V_{\min}$, and $V_{\max}$ (Tables 7.14 and 7.16), and for TDS of 450–900 mg/L ($r = 2$), $R_c = 0.333$. Any accidental or severe conditions such as heavy industrial activities, lithological causes, abandoned mines, or cumulative radioactive materials that might raise any highly hazardous substance minimum of ten times higher than the normal level in groundwater should be considered, and the r value is -0.001 instead of 0 (rejecting) in Eq. 7.16.

Step 2. In this step, rating factors (Q_i) of separate parameters are aggregated and then multiplied by the weight value and scoring ratio of a hazard class. The subindex value is calculated by Eq. 7.18:

$$S_i = \frac{s}{n} \times W_i \sum_{i=1}^{n} Q_i \tag{7.18}$$

where

S_i = subindex value of hazard class
s = scoring value of each class
n = the number of parameters included in a class

7.5.2.2.4 Aggregation of Subindices to Produce the Final Index

The two most traditional aggregation methods for the subindices are the multiplicative and additive methods. It should also be renowned that there are other adapted types of these two elementary methods. The additive and multiplicative methods still agonize from the eclipsing uncertainty problem. To avoid these difficulties, Liou et al. [35] projected a mixed aggregation method (an amalgamation of multiplicative and additive methods), but it was seen that this model is only for the indexing of drinking water. Hence, this study proposed a very simple additive technique exposed in Eq. 7.19:

$$\text{IIWQ}_{\text{Index}} = \sum_{i=1}^{n} S_i \tag{7.19}$$

7.5.2.3 Suitability of IIWQ$_{\text{Index}}$

This proposed index model explores some new ideas in irrigation water quality that show the influences of soil properties and crop yield. Previous indexing models were considered limited parameters, and the same water samples were given different classes of suitability when calculated by different index models. For example, if the dataset of water parameters is put in different models such as CCME [23], Simsek and Gunduz [1], Meireles et al. [2], and Maia et al. [31], then dissimilar outcomes are obtained. The calculated result of the Canadian WQI revealed that about 65% of samples ranged from 70 to 84 index values and fell into a 'good' category, and about 35% of samples were found within the 'fair' category. The mean value of this index was 71.52 with a standard deviation of ±4.56. In the cases of Simsek and Gunduz index, the same samples (97.5%) were considered in 'excellent' category, and the rest of the samples were in 'good' category. Besides, if the study followed the Meireles model, then the samples categorized as 32.5% are in 'no restriction' and 67.5% are in 'low restriction' categories. Lastly, 15.5% and 84.5% of samples fell under the 'excellent' and 'good' quality through the calculation of the Maia method. So, it is vital to formulate an appropriate equation in which all possible water quality parameters, all hazard classes, perfect ratings, scoring factors, etc., may be included to avoid these dissimilar results. In the present study, it was developed a best fitted indexing equation that involved the maximum number of parameters and hazard class to achieve the best results.

Another thoughtful incongruity was detected in the limit value of designated water quality parameters in present indexing models. The minimum and maximum values of irrigation water parameters differ with allowable authorities and places. Diverse index methods considered different limit values; as a result, the final index values are different from each other. To solve this problem, the present proposed model considers a uniform pattern of parameter ratings where the permissible range of all parameters is used in a modified form. Ayers and Westcot [12] proposed that the lowest rating value of parameters was one ($r = 1$), followed by other present index calculations for irrigation water quality. Nevertheless, the proposed method considered the lowest value of r to be zero ($r = 0$) for every parameter in the rejection category in water samples. If the r value of all taken parameters is zero, obviously the final value of $IIWQ_{Index}$ becomes zero. The highest value of $IIWQ_{Index}$ is unspecified because the extremely accepted value of parameters is undefined. The increased $IIWQ_{Index}$ value designates a better quality of water for irrigation practices (Table 7.15).

Table 7.15 Proposed irrigation water category according to $IIWQ_{Index}$

$IIWQ_{Index}$ value	Category/ suitability	Remarks
<40	Rejection	• Must be avoided this type of water is for irrigation practice in any condition. In the high sodic water, the penetrability index of soil must have very high (PI > 80), and to avoid saltation surplus excess amount of water should be used • The elevated SAR and water with low salt require lime or gypsum application in soil • Limited high salt tolerance crop tolerates this type of water
40 to <60	Poor	• This type of water may be used in sandy and porous soils with high penetrability • Heavy irrigation should be required with elevated EC and SAR. Moderate to high salt tolerance crops may be cultivated using special salinity control practices
60 to <70	Moderate	• May be used in soils with moderate to high penetration rates with low leakage of salts • Crops with moderate tolerance to salts may be yield
70 to <80	Good	• Soils with low clay content, moderate permeability rate, recommended salt leaching, and light texture are fit for irrigation with this type of water • Avoid very salt-sensitive crops
>80	Excellent	• Except for enormously low porousness in soils, water is used for all kinds of soils with a low possibility of causing salinity, sodicity, and toxicity problems • No hazard risk for most crops

In the case of any accidental or severe conditions, if any highly hazardous substance is found to be a minimum of tenfold higher than the typical level in groundwater, it should be considered that the *r* value is −0.001 as an alternative of 0 (rejection) in Eq. 7.17. Based on Table 7.16, if the average concentration of arsenic in groundwater samples was obtained at a minimum of 0.5 mg/L, then the computed Q_i value would be −364. In this condition, the S_i value of the toxicity category was −3.52 in its place of 13.83, i.e. around 17 points would be deducted from the final index value, and this water may fall in the 'poor' category.

The proposed $IIWQ_{Index}$ is not only for groundwater but may also be equally utilized for surface water. There are four classes of groundwater available for irrigation, i.e. calcite (Ca–HCO_3), calcite–dolomite (Ca–Mg–HCO_3), sodic (Na–Cl), and mixed type (Ca–Mg–Na–HCO_3–SO_4), and this method is similarly useful for these types of water. Now, the study applied the dataset of calcite and sodic water to assess irrigation water quality through the proposed method.

7.5.2.3.1 For Calcite Water

A total of 40 sampling sites in the Ganges River basin (Kushtia District) located in the middle-west part of Bangladesh (see Chap. 3) were selected for this investigation as an example of genuine calcite water. The concentrations of analysing geochemical parameters are stated in Table 4.1 of Chap. 4. Other parameters of irrigation water suitability, such as Na%, SAR, SAR_{adj}, SSP, MAR, RSC, RSBC, and PI, were considered from the traditional methods. A previous study (Chap. 4) demonstrated that the groundwater of the study zones was highly mineralized, and almost all of samples were found to be calcite-type. It was justified by Piper's and Chadha's diagram and other several statistical methods. The parameter rating factor (Q_i), hazard subindex (S_i), and integrated irrigation water quality index ($IIWQ_{Index}$) were computed utilizing Eqs. 7.17 to 7.19 and are reported in Table 7.16. The aggregated index value for this type of water was 75.77 which fell under the 'good' class of water for irrigation.

7.5.2.3.2 For Sodic Water

A total of 20 sampling stations in the coastal area (Chittagong District) located in the southeast zone of Bangladesh (Fig. 7.8) were considered sodic water types, and these water data were adopted from Majumder and Islam [36] and Islam and Mazumder [37]. In the coastal area of the country, the groundwater samples were found highly sodic with elevated levels of EC, Na^+, and Cl^- and relatively low Ca^{2+} and Mg^{2+} loads. The analysis results of the water samples of this area are revealed in Table 7.17. The parameter rating factor (Q_i), hazard subindex (S_i), and $IIWQ_{Index}$ were computed by the same process. The aggregated index value of this type of water is found to be 36.51, which fell under the 'rejected' category of water for irrigation.

Table 7.16 List of parameters used in each hazard class and calculation results of $IIWQ_{Index}$ for calcite water

Hazard class	Parameter[a]	Mean value	Rating limit ($r = 3$ to 0)	Q_i	S_i	Hazard class	Para.[a]	Mean value	Rating limit ($r = 3$ to 0)	Q_i	S_i
Salinity ($W_1 = 0.286$)	EC	806.8	<700–>3000	5.55	8.70	Soil structure changes ($W_5 = 0.095$)	SAR_{adj}	0.47	<8–>28	16.27	4.62
	TDS	488.5	<450–>2000	4.59			SSP	9.05	<20–>80	30.47	
Sodicity ($W_2 = 0.238$)	%Na	7.67	<20–>80	26.26	28.96		Na%	7.67	<20–>80	26.26	
	SSP	9.05	<20–>80	30.47		Miscellaneous effect ($W_6 = 0.048$) $\sum W_i = 1$	pH	7.83	>6.5–>8.5	715	4.53
	SAR_{adj}	0.47	<8–>28	16.27			NO_3	5.70	<2–>30	69.47	
Infiltration rates ($W_3 = 0.191$)	SAR_{adj}	0.47	<8–>28	16.27	15.13		SO_4	8.14	<10–>200	5.28	
	Na%	7.67	<20–>80	26.26			PO_4	1.05	<0.5–>20	33.06	
	PI	38.30	>90–<30	6.23			HCO_3	448.8	<50–>600	1.20	
	SSP	9.05	<20–>80	30.47			CO_3	1.64	<1–>15	86.65	

Toxicity to crop ($W_4 = 0.143$)	Na	15.56	<50–>400	1.92	13.83		Ca	114.4	>50–>400	3.70
	Cl	27.09	<30–>300	5.78			Mg	28.96	<10–>60	65.09
	B	2.1	<0.5–>2	0			TH	404.7	<75–>300	0
	K	3.21	<2–>35	31.34			RSC	−0.75	<0.5–>2.5	0
	Fe	4.11	<2.5–>30	52.15			RSBC	1.64	<1–>15	86.65
	Mn	1.6	<0.5–>20	49.08			MAR	29.67	<10–>50	88.92
	As	0.056	0.0– > 05	0		$\mathbf{IIQW_{Index}} = \sum_{i=1}^{n} \boldsymbol{S_i} = \mathbf{75.77}$				
	Cu	4.11	<2–>25	73.73						
	Zn	6.44	<3–>30	76.12						

[a]The unit of all metals, TDS, and ions is mg/L, EC in µS/cm, and other parameters are in traditional units

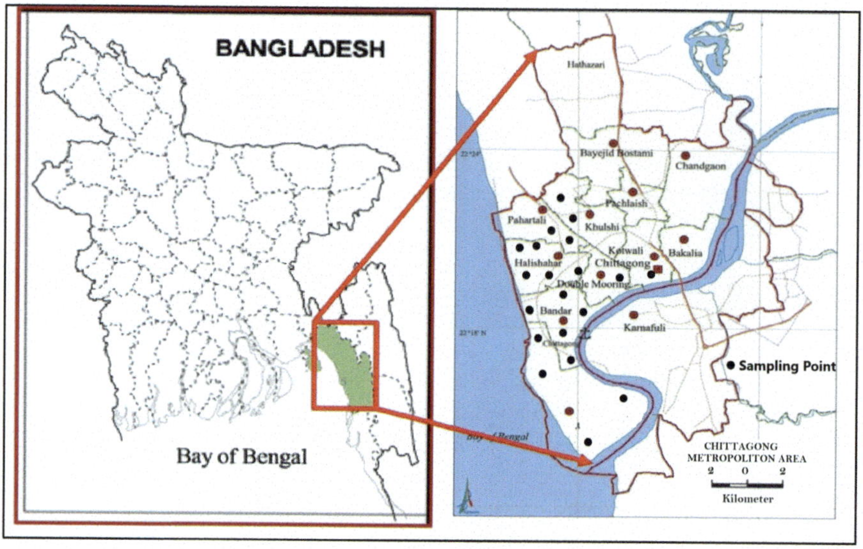

Fig. 7.8 Sampling locations of sodic-type groundwater

Table 7.5 shows that numerous water parameters crossed the maximum permissible limit. In this situation, the *r* value becomes zero, and obviously the Q_i value obtained zero. The higher level of Na^+ (767 mg/L) and less concentrations of Ca^{2+} and Mg^{2+} led to higher values of irrigation water parameters of Na%, SSP, and SAR, which lowered the index value. This kind of water is seriously harmful to both soil environment and crop yield, and hence this water is totally inadequate for irrigation practice.

7.5.2.4 Application of $IIWQ_{Index}$

With the present two case studies, eight (8) additional datasets of physical, geochemical, and irrigation water quality parameters from different countries are revealed in Table 7.18. The respective water types, electrical conductivity (EC, μS/cm), and sodium adsorption ratio (SAR) values are shown in this table. The results of the calculated $IIQW_{Index}$ value with the water category of the listed sample sources indicate the status of overall water quality for irrigation utility. The subindex values of sodicity and salinity hazard classes employed the lion's share of the aggregated value of $IIQW_{Index}$. Usually, these two classes depend mostly on the EC and SAR of water, respectively. EC measures the total dissolved salts of water such as Ca-, Mg-, and Na salt of Cl^-, SO_4^{2-}, CO_3^{2-}, HCO_3^-, NO_3^-, and PO_4^{3-}. Perfect calcite and sodic water exposed elevated levels of EC as well as TDS. A salinity hazard exists if salt gathers in the crop root zone to a level that causes a loss in production, as the roots of the plants are incapable of taking up enough water to keep the plant hydrated in saline soil. The unavailable water uptake in crops hampers

Table 7.17 List of parameters used in each hazard class and calculation results of $IIWQ_{Index}$ for sodic water

Hazard class	Parameter[a]	Mean value	Rating limit ($r = 3$ to 0)	Q_i	S_i	Hazard class	Para.[a]	Mean value	Rating limit ($r = 3$ to 0)	Q_i	S_i
Salinity ($W_1 = 0.286$)	EC	1407	<700–>3000	8.50	13.09	Soil structure changes ($W_5 = 0.095$)	SAR	25.22	<10–>30	0	0
	TDS	817	<450–>2000	6.75			SSP	92.6	<20–>80	0	
Sodicity ($W_2 = 0.238$)	%Na	93.77	<20–>80	0	0		Na%	92.45	<20–>80	0	
	SSP	93.96	<20–>80	0		Miscellaneous effect ($W_6 = 0.048$)	pH	8.6	>6.5–>8.5	0	1.95
	SAR	32.08	<10–>30	0		$\sum W_i = 1$	NO_3	9.69	<2–>30	106.4	
Infiltration rates ($W_3 = 0.191$)	SAR(EC)	32.08	<10–>30	0	4.85		SO_4	45.74	<10–>200	11.15	
	Na%	93.77	<20–>80	0			PO_4	3.67	<0.5–>20	102.8	
	PI	96.20	>90–<30	25.38			HCO_3	134. 9	<50–>600	2.04	
	SSP	93.96	<20–>80	0			CO_3	1.67	<1–>15	132.8	

(continued)

Table 7.17 (continued)

Hazard class	Parameter[a]	Mean value	Rating limit ($r = 3$ to 0)	Q_i	S_i	Hazard class	Para.[a]	Mean value	Rating limit ($r = 3$ to 0)	Q_i	S_i
Toxicity to crop ($W_4 = 0.143$)	Na	767	<50–>400	0	16.62		Ca	23.54	>50–>400	0.47	
	Cl	389.5	<30–>300	0			Mg	12.96	<10–>60	35.30	
	B	2.61	<0.5–>2	0			TH	112.7	<75–>300	3.06	
	K	42.5	<2–>35	0			RSC	−0.03	<0.5–>2.5	0	
	Fe	7.31	<2.5–>30	84.81			RSBC	1.2	<1–>15	65.11	
	Mn	3.51	<0.5–>20	100.64			MAR	47.85	<10–>50	31.68	
	As	0.06	0–>0.05	0		$\mathbf{IIQW_{Index}} = \sum_{i=1}^{n} S_i = \mathbf{36.51}$					
	Cu	4.81	<2–>25	84.20							
	Zn	6.74	<3–>30	79.00							

[a]The unit of all metals, TDS, and ions is mg/L, and other parameters are in usual units

Table 7.18 Calculated values of $IIWQ_{Index}$ of groundwater samples in different geographical places

Sample source	References	Water type	EC	SAR	No of parameter used	Calculated $IIWQ_{Index}$ value	Water category
Kodaikanal, Tamil Nadu, **India (1)** Shallow GW No. of samples: 15	Kalaivanan et al. [38]	Ca–Mg–HCO_3 (50%) Na–K–Cl–SO_4 (40%) Calcite–dolomite Sodic	1818	3.5	20	71.90	Good
Muktsar, Punjab, **India (2)** Deep and shallow No. of samples: 82	Kumar et al. [39]	Na–Cl–SO_4 Ca–Mg–HCO_3–SO_4 Sodic (major) Calcite–dolomite	1022	8.14	24	65.86	Moderate
Near-suburb area, **North China** Shallow GW No. of samples: 22	Xiao et al. [40]	Mg–Ca–HCO_3 (93%) Na–HCO_3 (7%) Dolomite–calcite Sodic	575	1.01	24	97.31	Excellent
Akure, Ondo State, **Nigeria** Shallow GW No. of samples: 60	Falowo et al. [41]	Ca–Mg–HCO_3 Calcite–dolomite	189.8	0.087	19	63.06	Moderate
Beni Mellal city, **Morocco** Shallow GW No. of samples: 51	Baghdadi et al. [42]	Ca–Mg–HCO_3–SO_4 Na–Cl (minor) Calcite–dolomite Sodic	778.5	0.68	23	90.37	Excellent
Talensi District, **Northern Ghana** Deep-26; Shallow-13 No. of samples: 39	Chegbeleh et al. [43]	Ca–Mg–Na–HCO_3 Calcite–dolomite Sodic	403.9	0.34	22	72.11	Good

(continued)

Table 7.18 (continued)

Sample source	References	Water type	EC	SAR	No of parameter used	Calculated $IIWQ_{Index}$ value	Water category
Calabria, **South Italy** No. of samples: 23 Shallow GW	Vespasiano et al. [44]	Ca–Mg–HCO_3 (65%) Na–Cl (27%) Calcite–dolomite Sodic	905	1.10	25	80.76	Excellent
Sargodha District, **Pakistan** No. of samples: 77 Shallow GW	Siddique et al. [45]	Na–HCO_3 (42%) Ca–Na–HCO_3 (37%) Sodic Calcite–dolomite	939	1.70	21	94.18	Excellent
Kushtia District, **Bangladesh (1)** No. of samples: 40 Shallow–semi-deep GW	This study	Ca–HCO_3 Calcite	806.8	0.47	27	75.77	Good
Chittagong coast, **Bangladesh (2)** No. of samples: 20 Shallow GW	This study	Na–Cl Sodic	1607	32.08	27	36.51	Rejection

the growth and production rate of crops [12]. SAR is a perfect parameter to assess the opportunity for Na–alkali hazards because it measures the soil capacity to adsorb Na^+ from irrigation water. Irrigation water with an elevated SAR value can abolish the soil structure by a cation exchange reaction among Na^+ in water and Ca^{2+} and Mg^{2+} in soil.

The $IIQW_{Index}$ analysis results presented that both very low and very high levels of water minerals in samples are harmful to the soil environment and plant health, which lowers the index value. For instance, in Nigeria and Ghana, groundwater samples carry a few mineral components (such as Ca, Mg, and Na) with less than 20 mg/L and have very low EC or TDS values. Consequently, the index value was comparatively lower than that in other countries, although this water is very fresh and non-sodic. The water types can also impact the index value, and the results displayed that the complete calcite- or dolomite-type water samples did not have the highest index value. The $IIWQ_{Index}$ values of different groundwaters in Table 7.18 shadowed the order sodic < calcite < calcite-dolomite < calcite–dolomite–sodic.

In Bangladesh (1), all water samples are calcite type (Ca–HCO_3), and the measured index value is not found to be at a satisfactory level (Table 7.18). Besides, the samples of the coastal areas of the country are the sodic type that showed very low index values and were not fit for irrigation uses. In China, Morocco, and Pakistan, the water samples are mixed type, i.e. dolomite–calcite–sodic type with low SAR and medium values of EC. The levels of Ca^{2+}, Mg^{2+}, and Na^+ in these waters were found to be almost similar to the medium level. This class of water is very good for irrigation uses. Due to the comparatively elevated level of sodicity, SAR, and EC, the samples of India (2) fell under the 'moderate' category with an index value of 65.86. The table reveals that the calculated index value is mostly inversely proportioned to EC and SAR (Fig. 7.9a, b) and is supported by several study findings [46–48]. Hence, the calculated water quality index values of different countries supported the acceptability of the proposed $IIWQ_{Index}$ model.

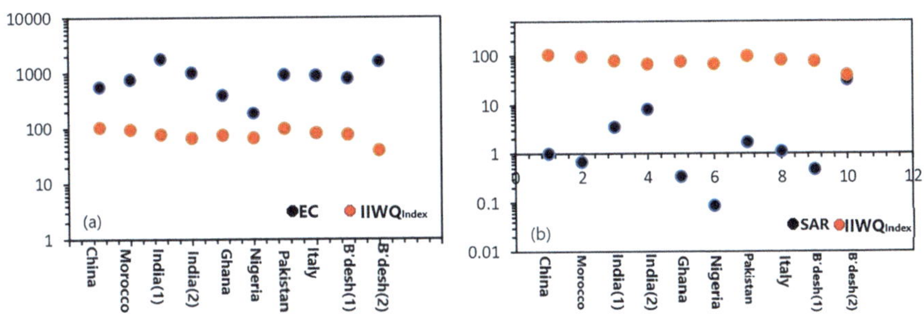

Fig. 7.9 (**a**) EC vs. $IIWQ_{Index}$ and (**b**) SAR vs. $IIWQ_{Index}$ of groundwater for different regions

7.6 Summary

Numerous geochemical and irrigation water suitability parameters of groundwater for the sampling period of PRM, MON, and POM were measured to evaluate the irrigation water quality in the Ganges River basin and delta areas of Bangladesh. The study utilized three irrigation water quality (IWQ) index models, viz. Simsek IWQ_{index}, Canadian WQI, and Meireles WQI, to assess the irrigation water quality. The calculated Simsek IWQ_{index} values ranged from 36.04 to 41.94 with an average of 38.89(±2.44) for the PRM, 35.74 to 39.21 with an average of 35.46(±2.04) for the MON, and 35.83 to 38.76 with an average of 34.12(±2.98) for the POM. According to the suitability range of the $SIWQ_{index}$, the irrigation water suitability in the areas degraded as PRM > MON > POM. The Meireles WQI (MWQI) model exhibited that the water quality of the study zone was designated as a 'low' and 'no' restriction (A and B class) in water use, whereas the Canadian model classified the water into 'good' to 'fair' (B to C) water quality. Statistically, the pair difference is very low, and the correlation matrix ($r = +0.71$) is strongly positive among the CWQ and MWQ indices. The results showed that the mean pair difference was −12.665 from the MWQI to CWQI. Also, the study found that there was no actual difference between the two methods. Finally, the study showed that the MWQI models were more suitable for assessing irrigation water quality. In addition, the study projected a new method, named the 'overall IWQ_{Index}', using the appropriate subindex and aggregated equations, which comprised the maximum number and kinds of water parameters to conveniently assess irrigation water quality.

The overall IWQ_{Index} method categorized irrigation water into six classes, i.e. rejection (<3.5), poor (3.5 to <6), fair (6 to <12), moderate (12 to <23), good (23 to <35), and excellent (>35). This method assessed the calcite facies (Ca–HCO_3) of water in the study zone to recognize their classes for irrigation practices. The calculated results presented that the index values were gotten at 34.57 and 29.65 in the sampling period of PRM and POM, respectively. The samples in the PRM period are of better quality with near the 'excellent' class than samples of the POM that were in the 'good' class. Also, to judge the irrigation water quality, different hydrogeochemical parameters (pH, EC, TDS, and TH) and several irrigation water quality parameters (Na%, SAR, SSP, RSBC, RSC, MAR, PI, and KR) of groundwater for three sampling seasons were investigated and reported. Without saline water intrusion in the study area, the average EC, TDS, and TH values were relatively higher than those in other similar parts of the country, which was the cause of the elevated concentrations of earth metals and bicarbonate in the samples. In this study, except for EC, TDS, and TH, the above-mentioned parameter values of the three sampling seasons revealed that the groundwater belongs to excellent to good categories for irrigation uses. The US salinity hazard diagram indicated that more than 70% of samples for the PRM season fell in the categories of C2S1 (medium salinity and low Na-sodicity) for irrigation practices. Approximately 90% of samples for the MON and POM sampling seasons fell in the categories of C3S1 (high salinity and low Na-sodicity). This finding indicated that the groundwaters of the study area were used for all categories of soil without the hazard of transportable Na^+ ions. According to the Wilcox diagram, almost all samples of the PRM period fell into the cluster of excellent to good categories. But about 75% of samples of the MON and POM seasons fell in the excellent to good categories, and 25% were in the good to acceptable types. Along with the permeability index diagram, all the samples are underclass I and II, which presented that the groundwater was of 'good' quality (25% to 75%) for irrigation in the study area with no soil penetration problem. Instead, the ratios among Mg:Ca and Na:Ca were substantial measures of irrigation water quality. The study area has very low Mg:Ca and Na:Ca ratios in groundwater compared to the maximum acceptable values (4 and 3, respectively), which showed no threat of permeability problems for soil. When the Na:Ca ratio crossed 3, higher level of Na^+ in water inspires soil dispersal and structural breakdown.

The irrigation water quality index (IWQI) is a tool that can assess the quality of irrigation waters concerning potential soil and crop problems. But every method has some limitations and drawbacks. The present study critically reviewed the present methods and recognized the causes and mode of those limitations. This study proposed a new integrated irrigation water quality index ($IIWQ_{Index}$) using the best-fitted subindex and aggregated equations that comprised the maximum number and types of water parameters to conveniently assess irrigation water. The $IIWQ_{Index}$ model categorized irrigation water into five classes, viz. rejection (<40), poor (40 to <60), moderate (60 to <70), good (70 to <80), and excellent (>80). This method assessed two types of water to recognize their classes for irrigation practices. The computed results exposed that the index values were 75.77 and 36.51, and the water category fell under 'good' and 'rejected' for the calcite (Ca–HCO_3) and sodic (Na–Cl) types of water, respectively. The designated water classification of this proposed $IIQW_{Index}$ model was justified through a diversified water quality dataset of eight different geographical locations, and satisfactory results were obtained. The proposed methodology is believed to be provided a simple analysis tool even for a nontechnical decision-maker and/or a farm manager. This indexing method is easy to calculate, delivers practical information on the appropriateness of irrigation water, and helps decrease crop and soil damage by using bad-quality groundwater.

References and Further Study

1. Simsek, C., & Gunduz, O. (2007). IWQ index: A GIS-Integrated technique to assess irrigation water quality. *Environmental Monitoring and Assessment, 128*, 277–300.
2. Meireles, A., Andrade, E. M., Chaves, L., Frischkorn, H., & Crisostomo, L. A. (2010). A new proposal of the classification of irrigation water. *Revista Ciencia Agronomica, 41*(3), 349–357.
3. Ashraf, M., Afzal, M., Ahmad, R., & Ali, S. (2011). Growth and yield components of wheat genotypes as influenced by potassium and farm yard manure on a saline sodic soil. *Soil and Environment, 30*, 115–121.

4. Romanelli, A., Lima, M. L., Londono, O. M. Q., Martı'nez, D. E., & Massone, H. E. (2012). A GIS-based assessment of groundwater suitability for irrigation purposes in flat areas of the wet pampa plain, Argentina. *Environmental Management, 50*, 490–503.
5. Bozdag, A. (2016). Assessment of the hydrogeochemical characteristics of groundwater in two aquifer systems in Çumra Plain, Central Anatolia. *Environmental Earth Sciences, 75*(674), 1–15.
6. Raghunath, H. M. (1987). *Groundwater*. Wiley Eastern.
7. Todd, D. K. (1980). *Groundwater hydrology*. Wiley.
8. Richards, L. A. (Ed.). (1968). *Soil and water conservation research branch. Agriculture handbook, no. 60*. Agricultural Research Service.
9. Gupta, S. K. (1983). Variations of water table in Yamuna drainage basin of Haryana implications and management strategies. In *Paper presented at the seminar on strategies for irrigation water management*, Patna.
10. Doneen, L. D. (1964). Notes on water quality in agriculture. Published as a Water Science and Engineering, Paper 4001, Department of Water Sciences and Engineering, University of California, California.
11. Kelley, W. P. (1985). Use of saline irrigation water. *Soil Science*. 1963, *95*, 355–391.
12. Ayers, R. S., & Westcot, D. W. *Water quality for agriculture, FAO Irrigation and Drainage Paper 29 Rev. I*. UN Food and Agriculture Organization.
13. UCCC. (1974). Guidelines for interpretations of water quality for irrigation. University of California Committee of Consultants (UCCC), USA. National field manual, US–Geological Survey.
14. Michael, A. M. (1978). *Irrigation theory and practice*. Vikas Publishing House.
15. Matthess, G. (1982). *The properties of ground water* (p. 397). John Wiley and Sons.
16. Hem, J. D. (1970). *Study and interpretation of the chemical characteristics of natural water* (2nd ed.). US Geo Survey, Water Supply, Paper 1473.
17. Fipps, G. (2003). *Irrigation water quality standards and salinity management strategies*. Texas A & M University.
18. Wilcox, L. V. (1955). Classification and use of irrigation water. U.S. Department of Agriculture, Circular No. 969, Washington, DC.
19. Gupta, U. C., & Gupta, S. C. (1998). Trace element toxicity relationships to crop production and livestock and human health: Implications for management. *Communications in Soil Science and Plant Analysis, 29*(11–14), 1491–1522.
20. Bauder, T. A., Waskom, R. M., Sutherland, P. L., & Davis, J. G. (2014). Irrigation water quality criteria. Colorado State University Extension Publication, Crop series/irrigation. Fact sheet no. 0.506, p. 4.
21. FAO. (1985). *Water quality for agriculture*. Food and Agriculture Organization of the United Nations.
22. Islam, M. S., & Mostafa, M. G. (2021). Groundwater suitability for irrigated agriculture in Alluvial Bengal delta plain: A review. *International Journal of Advances in Applied Sciences, 10*(2), 156–170.
23. CCME. (2001). Canadian water quality guidelines for the protection of aquatic life. Canadian Environmental Quality Guidelines CCME Water Quality Index 1.0 Technical Report. CCME, Winnipeg.
24. CCME. (2006). *A sensitivity analysis of the Canadian Water Quality Index*. Ontario CCME.
25. Mirza, A. T. M., Rahman, T., Rahman, S. H., & Majumder, R. K. (2012). Groundwater quality for irrigation of deep aquifer in southwestern zone of Bangladesh. *Songklanakarin Journal of Science and Technology, 34*(3), 345–352.
26. Roy, D. M., Md. Abdur Rahim, M. A., & Uddipon Roy, U. (2017). Assessment of groundwater quality and associated hazard identification in the Betagi Pauroshava, Barguna district. *Barisal University Journal Part 1, 4*(2), 361–376.

27. Bhuiyan, M. A. H., Bodrud-Doza, M., Islam, A. R. M. T., Rakib, M. A., Rahman, M. S., & Ramanathan, A. L. (2016). Assessment of groundwater quality of Lakshimpur district of Bangladesh using water quality indices, geostatistical methods, and multivariate analysis. *Environmental Earth Sciences, 75*(12), 1020.
28. Didar-ul, S. M. I., Bhuiyan, M. A. H., Rume, T., & Azam, G. (2017). Hydrogeochemical investigation of groundwater in shallow coastal aquifer of Khulna District, Bangladesh. *Applied Water Science, 7*, 4219–4236.
29. Eaton, F. M. (1950). Significance of carbonates in irrigation waters. *Soil and Science, 69*, 123–134.
30. Rhoades, J. D. (1972). Instrumental field methods of salinity appraisal. In G. C. Topp, W. D. Reynolds, & R. E. Green (Eds.), *Advances in measurements of soil physical properties: Bringing theory into practice* (pp. 231–248). SSSA Special Publication 30/ASA/CSSA/SSSA.
31. Maia, C. E., & Rodrigues, K. K. R. P. (2012). Proposal for an index to classify irrigation water quality: A case study in Northeastern Brazil. *Revista Brasileira de Ciência do Solo., 36*(36), 823–830.
32. Hussain, M. H., Salwan, S. A.-H., & Ayser, M. A.-S. (2018). Assessment of index for aquifer water quality for irrigation and livestock purposes of Dammam Aquifer in Najaf Area of Iraq. *Journal of Karbala University, 16*(3), 77–89.
33. Ayers, R. S. (1977). Quality of water for irrigation. *Journal of the Irrigation and Drainage Division, 103*(2), 135–154.
34. Abbasi, T., & Abbasi, S. A. (2011). Water quality indices based on bioassessment: The biotic indices. *J. Water and Health, 9*(2), 330–348.
35. Liou, S. M., Lo, S. L., & Wang, S. H. (2004). A generalized water quality index for Taiwan. *Environmental Monitoring And Assessment, 96*(1–3), 35–52.
36. Majumder, S. M. M. H., & Islam, M. S. (1995, April 25–28). Study of water quality of surface and underground water of Chittagong and Coz's Bazar Town. In *Proceedings of the international workshop on coastal aquaculture and environmental management*, Cox's Bazar, Bangladesh.
37. Islam, M. S., & Hoque, S. M. M. M. (2020). Alkalinity and hardness of natural waters in Chittagong City of Bangladesh. *International Journal of Science and Business, 4*(1), 137–150. https://doi.org/10.5281/zenodo.3606945
38. Kalaivanan, K., Gurugnanam, B., Pourghasemi, H. R., Suresh, M., & Kumaravel, S. (2017). Spatial assessment of groundwater quality using water quality index and hydrochemical indices in the Kodavanar sub-basin, Tamil Nadu, India. *Sustainable Water Resources Management, 4*, 627–641.
39. Kumar, M., Kumari, K., Ramanathan, A., & Saxena, R. (2007). A comparative evaluation of groundwater suitability for irrigation and drinking purposes in two intensively cultivated districts of Punjab, India. *Environmental Geology, 53*, 553–574.
40. Xiao, Y., Shiyang, Y., Qichen, H., Gu, X., Pei, Q., & Zhang, Y. (2020). Hydrogeochemical appraisal of groundwater quality and health risk in A near-suburb area of Northern China. *Journal of Water Supply Research and Technology-Aqua, 69*, 55–69.
41. Falowo, O., Amodu, M., Oluwasegunfunmi, V., Aliu, A., & Otuaga, M. (2019). Groundwater evolution, hydrochemical facies and quality evaluation for irrigation use in Akure, Ondo State, Nigeria. *Journal of Geoscience and Environmental Protection, 7*, 118–140.
42. Baghdadi, M. E., Zantar, I., Jouider, A., et al. (2019). Evaluation of hydrogeochemical quality parameters of groundwater under urban activities. Case of Beni Mellal city (Morocco). *Euro-Mediterranean Journal for Environmental Integration, 4*(6), 1–19.
43. Chegbeleh, L. P., Akurugu, B. A. & Yidana, S. M. (2020). Assessment of groundwater quality in the Talensi District, Northern Ghana. *The Scientific World Journal, 2020*, ID 8450860, 1–24.
44. Vespasiano, G., Muto, F., & Apollaro, C. (2021). Geochemical, geological and groundwater quality characterization of a complex geological framework: The case study of the Coreca Area (Calabria, South Italy). *Geosciences, 11*(3), 121.

45. Siddique, J., Jin, M., Shah, M. H., Shahab, A., Rehman, F., & Rasool, U. (2020). Integrated approach to hydrogeochemical appraisal and quality assessment of groundwater from Sargodha District, Pakistan. *Geofluids, 2*, 1–5.
46. Rhoades, J. D. (1992). Instrumental field methods of salinity appraisal. In G. C. Topp, W. D. Reynolds, & R. E. Green (Eds.), *Advances in measurements of soil physical properties: Bringing theory into practice* (pp. 231–248). SSSA Special Publication 30/ASA/CSSA/SSSA.
47. Hoffman, G. J. (2010). Water quality criteria for irrigation. University of Nebraska, Lincoln, Lincoln Extension Education Program. EC782.
48. Fipps, G. (2003). Irrigation water quality standards and salinity management. Fact Sheet B-1667. Texas Cooperative Extension. The Texas A & M University System, College Station, TX.

Chapter 8
Industrial Water Quality

Abbreviations

AI	Aggressiveness index
AWWA	American Water Works Association
CI	Corrosion index
CR	Concentration ratio
CSMR	Chloride–sulphate mass ratio
LI	Larson–Skold index
LSI	Langelier saturation index
PSI	Puckorius scaling index
RSI	Ryznar stability index

M. S. Islam, *Hydrogeochemical Evaluation and Groundwater Quality*,
https://doi.org/10.1007/978-3-031-44304-6_8

8.1 Industrial Water

Heavy water-using industries can include chemicals, food, paper, tannery, refined petroleum, or primary metals. As stated by the United Nations World Water Development Report, the industrial sector uses an average of 22% of all global water extractions. It ranges from 59% to 8% depending on the economic conditions of the countries. The percentage of water used in the industrial sector is usually higher in developed countries than in the least developed countries. It is not as much as is utilized by agricultural sectors, which account for approximately 50% of freshwater use. The industry tends to use freshwater, as salt water is inappropriate for the utmost applications because it rusts metal equipment used in machinery. A recent study showed that the industrial sector will consume approximately 24% of global freshwater withdrawals by 2025 [1]. Hence, wastewater needs to be treated before it returns to the water supply. A 20 shallow groundwater samples were collected from Kushtia Municipality area Fig. 8.1 in 3 periods viz. pre-monsoon (PRM), monsoon (MON), and Post-monsoon (POM) and analyzed these samples for the assessment of industrial water suitability.

The cooling systems in an industrial process consume large amounts of water, which is typically reused to save acquisition costs and water resources. Cooling water recycling may result in damage by encrusting, corrosion, or slime caused by the amount of natural soluble salts inside the flow systems owing to emission or evaporation. Thus, the quality assessment and regulation of cooling water is the most significant factor in the procedure of cooling water systems. There are two proposed ways to control its quality, i.e. the treatment of the circulated cooling water and makeup water. To maintain the quality of circulated cooling water, total dissolved solids (TDS) and pH adjustment, and corrosion inhibitors are applied. Furthermore, circulated water is partly discharged to avoid excessive amounts of soluble salts in water. The quality of makeup water depends on TDS removal, suspended solid removal, softening, pH control, dosing of the anti-corrosion agent, bacterial control, etc. Hence, this study attempted to assess the water quality of the cooling system in industrial plants.

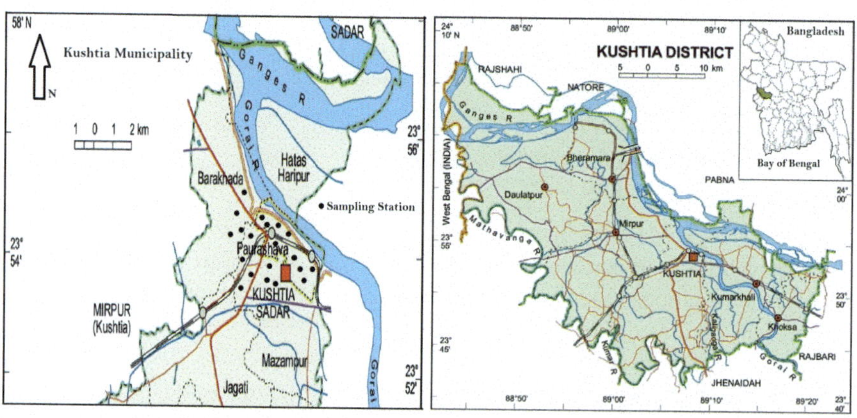

Fig. 8.1 Study area and sampling sites

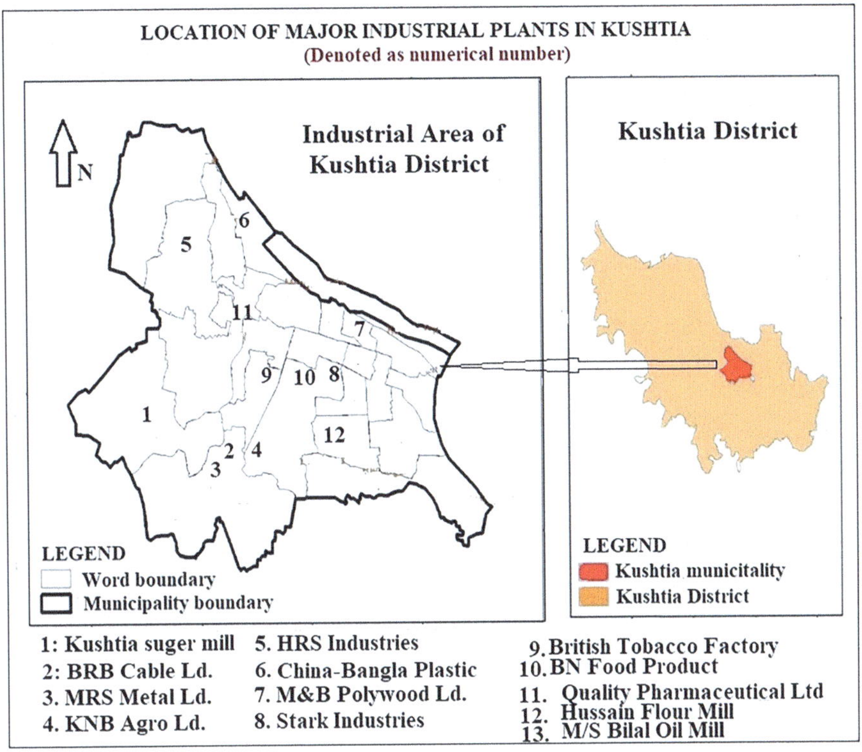

Fig. 8.2 Location of major industries in Kushtia (denoted by numerical numbers)

Various recently established medium to heavy industrial plants in the study zone use millions of gallons of shallow groundwater in this sector. The heavy and medium industrial plants in the study area are BRB Cable Ind., Kiam Metal Ind., M & B Polywood, MRS Industries Limited, KNB Agro-Industry, Quality Pharmaceuticals Ltd., Kushtia Sugar Mill, Sazzad Food Industries, etc., in the Kushtia District of Bangladesh (Fig. 8.2). Hence, it is crucial to monitor the suitability of groundwater resources for industrial use in those areas. In the study, where water hardness and a high level of iron are the main quality issues, there was not enough systematic information on the actual quality of water for industrial use. Highly corrosive and scale-forming water can cause industrial system failures. The longevity of industrial equipment and the quality of finished products mostly depend on supplied water quality.

Every industry has individual water quality standards to maintain the product quality. The standard values of water quality parameters for food, wood, sugar, leather, pulp and paper, synthetic polymer, frozen food industry, etc., are somewhat different. For example, TDS is not important for the pulp and paper and synthetic polymer industries, whereas it must be cautiously maintained in the sugar and food industries. The maximum acceptable values of the total hardness and Ca concentration are very high for the plywood industry, but the small amounts of these water parameters are harmful to the food industry. One of the aims of this study is to justify the water quality for each industry in the study zone.

Corrosion and scaling are the major issues of plant operation using hard water. Instead, scale formation inside water supply pipelines is the cause of dissolved toxic metals in the water that threaten public health. Investigations proved that some toxic and heavy metals dissolved through sorption processes on the surface of scales in supply pipes. Regular monitoring of water quality is the key factor for the operational management of an industry. The study attempted to assess the suitability of groundwater for industrial purposes for sustainable industrial management. The Langelier saturation index (LSI), chloride–sulphate mass ratio (CSMR), Puckorius scaling index (PSI), Ryznar stability index (RSI), aggressiveness index (AI), Larson–Skold index (LI), and corrosion index (CI) were calculated to assess the quality of the pipeline water supply and industrial purposes.

8.2 Industrial Water Quality Indices

The stability of industrial and pipeline supply water was determined through the following scaling and corrosive indices [2–7]. The guideline values for each index and water categorization for industrial water quality are presented in Table 8.1.

8.2.1 Puckorius Scaling Index (PSI)

The PSI tries to measure the connection among scale formation and saturation state by adding an estimate of the buffering capacity of the water sample into the index. Water with high Ca^{2+} but low buffering capacity and total alkalinity could have elevated the calcite ($CaCO_3$) saturation level. The high Ca^{2+} concentration in water increases the ion activity product. Due to the low buffering capacity, a diagram of ion activity product vs. precipitate for the water showed a rapid decrease in pH as Ca precipitated. Even a small decrease in the CO_3^{2-} level in the water would radically decrease the ion activity product due to the small amount present before the beginning of precipitation. This type of water might have a high tendency to create encrusting in the boiler. The degree of this tendency was measured by the index method of Puckorius, which was calculated from Eq. (8.1):

$$\mathrm{PSI} = 2\mathrm{pH}_{\mathrm{Eq}} - \mathrm{pH}_{\mathrm{S}} \tag{8.1}$$

where $\mathrm{pH}_{\mathrm{Eq}} = 1.465 \times \log(\text{Total alkalinity}) + 4.54$; $\mathrm{pH}_{\mathrm{S}} = (9.3 + A + B) - (C + D)$, in which $A = (\log[\mathrm{TDS}] - 1)/10$; $B = -13.12 \times \log(\mathrm{T}) + 34.55$; $C = \log[\mathrm{Ca}^{2+}] - 0.4$; and $D = \log[\text{Total alkalinity}]$.

Table 8.1 Guideline values for industrial water quality indices

Index	Value	Remarks
Puckorius scaling index (PSI)	<4.5	Water tends to encrust
	$4.5 \leq PSI \leq 6.5$	Water is set within an optimum range where no corrosion may happen
	>6.5	Significant corrosion tendencies
Langelier saturation index (LSI)	= 0	Water is thought to be neutral
	<0	Water tends to be corrosive
	>0	Water is supersaturated regarding $CaCO_3$ and encrust forming may happen
Ryznar stability index (RSI)	4.0–5.0	Strongly encrusting
	5.0–6.0	Slightly encrusting
	6.0–7.0	Slightly corrosive
	7.0–7.5	Pointedly corrosive
	7.5–9.0	Strongly corrosive
	≥9.0	Intolerably corrosive
Chloride–sulphate mass ratio (CSMR)	<0.5	Water has no potential for galvanic corrosion
	>0.5	Water with the potentiality of galvanic corrosion
Revelle index (RI)	<0.5	Equipment unaffected by salinization
	0.5–6.6	Equipment somewhat affected by salinization
	>6.6	Equipment strongly affected by salinization
Larson–Skold index (LI)	≪0.8	Probably Cl^- and SO_4^{2-} will not inhibit natural film formation in water
	$0.8 \ll LI \ll 1.2$	Cl^- and SO_4^{2-} may prevent film formation in water
	≫1.2	Tendency to high corrosion rates should be expected as the index increases
Aggressive index (AI)	<0.8	Water has a nonaggressive character
	$10 < AI < 12$	Water sample is moderately aggressive
	<10	Water presents a very aggressive nature
Corrosivity index (CI)	>1	Water has an unsuitable character
	<1	Water presents a suitable character

Here, pH values = pH at saturation in $CaCO_3$, pH_{Eq} = equilibrium pH, and T = Kelvin temperature of the water sample. In this equation, the total alkalinity is equal to ($[HCO_3^-] + 2[CO_3^{2-}] + [OH^-]$). The respective guideline values for PSI are given in Table 8.1.

8.2.2 *Langelier Saturation Index (LSI)*

The LSI is a water quality factor that is used to determine the tendency to form $CaCO_3$ scaling in water. This method estimates the degree of saturation and is measured by calculating the difference between the actual and theoretical pH of a water sample (Eq. 8.2):

$$LSI = pH_A - pH \quad (8.2)$$

where pH_A is the actual pH of the water and pH values are the pH at the saturation state in $CaCO_3$ (theoretical value).

8.2.3 *Ryznar Stability Index (RSI)*

The RSI is a tool used to predict the probability of $CaCO_3$ scale formation in a given sample of water. It attempts to provide the scientific community with a more accurate predictor of $CaCO_3$ scaling. The Ryznar index was calculated as follows:

$$RSI = 2pH_S - pH_A \quad (8.3)$$

Like LSI, pH_A is the actual pH of the sample, and pH values are the theoretical pH values.

8.2.4 *Chloride–Sulphate Mass Ratio (CSMR)*

The higher CSMR may increase the galvanic corrosion and related problems with lead solder water pipe joints. Values of CSMR exceeding about 0.6 were connected with accelerated lead solder corrosion. The CSMR was calculated using Eq. (8.4), and the respective guideline values of this index are shown in Table 8.1:

$$CSMR = [Cl^-]/[SO_4^{2-}] \quad \text{(all ions were measured in the unit of mg/L)} \quad (8.4)$$

8.2.5 *Revelle Index (RI)*

Revelle (1941) projected the following method for calculating the RI, which was used to evaluate groundwater contamination from seawater intrusion:

$$RI = [Cl^-] - \{[HCO_3^-] + [CO_3^{2-}]\} \quad (8.5)$$

where $[Cl^-]$ is the concentration of chloride, $[HCO_3^-]$ is the concentration of bicarbonate, and $[CO_3^{2-}]$ is the concentration of carbonate in the water sample.

8.2.6 Larson–Skold Index (LI)

The LI was established for the measurement of corrosion rates of Great Lakes (USA–Canada) waters. This method extrapolates to another type of water, such as those of less alkalinity or high alkalinity, and goes outside the range of the original data. The index is recognized as a suitable tool for forecasting the aggressiveness of cooling waters in industry and was calculated using Eq. (8.6):

$$\mathrm{LI} = \left([\mathrm{Cl}^-] + [\mathrm{SO_4}^{2-}]\right) - \left([\mathrm{HCO_3}^-] + [\mathrm{CO_3}^{2-}]\right) \quad \text{(all concentrations are in mEq/L)} \tag{8.6}$$

8.2.7 Aggressive Index (AI)

The aggressive index is a part of standard C-400 recognized by the American Water Works Association (AWWA). It narrates the corrosivity of water and its effect on asbestos cement pipes. This index uses a formulation that is computed from the water pH, whether the water has more pH or less pH, the water hardness, how much Ca the water contains, and the value of the total alkalinity in the water:

$$\mathrm{AI} = \mathrm{pH_A} + C + D \tag{8.7}$$

where $\mathrm{pH_A}$ = actual pH of the sample; $C = \log[\mathrm{Ca}^{2+}] - 0.4$; and $D = \log[\text{Total alkalinity}]$.

8.2.8 Corrosivity Index (CI)

The corrosivity index is a degree of water quality that is used to determine whether the water will cause corrosion in the water supply pipes or will form a scale that can create blockages. It is measured using Eq. (8.8):

$$\mathrm{CI} = \left(\mathrm{Cl}^-/35.5 + 2\mathrm{SO_4}^{2-}/96\right) - \left[\left(\mathrm{HCO_3}^- + \mathrm{CO_3}^{2-}\right)/100\right] \tag{8.8}$$

8.3 Groundwater Characterization

The values of detected water parameters for three sampling periods are included in Chap. 4. Here, descriptive statistical analyses of some selected physicochemical water parameters of two sampling periods are displayed in Table 8.2. The analysis results illustrated that the pH was neutral to slightly basic, and higher levels of TDS, alkalinity, Ca^{2+}, and HCO_3^- and lower values of Cl^-, SO_4^{2-}, and CO_3^{2-} were the main features of the groundwater chemistry in the study. These water parameters were used to calculate the industrial water quality indices and justify the cooling water suitability through a chemical approach. The study showed that temporary water hardness and alkalinity have the potential for scale formation in boilers and supply pipes.

The detailed dataset of the saturation index (SI), which is massively illustrated in Chap. 4, revealed that in approximately 90%, 20%, and 10% of the samples in the PRM sampling period, the index values for calcite (polymorph $CaCO_3$), dolomite ($CaMg(CO_3)_2$), and aragonite (crystal $CaCO_3$) were more than 0, which indicates that the samples are oversaturated in these minerals and thus precipitated. Accordingly, the results designated that maximum of the samples were supersaturated with calcite minerals, which led to elevated levels of Ca^{2+} and HCO_3^- in the groundwater of the three sampling rounds in the study area. The elevated value of *SI* for calcite focused on the likelihood of a further increase in the concentration of Ca^{2+}, Mg^{2+}, and HCO_3^- in the samples due to additional dissolution of this type of mineral. Besides, a higher level of aquifer CO_2 pressure encourages carbonate mineral

Table 8.2 Descriptive statistics of physicochemical parameters in groundwater samples during the PRM and POM sampling rounds

	Pre-monsoon, PRM (n = 18)				Post-monsoon, POM (n = 18)			
Parameter	Mean	Min.	Max.	±SD	Mean	Min.	Max.	±SD
Temperature	32	27	36	1.88	29	24	33	2.12
pH	7.02	6.65	7.80	0.218	7.83	7.0	8.91	0.404
TDS	413.15	219	675	113.21	601.5	450.5	1109	156.2
Total hardness (TH)	362.8	121.5	564.0	93.66	404.8	225.7	613.3	93.8
Total alkalinity (T. alk.)	418.6	271.5	703	110.81	451.56	251.62	819.65	124.27
Ca^{2+}	95.64	56.6	151.4	22.46	114.4	67.2	187.8	26.87
Cl^-	31.19	12.0	562	9.292	27.09	12.9	41.8	7.60
HCO_3^-	418.6	271.5	703	110.81	448.78	248.5	817	124.45
CO_3^{2-}	0	0	0	0	3.12	0	5.33	0.43
SO_4^{2-}	16.46	2.91	45.7	9.292	15.14	2.95	41.7	9.25
NO_3^-	4.14	0.82	14.30	3.81	3.71	0.87	18.38	3.64
Fe	8.02	1.14	12.62	3.16	8.784	1.985	14.44	3.36
Cu	1.69	0.09	3.27	0.85	1.71	0.09	3.39	0.85

Note: The unit of all water parameters is mg/L, except temperature in °C and pH

weathering. Thus, the groundwater facies of the study area are Ca–HCO_3 (calcite type) classes, and this factor mostly regulates the corrosivity and scale-forming tendency of the water for industrial uses.

8.4 Groundwater Suitability for Industrial Uses

Evaluation of industrial water quality is performed through two methods:

- Chemical approaches
- Industrial water quality indices

8.4.1 *Storability and Corrosivity Assessment:* Chemical Approaches

Most of the industrial plants in Kushtia District are situated in the municipality area. Plastic and polymer, metal, food processing, plywood, oil, agro-based, pulp and paper, leather, sugar, etc., are the main types of industries in the area. The water supply from shallow aquifers is the major source of water for these industries to use in their boilers, coolers, and other production processing units. Water quality highly influences the finished product and the efficiency of machinery. For example, very hard water can cause scale and crust formation in the boiler, resulting in the lifetime of the equipment being decreased. The locations of major industrial plants in Kushtia are presented in Fig. 8.2.

Water storage and distribution systems for domestic, agricultural, and industrial purposes are often marred by scaling and encrustation phenomena. These issues were influenced by variations in the physicochemical parameters of the used water. These most common parameters include temperature, pH, EC, TDS, Ca^{2+}, Mg^{2+}, Fe, Cu, Cl^-, HCO_3^-, CO_3^{2-}, and SO_4^{2-}. However, the combined effect of these water parameters on corrosion and scale formation can be more complex. The chemical features of the water flowing through a supplying pipe will affect whether the water is stable and will also influence the corrosive reactions. The main factors include pH, alkalinity, and water hardness, but oxidizing agents, saturated CO_2, dissolved oxygen (DO), and TDS can also influence corrosion [7]. Usually, at low pH, acidic waters have many H^+ ions to react with the electron at the cathode, so corrosion is boosted. Water with a higher pH (alkaline) decreases the solubility of $CaCO_3$ so that it is more likely to precipitate out as scale. On the other hand, hard water contains more Ca salts, which can precipitate out as $CaCO_3$ forms scale. However, if the hardness is noncarbonate, Cl^- and SO_4^{2-} will tend to keep Ca^{2+} in the solution and will prevent precipitation. Another water parameter, alkalinity, is the degree to which the water pH can be effortlessly changed, so it can have a mitigating effect on pH. High alkalinity in water is more likely to be scale-forming even at a comparatively low pH. Besides, waters with a low alkalinity deficiency have the buffering capacity to deal with acids, so these types of water can easily become acidic and corrosive. Therefore, a high pH with low alkalinity in water led to corrosivity, and an inverse situation led to scale formation. The main features of groundwater chemistry in the study area were higher TDS, alkalinity, TH, Ca^{2+}, HCO_3^-, Cu, and Fe and lower values of pH (mostly in PRM), SO_4^{2-}, and Cl^-. Very high carbonate hardness and bicarbonate alkalinity at relatively lower water pH in the samples create a complex situation regarding corrosion and water stability. Water with high alkalinity and hardness is more likely to be scale-forming even at a comparatively low pH. On the other hand, in this situation, water promotes the buffering capacity, so it can easily become less acidic and less corrosive. The average pH value of the samples in the study zone is 7.4 in both sampling periods, which is not sufficiently low or high to control the potentiality of corrosion and scale formation. Therefore, the higher alkalinity and hardness of calcite-type water are the major regulating factors that create corrosion and precipitation in industrial equipment. However, the standard chemical characteristics of water vary with the type of industry and finished products.

The quality requirements for industrial water supplies range widely, and almost every industrial application has set its standards (Table 8.3). The contaminated water used in industrial production seriously impacts the quality of products.

After comparing the parameters of water samples with industrial standards given in Table 8.2, 100% of the studied samples are unsuitable for the wood, food, paper, and sugar industry (Fig. 8.3a–c, e). However, almost 89% of the studied samples are considered inappropriate for the polymer industry (Fig. 8.3d). In addition, the canned and frozen fruit diagram shows that nearly 84% of the water samples are nonpermissible (Fig. 8.3f). The absence values in Table 8.3 indicate either that there is no limit for the component or that the component cannot attain offensive levels if the water meets the other conditions.

Table 8.3 Industrial water quality requirements, constructed by Hem [8], Magara [9], and Zeid et al. [10]

Water parameter	Wood industry	Food industries	Pulp and paper	Synthetic polymer	Sugar mill	Canned and frozen fruits and vegetables
pH	6.5–8.0	6.5–7.5	6.0–10.0	6.2–8.3	6.5–7.5	6.5–8.5
TDS	1000	500	–	–	<500	500
TH	900	100	100	350	100	250
Ca^{2+}	100	20	20	80	<50	–
Mg^{2+}	50	10	12	36	–	–
HCO_3^-	250	200	–	–	250	–
NO_3^-	5	5	–	–	–	10
Cl^-	500	200	200	–	200	250
SO_4^{2-}	100	50	–	–	–	250

Note: The unit of all water variables is mg/L, except pH

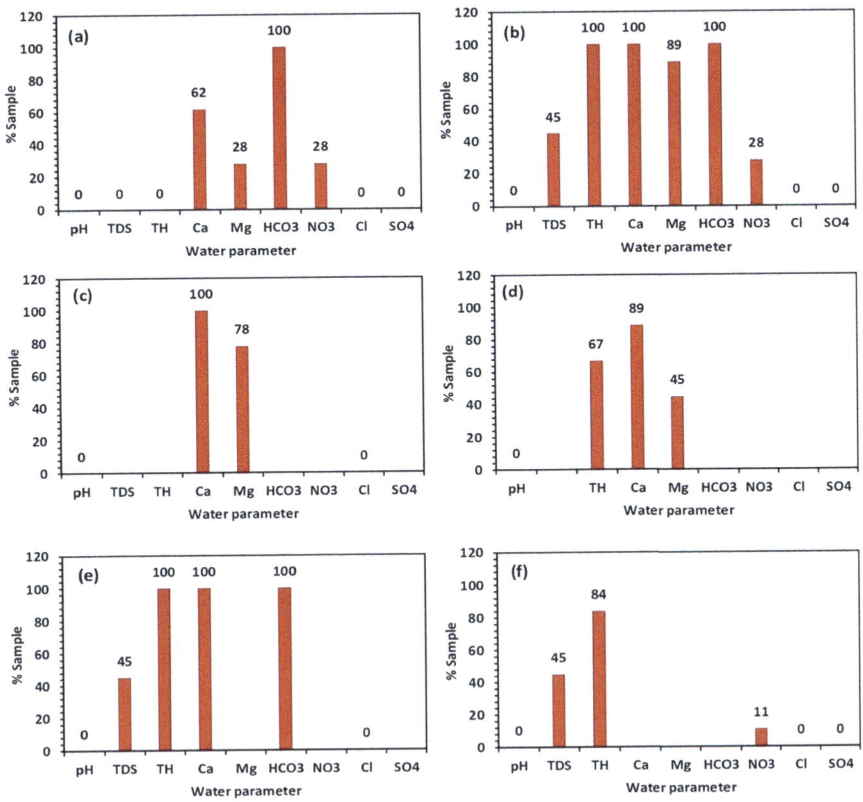

Fig. 8.3 Percentage of samples exceeding various water quality parameters: (**a**) wood, (**b**) food, (**c**) paper, (**d**) polymer, (**e**) sugar, and (**f**) canned fruit industries

8.4.2 Water Quality Management in Cooling Systems

Cooling water includes a major part of the feeding of industrial water. The standards of cooling water quality are represented in Table 8.4. Circulated water is repetitively used in cooling operation systems. The quality of circulated water should be checked and controlled while it is in direct contact with cooling equipment and piping systems in the industry. Water added to circulation arrangements to compensate for water loss due to evaporation, emission, or blowdown is called makeup water. The water quality standard and reference values for both types of water are quite different (Table 8.4). The concentration ratio (*CR*) is the fraction of the circulated water quality to that of makeup water, as revealed in Eq. (8.9) [9]. The standards in Table 8.4 take up a common *CR* of 2–5 in cooling systems:

Table 8.4 Cooling water standard for industrial purposes

		Cooling water			Consideration	
	Water parameter	Circulated water	Makeup water	CR	Corrosion	Scale formation
Standard	pH	6.5–8.2	6.0–8.0	–	√	√
	EC	≤8000	≤3000	2.67	√	√
	TH	≤200	≤70	2.86		√
	T. alkalinity	≤100	≤50	2.0		√
	Cl^-	≤200	≤50	4.0	√	
	SO_4^{2-}	≤200	≤50	4.0	√	
	CO_3^{2-}	≤50	≤10	5.0		√
	HCO_3^-	≤250	≤100	2.50		
Reference	Fe	≤1.0	≤0.3	–	√	
	Cu	≤0.3	≤0.1	–	√	
	SO_4^{2-}	UDL	UDL	–	√	
	Residual Cl^-	≤0.3	≤0.3	–	√	
	CO_3^{2-}	≤4.0	≤4.0	–	√	

Source: JRA GL02 [11]; ISO 16784 [12]; JIS K 0101 [13]
Note: "Standard" denotes critical parameters that directly influence the performance of facilities
"References" are suggested values to avoid nuisance by water-borne diseases or algal growing
"√" marks in the table show the factor relevant to corrosion or scale generation
UDL under detection limit

$$\text{Concentration ratio (CR)} = \frac{\text{Quality of circulated water}}{\text{Quality of makeup water}} \tag{8.9}$$

In cooling systems, metal parts may suffer from corrosion at a lower pH value, whereas scales may form at a higher pH value. Hence, the pH or acidity of the cooling water should be controlled in the range of 6.5–8.0. Although the corrosion tendencies cannot be evaluated by pH values alone, the other water parameters should also be considered. The pH value found in the groundwater samples varies from 6.65 to 7.80 in the pre-monsoon (PRM) and 7.0 to 8.91 in the post-monsoon (POM) sampling periods, which falls within the standard value for both circulated and makeup water (Tables 8.2 and 8.4).

Electrical conductivity (EC) illustrates the total concentration of soluble salts in water. In industrial water, soluble salts are a factor of damage, such as scale formation and corrosion, as described above. The high EC in water is likely to cause such damage. The lower value of EC in makeup water increases the *CR* (Eq. 8.9) of the circulated cooling water and diminishes the loss of water because the quality of makeup water is inversely proportional to the *CR*. The EC and pH values of water are easily counted and hence readily used in routine quality control. As with the pH value, the EC value of the samples is within the standard value of circulated and makeup water for the cooling system (Table 8.4).

The scale is formed on the heat exchange conductors in the cooling system and therefore causes impairment such as reduced heat conductivity. The mean concentration of Ca^{2+} and HCO_3^- and the value of total hardness (TH) in both sampling rounds are higher than the standard value of circulated and makeup water in the cooling system (Tables 8.2 and 8.4). So, there are great possibilities to form scale precipitation in cooling equipment in the industries of the study area. Also, it was found that the measured concentration of iron (Fe) in all the samples was very higher than the reference value for both circulated and makeup water, resulting in great prospects for corrosion in industrial equipment. The study showed that except for CO_3^{2-}, the levels of other items, such as residual Cl^-, Cu, and SO_4^{2-} in groundwater samples crossed the reference value, which encourages algal growth and water-borne diseases. Chloride (Cl^-) and sulphate (SO_4^{2-}) ions are important factors for metal corrosion [7]. Water with a high level of SO_4^{2-} (as H_2SO_4) is not appropriate for cooling water. Normally, the concentration of SO_4^{2-} is lower than that of Cl^- in any water class. If the water contains a high level of SO_4^{2-}, the water must be oxidized with aeration to strip SO_2 gas into the air.

8.4.3 *Evaluation of Industrial Water Quality:* Using Index Methods

High-scale-forming and corrosive water can cause system failures in equipment or result in public health problems because of dissolved harmful components. The durability of industrial equipment and the quality of finished products depend on the quality of supplied water. Thus, corrosion and scaling are the major threats to plant operation. Regular monitoring of water quality is a vital operational tool for the good governance of an industry. Several industrial water quality indexing methods were computed using the values of detected water parameters of different sampling seasons in the study area (Table 8.2), and several industrial water quality indexing methods (described in the methods section) were computed. Such indexing methods are Puckorius (PSI), Langelier (LSI), chloride–sulphate mass ratio (CSMR), Ryznar (RSI), Revelle (RI), aggressive (AI), Larson–Skold (LI), and corrosive index (CI). The results for both sampling periods with some statistical values of those calculations are included in Table 8.4 and Fig. 8.3. According to the results obtained, seasonal variation was observed. The pH, total alkalinity, Ca, and HCO_3^- loads of groundwater in the post-monsoon (POM) season were higher than those in the pre-monsoon (PRM) season, which is the main cause of this variation (Table 8.2). The findings from the index calculation are discussed below:

8.4.3.1 Puckorius Scaling Index (PSI) and Chloride–Sulphate Mass Ratio (CSMR)

The precipitation features of water to achieve an equilibrium state and the buffering capacity of the aqueous medium can be effectively measured using PSI. This is attained by establishing the connotation among the acidic and alkaline nature of water. Knowing the pH or alkalinity/acidity level of a certain water supply system helps in measuring and categorizing the corrosivity and scaling tendency of domestic and industrial water supply systems. The PSI values fluctuated from 8.46 to 10.79 (PRM) and 8.48 to 10.86 (POM), with mean values of 9.59 (PRM) and 9.57 (POM) (Table 8.4 and Fig. 8.4). Following the PSI classification criteria of the US-EPA [14], substantial corrosion tendencies (PSI > 6.5) are observed in 100% of the water samples in both sampling periods. This result suggests that the calcite groundwater samples in the study zone have less encrusting than corrosive tendencies. This result is in agreement with that obtained for the Larson–Skold index (LI) and Langelier saturation index (LSI).

The level of Cl^- and SO_4^{2-} can have a direct bearing on the corrosivity of water, though the higher level of Cl^- in water is known to have a greater influence on its corrosive potentiality. The elevated level of CSMR (>0.5) found for the studied groundwater samples (94.44%) suggests that the water samples have galvanic corrosion potential and can corrode leaded supply systems (Table 8.1). However, only one (5.66%) out of the total samples ($n = 18$) recorded CSMR values <0.5 and thus had no potential for galvanic corrosion in both sampling seasons (Tables 8.1 and 8.5).

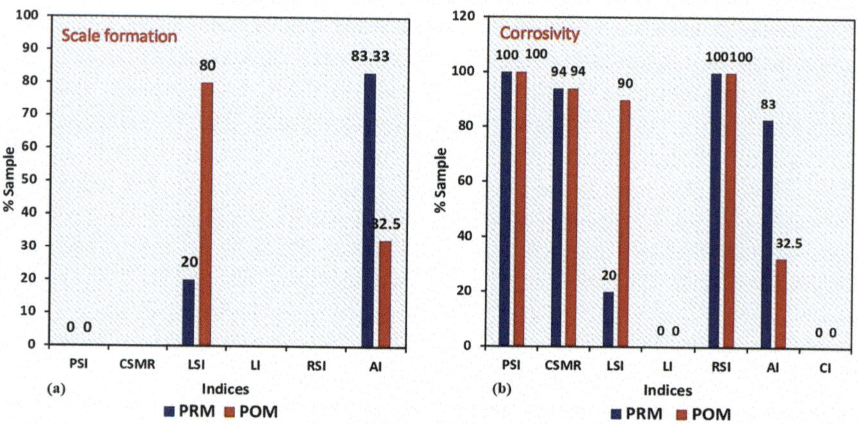

Fig. 8.4 Percentage of the samples crossing the (**a**) scale formation and (**b**) corrosive tendency in the pre-monsoon (PRM) and post-monsoon (POM) sampling seasons

Table 8.5 Statistical illustration of water quality indices for industrial purposes

Index	Pre-monsoon (PRM)			Post-monsoon (POM)		
	Range	Mean	Result (%sample)	Mean	Range	Result (%sample)
Puckorius scaling index (PSI)	8.46–10.79	9.59	High corrosion (100%) Dissolve scale (100%)	8.48–10.86	9.57	High corrosion (100%) Dissolve scale (100%)
Chloride–sulphate mass ratio (CSMR)	0.36–4.08	1.88	Galvanic corrosion (94.44%)	0.44–3.17	1.61	Galvanic corrosion (94.44%)
Langelier saturation index (LSI)	−0.68–0.97	−0.197	No scale (80%) Encrusting (20%)	−0.24–1.44	0.590	No scale (10%) Encrusting (90%)
Revelle index (RI)	0.03–0.12	0.07	Free from salinization (100%)	0.03–0.10	0.06	Free from salinization (100%)
Larson–Skold index (LI)	0.090–0.265	0.158	No corrosion (100%)	0.070–0.251	0.131	No corrosion (100%)
Ryznar stability index (RSI)	6.06–8.21	7.408	Slightly corrosive (15%) Strongly corrosive (85%)	5.82–7.83	6.651	Slightly encrusting (5%) Slightly corrosive (75%) Strongly corrosive (20%)
Aggressive index (AI)	10.71–12.39	11.22	Nonaggressive and scaling tendencies (16.67%) Moderately aggressive (83.33%)	11.28–12.97	12.12	Nonaggressive (67.5%) Moderately aggressive (32.5%)
Corrosivity index (CI)	0.187–4.090	0.300	Suitable (100%)	0.123–3.651	0.252	Suitable (100%)

8.4.3.2 Larson–Skold Index (LI) and Langelier Saturation Index (LSI)

To determine the corrosivity nature of water, the relation between weak (CO_3^{2-} and HCO_3^-) and strong (Cl^- and SO_4^{2-}) conjugated acidic anions is typically measured using the Larson–Skold index (LI). The higher concentrations of SO_4^{2-} and HCO_3^- are known to stimulate the scaling and corrosivity tendencies of water. This is due to the conversion of SO_4^{2-} to S^{2-} ions by sulphate-reducing bacteria present in water under anaerobic conditions. The sulphate-reducing bacteria that exist in dissolved organic matter act as electron donors and help convert SO_4^{2-} to S^{2-}, thereby enhancing encrustation. This process is shown in the following reaction (Eq. 8.10):

$$SO_4^{2-}\,(\text{soluble}) + \text{organic matter} \rightarrow S^{2-}\,(\text{precipitate}) + CO_2 + H_2O \qquad (8.10)$$

The acidic pH (<7) of water may be impacted by the presence of bicarbonate (HCO_3^-), which reacts with metal sulphide (MS) minerals in the primary rock and can result in the formation of strong acid (H_2SO_4), thereby increasing the corrosive potential of the water. In this study, HCO_3^- was observed to be the dominant anion among all water samples (Table 8.2).

An observation of Table 8.5 reveals that the Larson–Skold index (LI) results varied from 0.090–0.265 (PRM) to 0.070–0.251 (POM), averaging 0.158 (PRM) and 0.131 (POM). Following the LI classification standards, no scaling or corrosion tendencies were observed in any of the samples at either sampling period.

The Langelier saturation index (LSI) results varied from −0.68–0.97 (PRM) to −0.24–1.44 (POM) with recorded mean values of −0.197 (PRM) and +0.590 (POM) (Table 8.5). An LSI value greater than 0 suggests that the water has a higher scaling possibility (supersaturated with $CaCO_3$); an LSI value equal to 0 implies that there is no scaling potential in water (saturation is in an equilibrium state); nevertheless, an LSI value undersaturated with $CaCO_3$ suggests that the water has higher corroding potential but lower scaling tendency (Table 8.1). Table 8.5 shows that in the PRM season, 20% of the samples have a serious encrusting problem, whereas, due to a relatively higher concentration of HCO_3^- in water, 90% of the samples of the POM season have a great chance to form encrust on industrial equipment. The high scaling tendency detected could be attributed to the geological formation of the study area. As stated by Egbueri [15], water saturation to $CaCO_3$ is mostly controlled by the number of carbonate rocks or minerals (e.g. dolomite and calcite) in the aquifer basement. This study confirmed that the groundwater facies of the study area is mainly the Ca–HCO_3 type and that the dominant water–rock interaction is calcite dissolution. The high content of carbonate minerals reported (from the geological setting in the study area) within the primary mineral is in direct correlation with the high saturation of $CaCO_3$ in the water. If carbonate minerals had been in low abundance in the fundamental geology, lower levels of Ca^{2+} and Mg^{2+} would have been expected. It can consequently be hypothesized that the comparatively high weathering rate of carbonate minerals in the study area enhanced the scaling tendency of water.

8.4.3.3 Ryznar Stability Index (RSI) and Revelle Index (RI)

The stability index was initially proposed by Ryznar (1944) for the evaluation of water stability. This index is found to be convenient as a projecting method for corrosivity and scale formation tendency in water delivery systems. Following the RSI cataloguing standards, almost 100% of the samples are indicative of a slight to strong corrosive tendency (RSI ≥6), with a very low scaling tendency (Tables 8.1 and 8.5). This remark correlates effortlessly with the results obtained for the Langelier saturation index (LSI). Along with these findings, it can be assumed that the low scaling tendency detected among the water samples is accompanied by low contributions of Ca^{2+} and Mg^{2+} from the primary aquifer basements. However, the high corrosivity tendency is attributed to the low pH (acidic) of the water. The Revelle index (RI) was utilized to regulate the influence of salinization on the equilibrium chemistry, corrosion, and scaling level of given water. Because of the low Cl^- and elevated HCO_3^- levels, approximately 100% of the total water samples seem to be unaffected by salinization (RI < 0.5) (Table 8.4). It can be assumed from this result that the source of salinity-forming salts (mainly Na^+ and Cl^-) was not halite-salt dissolution or seawater intrusion. These may come from any mixed rocks in the groundwater systems.

8.4.3.4 Aggressive Index (AI)

The aggressive index was initially established to monitor water in asbestos pipes. It is occasionally substituted for the Langelier saturation index (LSI) as an indicator of the corrosive potential of water. The AI varied from 10.71 to 12.39 with an average of 11.22 in the PRM period and varied from 11.28 to 12.97 with an average of 12.12 in the PRM period (Table 8.5 and Fig. 8.3). The AI analysis results illustrated that about 16.67% of the samples had to scale tendencies, but they were nonaggressive. Approximately 83.33% of the samples were of a moderately aggressive/corrosive nature in the PRM season, whereas samples were of a relatively less aggressive nature in the POM season. The AI is dependent on the total hardness (TH), pH, and total alkalinity (HCO_3^-). The experimental results are consistent with other corrosivity and scaling index results from this investigation.

8.4.3.5 Corrosivity Index (CI)

The CI varied from 0.187 to 4.090 with an average of 0.30 in the PRM period and from 0.123 to 3.651 with an average of 0.52 in the PRM period (Table 8.5). According to the CI analysis results, 100% of the samples were within the safe ranges, i.e. no corrosion occurred. If CI >1, the water was an unsuitable character, and when CI <1, the water presented suitable characteristics regarding scale formation and the corrosive nature of water. Due to the very high concentration of HCO_3^- compared to Cl^- and SO_4^{2-} in the samples, it is expected that the CI value would be very low.

8.5 Summary

The study attempted to evaluate the suitability of calcite-type groundwater in the middle-west part of Bangladesh for industrial uses. Samples were collected from the shallow aquifer twice a year, viz. the pre-monsoon (PRM) and post-monsoon seasons. The analytical results showed that an almost neutral pH; higher values of TDS, total hardness, Ca^{2+}, Fe, Cu, and HCO_3^-; and lower concentrations of Cl^-, SO_4^{2-}, NO_3^-, and CO_3^{2-} are the main water chemistry in the study area. Cooling water comprises the majority of industrial water consumption. The study illustrated that the circulated and makeup cooling water was less corrosive but had a higher tendency to scale formation. In addition, approximately 100% of the samples are less suitable for the wood, food, paper, and sugar industries, and approximately 85% of the samples are not appropriate for the canned and polymer plastic industry. Various indexing models were calculated, and the results were included methodically for industrial water quality assessment. The results showed that water sources were found to be noncorrosive concerning the Larson–Skold index (LI) and the corrosivity index (CI). However, low and strong corrosion and aggressive nature were observed by the Ryznar stability index (RSI) and aggressive index (AI) of the water for both seasons. Furthermore, according to the Puckorius scaling index (PSI), all samples have no scale-forming nature, but 100% of the samples have high corrosivity. For the Langelier saturation index (LSI), 90% of the samples in the post-monsoon period have an encrusting trend and have a low tendency to scale formation. Generally, the LSI, RSI, and AI values depend on the actual pH of water samples. The actual pH value (pHA) of water samples in the POM ($\approx$ 8) is greater than in the PRM ($\approx$ 7) period, and for this reason, the mean values of the LI and AI obtained were higher and the RSI obtained were lower in the POM than in the PRM seasons. The high CSMR values recorded for the analysed groundwater samples (94.44%) suggest that the water has galvanic corrosion potential. In addition, the RI values showed that the samples were free from sodicity and that no seawater intrusion occurred in the study area. The results of the incorporation of multiple storabilities and corrosivity assessment indices showed that the groundwater samples have low corrosion and high scaling tendency, resulting in a high risk of corrosion and encrustation to domestic, industrial, municipal, and agricultural water supply and storage facilities. However, there is a higher risk of scale formation in boiling pools at high temperatures due to the higher concentrations of HCO_3^- and Ca^{2+} in the groundwater of the study area. Hence, water should be used in industrial plants after necessary treatment.

References and Further Study

1. Islam, M. S., & Mostafa, M. G. (2023). Suitability assessment of shallow groundwater for industrial uses in terms of storability and corrosivity. *Journal of Pollution and Effects on Community Health, 1*(1), 1–20. https://doi.org/10.58489/JPECH.007

2. Abbasnia, A., Alimohammadi, M., Mahvi, A. H., et al. (2018). Assessment of groundwater quality and evaluation of scaling and corrosiveness potential of drinking water samples in villages of Chabahar city, Sistan and Baluchistan province in Iran. *Data in Brief, 16*, 182–192. https://doi.org/10.1016/j.dib.2017.11.003
3. Aghazadeh, N., Chitsazan, M., & Golestan, Y. (2016). Hydrochemistry and quality assessment of groundwater in the Ardabil area, Iran. *Applied Water Science, 7*, 3599–3616. https://doi.org/10.1007/s13201-016-0498-9
4. Asghari, F. B., Jaafari, J., Yousefi, M., Mohammadi, A. K., & Dehghanzadeh, R. (2018). Evaluation of water corrosion, scaling extent and heterotrophic plate count bacteria in asbestos and polyethylene pipes in drinking water distribution system. *Human and Ecological Risk Assessment: An International Journal, 24*, 1138–1149.
5. Mirzabeygi, M., Naji, M., Yousefi, N., et al. (2016). Evaluation of corrosion and scaling tendency indices in water distribution system: A case study of Torbat Heydariye, Iran. *Desalination and Water Treatment, 57*, 18–26. https://doi.org/10.2166/wh.2014.157
6. Kumar, P. J. (2019). Assessment of corrosion and scaling potential of the groundwater in the Thanjavur district using hydrogeochemical analysis and spatial modelling techniques. *SN Applied Sciences, 1*(5), 395. https://doi.org/10.1007/s42452-019-0423-6
7. Omeka, M. E., Egbueri, J. C., & Unigwe, C. O. (2022). Investigating the hydrogeochemistry, corrosivity and scaling tendencies of groundwater in an agrarian area (Nigeria) using graphical, indexical and statistical modelling. *Arabian Journal of Geosciences, 15*, 1233–1256. https://doi.org/10.1007/s12517-022-10514-7
8. Hem, J. D. (1985). *Study and interpretation of the chemical characteristics of natural water* (3rd ed.). U.S. Geological Survey.
9. Magara, Y. (2014). Industrial water. In *E-book: Water quality standard* (Vol. 1, p. 23). EOLSS.
10. Zeid, S. A. M., Seleem, E. M., Salman, S. A., & Abdel-Hafiz, M. A. (2018). Water quality index of shallow groundwater and assessment for different usages in El-Obour city, Egypt. *Journal of Materials and Environmental Science, 9*(7), 1957–1968.
11. JRA-GL-02. (1994). *Japan Refrigeration and Air Conditioning Industry Association Standard.* http://www.mech.co.jp
12. ISO 16784-1. (2006). *Corrosion of metals and alloys – Corrosion and fouling in industrial cooling water systems – Part 1: Guidelines for conducting pilot-scale evaluation of corrosion and fouling control additives for open recirculating cooling water systems* (1st ed., pp. 1–14). ISO.
13. JIS K 0101. (2012). *Testing methods for industrial water* (98th ed., p. 488). Japanese Testing Association.
14. US-EPA. (1984). *Corrosion manual for internal corrosion of water distribution systems.* U.S. Environmental Protection Agency.
15. Egbueri, J. C. (2020). Groundwater quality assessment using pollution index of groundwater (PIG), ecological risk index (ERI) and hierarchical cluster analysis (HCA): A case study. *Groundwater for Sustainable Development, 10*, 100292. https://doi.org/10.1016/j.gsd.2019.100292

Chapter 9
Climate Change and Groundwater Management

Abbreviations

IPCC	Intergovernmental Panel on Climate Change
NEMAP	National Environmental Management Action Plan
NGO	Non-governmental Organization
NWMP	National Water Management Plan

9.1 Climate Change and Groundwater

As stated by the Intergovernmental Panel on Climate Change (IPCC), owing to natural internal courses or external forcing, or persistent human-caused changes in the composition of the air or land use, climate change is the change in the state of the climate that can be recognized by changes in its properties and that persists for an

M. S. Islam, *Hydrogeochemical Evaluation and Groundwater Quality*,
https://doi.org/10.1007/978-3-031-44304-6_9

extended time period, usually decades or longer. It arises when differences in the Earth's climate arrangement result in new weather patterns that keep on in place for an elongated period. Scientists have recognized many incidents of climate change throughout the geological history of the Earth; more recently since the industrial revolution, the climate has gradually been affected by human events driving global warming, and the terms are usually used interchangeably in that framework. It is well established that the global average temperature, as well as climate change, is growing up in an alarming position and cautions that the world has already warmed by 1 °C since the mid-nineteenth century and could reach 1.5 °C before the middle of this century at the present rate of warming due to the documented increase in human-caused greenhouse gas concentrations. Therefore, our planet's climate and weather designs are changing and will continue to shift and changes the typical hydrological balance and ecosystems leading to a decline in total water chemistry and quality. Understanding the probable effects of climate change and variability on groundwater is more complex and complicated than that on surface water. Except for extreme weather events, several anthropogenic factors, such as the population growth rate, urbanization, deforestation, crop diversity, and changes in land use patterns, also affect groundwater resources.

9.1.1 Impact of Climate Change on Groundwater Availability

The possible impact of climate change on the availability of groundwater is poorly understood. The recharge processes and water flow mostly depend on the aquifer and lithological conditions with some consequence of climatic factors. Many water supply programmes depend on groundwater, predominantly in rural locations, so developing a better understanding of climate–groundwater links is important. Climate change is possible to modify groundwater recharge patterns, as changes in rainfall and evaporation translate directly to shifts in soil moisture shortages and surface water runoff. Climatic and other issues, such as warming, sea level rise, changes in precipitation rate and distribution, rain–snow ratio, rate of evapotranspiration, overexploitation, and recharge–withdrawn ratio, are the main cause of global water stress. Precipitation change may directly impact recharge to groundwater systems, while changes in several other climatic variables may separately or jointly influence the various mechanisms of groundwater discharge: a 20% decrease in average yearly rainfall volumes could translate to an 80% deterioration in recharge, with no change in temperature, for areas that presently receive 500 mm rainfall per annum or less. Besides, Green et al. [1] and Taylor et al. [2] stated that under higher air CO_2 levels, terrestrial plants open their stomata less, and this response is predictable to reduce evapotranspiration and increase continental runoff.

Globally, in subhumid to arid areas, the total groundwater depletion increased from 126 (±32) km3/y to 283 (±40) km3/y from 1960 to 2000 [3]. Country estimates of the USA (32 ± 7 km3/y) and Saudi Arabia (10 ± 2 km3/y) are similar to estimates of 35 km3/y and 16 km3/y, respectively. Figure 9.1 displays the rate of (1960–2010)

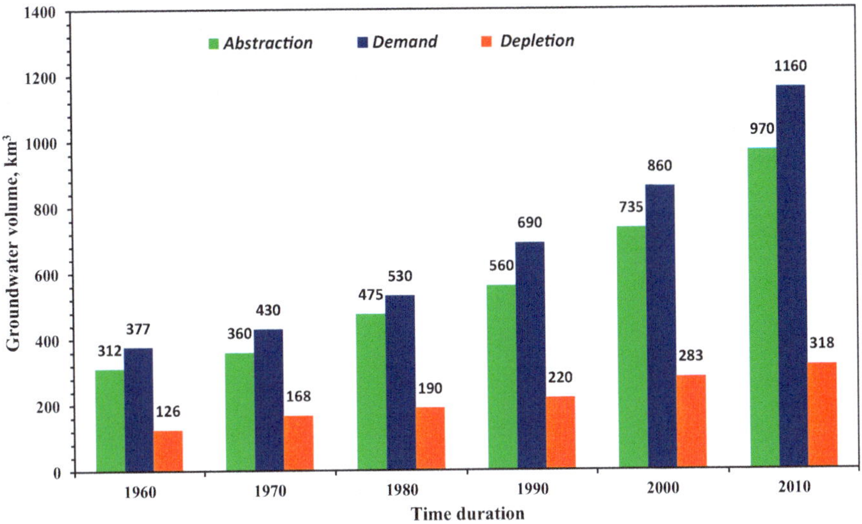

Fig. 9.1 Rate of (1960–2010) worldwide groundwater demand, abstraction, and water table depletion

worldwide groundwater demand, abstraction, and water table depletion from 1960 to 2010. On the other hand, Table 9.1 shows the groundwater stock, rate of abstraction, depletion, and water level decline of the most vulnerable countries and regions (Fig. 9.1 and Table 9.1).

9.1.2 *Impact of Climate Change on Groundwater Quality*

Global change may impact the suitability of groundwater in many complex ways through contaminant transport, but this is a minor cause compared to other causes, such as overwater mining, reduced recharge, contamination, and changes in the soil profile. Hot weather, precipitation variation, drought, and sea level rise are the major causes of climatic impacts on groundwater quality. Sea level rise, a notorious outcome of climate change, may lead to saltwater intrusion into coastal aquifers, affecting groundwater quality and making it unfit for drinking and irrigational purposes [7]. The countries in the Middle East and Central and South Asia, including India, China, Israel, the Nile Delta in Egypt, and Bangladesh, already suffered from this hazard, and the sea level change of 0.4–1.0 m by 2100 could induce saltwater intrusions to ingress 1–3 km inland [8]. Groundwater recharge throughout comparatively dry seasons may have higher concentrations of salt ions, TDS, and some metals, though recharge during comparatively wet seasons may have relatively lower concentrations of those components. Toxic heavy metals are a great threat to groundwater quality, but no study has been conducted on the long-term impact of

Table 9.1 Groundwater stock, rate of abstraction, depletion, and water level decline of most vulnerable countries and regions (comprises of Wada et al. [3], Döll et al. [4], Moseki [5], and Bierkens and Wada [6])

Country/ region	Total volume (Km3)	GW abstraction (Km3)				Total GW depletion (Km3)		GW level decline (depth, m)		
		1950	1980	2000	2010	1900–2000	1900–2008	1950	1970	2000
India and adjacent	–	15	95	205	240	937.5	1361	–	20	40
USA (total)	125,040	52	110	110	105	799.2	1003	20	40	68
North China Plain	–	10	45	98	102	130.3	170.3	–	12	33
Pakistan	5700	10	27	50	57	–	–	–	–	–
Iran	–	9	27	48	55	–	–	–	12	76
Mexico	–	10	17	22	25	–	–	–	70	130
Nubian region	542,180	–	–	7.5	12.5	79.5	98.4	–	–	–
S. Arabia	2185	5.0	8.0	17	23	358.6	467.7	20	–	40
Bangladesh	2570.5	–	–	10	35	–	23.5	–	10.5	41.5
Indonesia	–	–	–	–	15.0	–	–	10	14	28
Australia	20,000	–	–	–	5.8	25	31	–	–	–
North-west Sahara	60,000	–	–	2.5	3.8	52.7	70.3	–	–	–
Libya	35,000	–	–	3.2	4.0	–	–	18	20	105
S. Africa	750	–	1.6	2.0	2.7	–	–	–	–	–
Italy	–	–	20	25	27	–	–	5	30	45
Greece	10.3		0.4	2.6	3.7	–	–	–	–	–
France	2000	4.0	4.9	5.5	6.0	–	–	10	46	65
UK	–	–	–	2.2	2.4	–	–	–	–	–
World	60 × 106	310	475	737	970	3371	4534	–	–	–

global change on this issue. Some studies have been performed in Indian subcontinent countries on the variation in arsenic concentration with climatic and lithological conditions. As air temperature increases, the river-fed and precipitation-fed groundwater could warm up to 3.5 °C and 1 °C (CH2014 Impact) by the end of the twenty-first century, and it is expected that the maximum water quality parameter will be affected [9]. During drought periods, due to higher evapotranspiration and reduced nutrient uptake by plants under water stress, nitrates and DOC may accumulate in the topsoil and can be mobilized and transported into aquifers after abundant rainfall, resulting in a change in the chemical pulse in groundwater. In the extremely hot year 2003, there was a strong increase in nitrate concentrations (10–40 mg/L) in groundwater near the Swiss Thur River basin, and according to the results of the NAQUA National Groundwater Monitoring programme, the NO3- concentration in all over Swiss groundwater increased significantly from 2003 to 2006, and an increase in EC was observed during low groundwater levels in the summer of 2015. Besides, in the same aquifer, reductions in Cl- and SO42- concentrations were also observed, and an increase in Ca and CO32- also occurred over the last four decades due to higher temperatures and increased biological activity in the soil. On the other hand, Jeannin et al. [11] suggest that carbonate dissolution quickly increases in curst groundwater in Jura Mountain because of climate warming and the presence of high partial pressure of CO2 in soils at the top of the bedrock.

9.2 Challenges in Groundwater Management

Over the past three decades, the Bangladesh government has attempted to introduce strategies and plans to regulate and monitor groundwater resources, but sustainable and practical solutions have proven difficult. In Bangladesh, current water resources laws and regulations do not adequately cover areas such as the powers, rights, and obligations of private users and the government. However, in 1985, Bangladesh government broadcasted an ordinance, Ground Water Management Ordinance 1985, only for the management of farming groundwater resources. The ordinance permits were introduced to regulate the fitting of privately owned hand tube wells in rural areas, where the groundwater was declining at rapid rates and where groundwater quality was failing. Subsequent regulations such as Irrigation Act, 1876; Irrigation Water Ordinance, 1983; the National Environmental Policy, 1992; the Water Resources Planning Act, 1992; National Policy for Safe Water and Sanitation, 1998; National Water Policy, 1999; and National Policy of Arsenic Mitigation in Bangladesh, 2004, maintain on the need for the preservation of groundwater resources. Finally, the Bangladesh Water Act 2013, which aims to manage, integrate, develop, use, distribute, extract, conserve, and protect water resources, has been accepted by the Government of Bangladesh. According to this law, any individual or non-governmental organization (NGO) must obtain a permit or license to extract groundwater on a large scale out there domestic and agricultural uses. Many of these laws and policies require modern extensions.

A milestone in the planning and development of water authority policy was the 1999 National Water Policy (NWP) passage. This is done in discussion with NGOs, civil society, advocacy groups, social activists, and key donors. Significant water resource management and orderliness issues addressed by the NWP are water rights and provisions, river basin management, water supply and sanitation, public and private ventures, water and agriculture, stakeholder participation, water for the environment, regulative framework, etc. The main purpose of the NWP is to promote water resource monitoring and environmental protection and to provide guidance to all delegations working in the water sector on how to achieve their objectives. The need for joint use of surface water and groundwater was also emphasized in his NWP, and this law established overarching principles for the development and rational use of water capital under various restrictions. This law considers actions to restore, protect, and enhance the environment through the National Water

Management Plan (NWMP) and the National Environmental Management Action Plan (NEMAP).

Several laws, regulations, and laws are recognized in Bangladesh regarding sustainable management of groundwater and its utilization, protection, and management. Nevertheless, the introduction of licensing systems, law enforcement, and tradable property rights have all been largely unsuccessful and ineffective. This is probably due to the complexities in responsibility and implementation that are especially prevalent in rural areas [10]. Most legislation, including the Groundwater Management Ordinance (1985), addresses groundwater management at the rural level (subdistrict). But underground oversight beyond official requirements is neglected. An important lesson learned about what does not work in the evolution of groundwater laws comes from knowledge of regulations in adjacent countries such as China, Pakistan, and India, where licensing systems also have the necessary consequences. In addition, the sheer number of tube well operators, less spread to remote areas of the state, and lack of accountability and enforcement are major causes of failure of licensing strategies in these countries. In these countries, the population of the shallow pump and socioeconomic situation bears great resemblances with Bangladesh. Outdated groundwater regulation through the inventiveness of licensing systems and groundwater usage rights, direct and unnoticeable pricing, and offering water on a volumetric basis is thus not alike to attain in Bangladesh.

Bangladesh is presently facing a crisis of fresh, arsenic-free groundwater due to overexploitation and overpressure of shallow groundwater, as well as insufficient knowledge on water management and failure to influence groundwater-related behaviours. It is therefore important to raise awareness through education and training programmes for all stakeholders and work to improve programmes to produce alternative crops and create strong production markets for these alternative crops. For the improvement of groundwater management, it should be privalized on significant investments in modern agricultural technologies and water, e.g., charging water on a volumetric base or definite crop pricing, simplifying markets for non-price crops, fixing allocations for groundwater withdrawal for different users, upgrading alternating cropping patterns, and extra backing through a financial subsidy for cultivators, shifting to less water demanding crops.

9.3 Recommendations

Long-term pumping of groundwater for recharge and water quality monitoring is the main goal of sustainable development of groundwater resources. Maintaining a balance between water extraction and replenishment is critical to managing the human impact on the environment and water resources. The groundwater quality in relation to drinking water, irrigation water and industrial water. Through forecasting the potential for the next socioeconomic development and environmental impact assessment, the management of groundwater resources can achieve the following characteristics:

- Obsessive extraction of groundwater for domestic, irrigation, and industrial uses needs to be controlled.
- Should the chemical and bacteriological characteristics of countrywide groundwater be frequently monitored.
- Because of the excessive demand for water and to reduce dependence on confined groundwater resources, consumption of obtainable surface water and cohesive use should be stressed, according to NWP and other strategies of the government, which will reduce the seasonal variation rate of the water table and minimize stress on groundwater stock.
- Assessment of the most valuable exploitation of groundwater resources by developing priority for long-time use considering extensive droughts, shifting inhabitants, and agricultural growth to minimize the increasing pressure on groundwater stock in an area. Evaluate groundwater contamination and substitute measures of keeping the resource in the future and protect public health.
- Regional geologic modelling of groundwater resources must be developed for effective water management to plan agricultural and domestic water supplies and to forecast the groundwater circumstances in advance for dry seasons.
- Improved maintenance and operation of the shallow tube wells (STWs), crop diversification, buildup management efficiency, electrification of deep tube wells (DTWs), and installation of new DTWs in potential zones may increase crop production.

9.4 Summary

Groundwater resources reply more slowly to changes in climate, but the route and magnitude of lagged effects are uncertain. In part, this reflects the deficiency of data on groundwater; monitoring information on groundwater situations is patchy or non-existent in most developing countries, and relatively little is known about available groundwater in many regions. It was found that the absence of contemporary water treatment and supply services and suitability monitoring and the reluctance of water law execution are the key challenges to a safe water supply in the country. Besides, public consciousness and publicity programmes towards groundwater pollution are immediately needed to ensure safe water for all. Furthermore, advanced research and survey-based investigations are too improved to obtain better scenarios of safe water status and water-related problems in Bangladesh.

References and Further Study

1. Green, T. R., Bates, B. C., Charles, S. P., & Fleming, P. M. (2007). Physically based simulation of potential effects of carbon dioxide-altered climates on groundwater recharge. *Vadose Zone, 6*(3), 597–609.
2. Taylor, R. G., Koussis, A. D., & Tindimugaya, C. (2009). Groundwater and climate in Africa – A review. *Hydrological Sciences Journal, 54*(4), 655–664. https://doi.org/10.1623/hysj.54.4.655
3. Wada, Y., van Beek, L. P. H., van Kempen, C. M., et al. (2010). Global depletion of groundwater resources. *Geophysical Research Letters, 37*(20), 1–5. https://doi.org/10.1029/2010GL044571
4. Döll, P., & Fiedler, K. (2008). Global-scale modelling of groundwater recharge. *Hydrology and Earth System Science, 12*, 863–885. https://doi.org/10.5194/hess-12-863-2008
5. Moseki, M. C. (2017). Climate change impacts on groundwater: Literature review. *Environmental Risk Assessment and Remediation, 2*(1), 16–20.
6. Bierkens, M. F. B., & Wada, Y. (2019). Non-renewable groundwater use and groundwater depletion: A review. *Environmental Research Letters, 14*, 063002. https://doi.org/10.1088/1748-9326/ab1a5f
7. IPCC. (2007). Climate change 2007: Impacts, adaptation and vulnerability. In M. L. Parry, O. F. Canziani, J. P. Palutikof, P. J. van der Linden, & C. E. Hanson (Eds.), *Contribution of Working Group II to the* 4th *assessment report of the intergovernmental panel on climate change* (p. 1000). Cambridge University Press.
8. Yechieli, Y., Shalev, E., Wollman, S., Kiro, Y., & Kafri, U. (2010). Response of the Mediterranean and Dead Sea coastal aquifers to sea level variations. *Water Resources Research, 46*, W12550. https://doi.org/10.1029/2009WR008708
9. Benateau, S., Gaudard, A., Stamm, C., & Altermatt, F. (2019). *Climate change and freshwater ecosystems: Impacts on water quality and ecological status. Hydro-CH2018 project* (p. 110). Federal Office for the Environment (FOEN).
10. CSISA-MI. (2015). *Groundwater management in Bangladesh: An analysis of problems and opportunities*. Central System Initiative for South Asia – mechanization and Irrigation, USAID, Research Report No. 2. Dhaka. https://doi.org/10.13140/2.1.2381.9040
11. Jeannin, P.-Y., Hessenauer, M., Malard, A., & Chapuis, V. (2016). Impact of global change on karst groundwater mineralization in the Jura Mountains. *Science of the Total Environment, 541*, 1208–1221. https://doi.org/10.1016/j.scitotenv.2015.10.008

Appendices

Appendix I: Average Value of Geochemical Data of Groundwater for Pre-monsoon (PRM) Period in 2 Years

S. ID	pH	EC	TDS	Ca	Mg	Na	K	TH	Cl	SO_4	PO_4	NO_3	H_2CO_3	Depth
S1	6.63	1043.5	663.4	125.7	31.3	43.40	2.20	444.7	51.00	34.70	1.01	7.80	643.9	45
S2	6.86	752.5	421.5	106.7	54.5	22.25	1.87	487.2	28.50	17.08	0.95	1.95	447.7	125
S3	7.06	811.0	530.1	122.3	61.1	29.60	2.80	121.5	29.80	18.12	0.35	0.93	413.9	26
S4	6.72	1013.0	676.5	155.2	25.5	18.95	1.91	483.9	35.15	45.74	1.23	13.86	674.5	25
S5	6.75	785.5	482.3	104.4	34.8	26.20	2.32	411.3	22.45	12.99	2.71	1.42	410.4	102
S6	6.86	925.5	589.5	127.8	39.4	10.55	1.32	440.1	39.25	22.15	0.64	0.81	675.5	28
S7	6.88	770.0	487.5	111.5	52.6	13.35	0.81	494.4	29.30	19.33	0.42	2.13	508.4	92
S8	7.19	866.0	533.0	116.4	36.8	13.00	0.69	440.2	39.55	12.68	0.79	0.94	555.8	22
S9	6.93	717.5	446.9	101.7	30.7	11.30	0.50	373.5	38.60	27.71	1.24	1.82	345.4	79
S10	6.87	772.0	474.4	112.0	63.8	12.75	1.93	544.8	60.70	19.31	2.15	8.84	602.1	23
S11	7.12	928.0	596.6	125.3	40.3	10.90	2.60	474.3	37.80	31.81	1.17	5.33	708.3	22
S12	7.07	975.5	618.5	134.4	57.4	11.60	1.64	564.0	14.35	38.47	0.43	1.05	542.7	93
S13	6.85	367.5	220.3	57.5	13.4	12.10	1.54	195.3	15.80	14.25	1.84	1.54	316.0	30
S14	7.03	508.0	316.2	81.7	22.4	8.50	1.29	301.2	23.45	17.93	0.72	2.11	350.6	105
S15	7.09	571.0	366.1	88.1	17.5	7.25	0.82	289.7	26.25	9.75	0.55	0.81	510.9	31
S16	7.16	413.0	240.9	56.4	21.6	11.45	0.97	232.5	29.55	7.35	1.31	1.87	344.0	28
S17	7.16	562.0	360.0	88.4	13.9	18.45	1.20	276.1	30.90	8.94	1.70	2.12	308.3	70
S18	6.90	871.0	555.1	97.8	25.8	71.00	2.96	300.3	30.50	20.98	0.35	11.27	402.7	40
S19	6.81	719.0	455.0	88.6	26.1	19.35	1.84	332.7	29.40	18.20	0.62	5.42	403.1	25
S20	7.24	412.0	239.6	77.4	19.7	16.30	0.31	241.8	19.15	5.83	0.23	1.20	401.5	93
S21	6.93	542.0	321.2	123.4	23.9	7.55	0.62	464.5	22.55	6.92	0.62	1.73	473.2	40

(continued)

M. S. Islam, *Hydrogeochemical Evaluation and Groundwater Quality*,
https://doi.org/10.1007/978-3-031-44304-6

S. ID	pH	EC	TDS	Ca	Mg	Na	K	TH	Cl	SO_4	PO_4	NO_3	H_2CO_3	Depth
S22	7.01	641.0	410.6	93.1	28.6	8.45	0.52	353.7	21.20	12.06	0.95	1.94	387.4	29
S23	7.09	680.0	403.8	93.8	32.2	7.15	0.45	361.2	23.40	17.15	1.35	2.42	324.8	31
S24	7.11	675.5	412.3	109.6	22.1	5.85	0.47	364.5	25.30	11.59	0.92	11.05	362.3	28
S25	7.09	593.5	373.4	95.9	20.2	5.25	0.41	322.7	39.15	12.52	0.35	8.96	326.2	32
S26	6.97	522.5	321.0	77.1	24.5	7.70	0.44	291.4	36.70	9.11	0.51	2.41	312.8	28
S27	6.94	425.5	273.0	81.9	21.6	7.30	0.87	295.6	27.40	2.77	0.84	0.83	363.7	31
S28	6.82	495.0	320.3	70.6	17.9	6.10	1.03	284.4	25.40	5.50	0.46	0.85	352.2	102
S29	7.25	606.5	371.2	105.7	41.4	8.35	2.22	403.3	41.00	11.35	1.14	2.50	361.9	27
S30	7.13	591.5	357.2	114.8	44.3	12.95	1.62	415.4	28.55	19.52	0.74	1.17	373.5	29
S31	6.98	506.0	324.2	89.3	32.9	17.65	0.87	399.3	28.65	18.23	0.70	6.54	337.2	100
S32	7.28	653.5	409.1	75.3	34.9	12.65	1.28	381.2	29.40	9.13	0.32	10.82	388.7	34
S33	7.21	565.5	364.1	69.0	36.5	7.70	1.07	399.4	28.85	10.64	0.74	4.18	364.0	31
S34	6.96	628.5	307.8	81.4	34.5	8.15	0.97	300.6	48.20	5.23	1.06	1.96	341.0	103
S35	8.04	875.0	421.0	111.5	37.0	12.80	1.95	401.7	43.80	19.30	2.58	2.69	403.6	23
S36	7.07	658.0	456.3	78.8	33.4	9.30	0.90	333.2	22.80	22.69	0.73	7.84	332.2	43
S37	6.94	543.5	321.1	99.2	34.9	11.80	1.15	390.1	34.50	18.59	1.42	10.65	270.8	99
S38	7.30	517.0	356.9	66.9	41.9	7.50	0.83	333.0	31.30	11.83	1.31	3.91	360.7	28
S39	7.13	544.5	307.9	54.1	28.7	5.70	0.68	260.5	22.25	9.63	0.99	0.91	411.1	32
S40	6.74	752.5	430.0	73.8	29.0	12.60	1.09	305.7	35.45	19.61	1.65	7.90	372.7	27

Appendix II: Average Value of Geochemical Data of Groundwater for Monsoon (MON) Period in 2 Years

S. ID	pH	EC	TDS	Ca	Mg	Na	K	TH	Cl	SO_4	PO_4	NO_3	HCO_3	Depth
S1	7.06	1237.0	802.3	147.6	29.0	40.82	2.09	484.3	44.61	34.66	1.12	7.87	667.1	45
S2	7.21	927.0	554.3	121.1	52.5	21.15	1.68	520.8	25.76	16.70	0.83	1.95	474.1	125
S3	7.46	1034.0	675.0	132.5	50.7	27.81	2.77	157.1	26.44	17.71	0.23	0.72	433.9	26
S4	7.07	1193.0	815.9	181.8	23.5	17.49	1.76	524.7	31.81	43.95	1.20	13.33	690.8	25
S5	7.18	976.5	624.5	125.8	33.3	21.52	2.19	446.6	20.30	13.61	2.81	1.21	432.2	102
S6	7.41	1119.5	734.7	137.3	38.0	8.67	1.17	467.8	35.60	21.80	0.50	0.83	693.1	28
S7	7.29	966.5	633.2	120.0	46.8	10.80	0.67	521.0	26.52	18.50	0.30	2.63	527.8	92
S8	7.50	1045.5	668.6	129.3	35.3	11.10	0.43	472.6	37.10	12.22	0.71	0.99	581.1	22
S9	7.30	906.5	589.6	112.7	28.4	10.06	0.35	406.5	37.69	28.91	1.06	1.45	363.0	79
S10	7.23	963.0	616.3	137.9	65.8	11.39	1.72	586.2	57.33	18.79	2.04	8.78	622.4	23
S11	7.49	1209.5	744.5	139.3	38.2	9.31	2.62	516.4	35.60	30.55	1.01	4.24	722.7	22
S12	7.46	1172.0	758.7	145.9	55.4	10.35	1.53	594.4	11.04	38.36	0.36	0.92	565.4	93
S13	7.29	560.5	369.0	61.5	11.4	10.35	1.12	228.6	13.00	13.59	1.55	1.70	338.2	30
S14	7.47	706.0	461.7	93.9	21.4	7.06	1.10	331.3	21.01	16.10	0.64	2.18	371.7	105
S15	7.51	753.5	503.7	99.6	20.4	5.58	0.63	321.4	24.90	8.88	0.46	0.85	531.6	31
S16	7.51	600.5	382.5	63.5	19.8	10.59	0.76	266.4	27.33	6.48	1.05	1.90	366.8	28
S17	7.59	745.0	498.4	113.6	19.2	16.86	1.22	308.7	27.81	7.58	1.61	1.92	325.5	70
S18	7.30	1154.5	747.8	113.6	23.8	70.14	2.68	332.3	28.29	21.32	0.28	11.14	420.4	40
S19	7.24	909.5	591.4	101.5	24.7	17.57	1.62	364.0	26.03	17.59	0.54	5.50	422.1	25
S20	7.61	596.5	384.5	92.2	17.8	15.34	0.26	273.4	17.31	4.68	0.10	1.16	420.3	93
S21	7.27	725.0	461.2	139.0	22.0	6.01	0.42	488.4	20.22	6.08	0.53	1.73	489.6	40
S22	7.42	828.0	550.7	108.9	27.1	7.44	0.29	380.5	18.13	11.20	0.80	1.94	405.5	29
S23	7.44	862.5	542.0	104.0	30.5	5.67	0.23	397.0	20.51	16.90	1.31	2.55	345.7	31
S24	7.52	865.0	551.5	123.5	20.3	4.43	0.23	396.6	22.14	10.20	0.85	13.83	385.3	28
S25	7.53	787.5	511.3	110.4	18.2	5.79	0.33	354.7	37.24	11.81	0.21	9.87	346.2	32
S26	7.39	722.0	473.3	92.3	22.5	6.43	0.23	323.7	32.11	8.18	0.64	3.00	329.1	28
S27	7.36	635.0	421.1	97.3	19.8	6.57	0.72	328.0	23.51	2.08	0.61	0.74	381.4	31
S28	7.28	681.0	463.8	88.1	15.1	4.76	0.93	337.6	24.02	4.28	0.32	0.80	371.4	102
S29	7.61	802.5	511.5	119.8	39.7	4.68	2.05	435.5	38.41	11.30	1.04	1.86	381.7	27
S30	7.55	777.5	496.8	124.2	42.4	11.29	1.43	441.3	26.32	18.69	0.75	1.19	398.4	29
S31	7.37	685.0	462.4	104.4	30.2	16.57	0.74	435.2	27.24	17.36	0.52	6.50	359.5	100
S32	7.68	844.0	550.7	85.4	32.8	10.68	1.03	410.8	26.41	8.23	0.21	10.44	409.9	34
S33	7.62	856.5	587.1	85.1	34.0	4.70	0.83	431.4	26.42	10.80	0.69	4.19	384.6	31
S34	7.41	811.0	442.4	96.6	38.7	6.68	0.73	344.0	45.55	3.98	1.07	1.90	363.9	103
S35	8.41	1072.0	568.5	124.5	35.5	11.64	1.73	433.7	41.54	17.96	2.52	2.69	428.7	23
S36	7.53	842.0	598.1	93.7	37.0	7.71	0.75	364.4	20.37	21.73	0.71	7.99	346.0	43
S37	7.34	744.0	467.7	107.1	33.3	10.53	0.93	421.6	31.53	17.69	1.46	10.92	292.6	99
S38	7.72	702.5	492.8	80.6	35.0	5.40	0.60	367.4	28.64	11.52	1.21	4.90	374.1	28
S39	7.39	734.0	443.9	70.7	26.7	7.32	0.63	291.3	21.79	8.76	0.81	1.22	434.7	32
S40	7.15	945.0	576.2	89.7	27.1	10.67	0.98	338.1	33.06	18.45	1.51	7.82	396.8	27

Appendix III: Average Value of Geochemical Data of Groundwater for Post-monsoon (POM) Period in 2 Years

S. ID	pH	EC	TDS	Ca	Mg	Na	K	TH	Cl	SO_4	PO_4	NO_3	HCO_3	Depth
S1	7.44	1410.0	891.2	184.3	32.3	32.66	2.02	603.4	41.10	31.71	1.01	7.09	815.6	45
S2	7.61	1152.5	744.3	143.6	43.8	13.00	1.73	545.0	18.90	18.22	0.98	1.71	625.7	125
S3	8.14	1114.0	690.8	159.5	50.9	26.62	1.89	613.3	27.30	13.09	0.32	1.12	559.9	26
S4	7.62	1710.0	1108.1	142.6	24.0	12.36	0.91	453.3	39.30	41.71	1.31	18.34	546.0	25
S5	7.15	885.5	550.8	126.9	36.6	23.76	2.32	471.0	21.10	9.99	1.91	0.92	446.4	102
S6	7.76	995.5	624.9	144.6	39.0	7.19	0.41	528.0	36.00	21.13	0.65	0.82	594.0	28
S7	7.47	1066.5	683.7	131.1	52.6	13.76	0.89	543.3	27.50	18.09	0.42	2.45	457.6	92
S8	7.19	1061.5	594.2	137.3	32.1	11.46	0.55	465.3	29.80	10.12	0.80	1.88	670.8	22
S9	7.85	995.5	658.3	128.6	29.2	9.27	0.43	439.5	19.80	32.73	1.20	1.01	445.8	79
S10	7.57	975.5	625.0	133.2	33.5	9.74	1.20	470.2	23.70	20.87	2.14	4.22	516.2	23
S11	7.15	1231.0	783.7	164.0	39.4	11.38	2.11	570.3	35.00	30.38	1.17	3.13	610.1	22
S12	7.14	1182.5	757.2	151.2	46.9	10.93	1.58	571.2	18.00	41.10	0.40	0.92	594.8	93
S13	7.36	663.0	466.0	67.3	14.0	10.81	0.71	225.8	15.90	19.80	1.67	1.75	317.0	30
S14	8.04	811.0	511.7	92.4	20.9	6.66	0.23	318.5	23.60	18.76	0.61	2.15	338.3	105
S15	8.19	665.5	449.0	125.0	19.2	9.00	0.58	396.8	23.70	8.61	0.37	0.80	618.3	31
S16	8.21	915.5	567.8	78.5	22.4	8.90	0.31	286.6	19.80	9.32	1.17	1.05	249.1	28
S17	8.06	664.0	453.3	99.0	18.4	16.20	1.16	322.9	31.40	9.92	1.70	1.99	307.2	70
S18	7.39	1166.5	763.4	126.8	19.3	50.25	2.56	397.0	35.50	20.64	0.31	8.71	334.5	40
S19	7.71	815.0	559.8	99.9	18.9	18.59	1.66	325.6	27.00	17.36	0.63	5.51	513.5	25
S20	8.29	806.5	500.2	87.9	17.8	6.63	0.24	292.6	12.90	6.58	0.25	1.85	321.9	93
S21	7.81	838.5	534.2	114.4	21.9	5.31	0.53	372.7	24.40	5.90	0.65	1.75	518.5	40
S22	8.10	741.5	500.4	112.7	20.8	4.81	0.45	368.6	22.70	9.87	0.37	1.96	433.1	29
S23	8.19	775.5	501.8	83.2	28.5	6.10	0.47	321.8	28.80	19.21	1.39	2.53	374.4	31
S24	8.18	1073.0	685.8	129.5	20.1	6.02	0.22	406.5	15.80	10.51	1.11	10.95	412.4	28
S25	8.09	797.5	503.9	105.9	21.4	5.14	0.52	346.3	31.50	9.67	0.36	7.95	377.2	32
S26	8.91	827.0	540.3	97.3	21.0	3.91	0.39	329.8	39.50	9.91	0.81	2.01	282.9	28
S27	7.92	827.0	528.9	92.4	20.7	7.76	0.90	316.5	19.80	2.95	0.81	0.81	411.5	31
S28	7.67	1192.0	755.8	90.6	19.4	6.51	1.11	304.2	26.70	5.18	0.47	0.84	266.6	102
S29	8.12	812.0	520.5	115.9	31.5	7.10	2.03	418.6	39.90	10.67	0.86	1.97	461.5	27
S30	7.39	788.0	499.7	121.0	40.2	10.85	1.65	466.3	29.90	17.90	0.80	1.10	427.5	29
S31	7.94	891.0	546.8	78.9	33.2	8.14	0.95	333.0	19.80	17.14	0.60	4.50	437.5	100
S32	8.06	854.0	555.5	81.5	31.8	12.64	1.12	334.7	22.10	10.04	0.37	8.44	486.8	34
S33	8.11	767.0	489.6	99.1	35.4	6.66	0.82	390.8	26.90	10.14	0.71	4.17	462.9	31
S34	7.59	934.0	600.6	101.5	31.9	7.19	0.89	377.1	38.40	3.87	1.14	1.91	274.6	103
S35	8.21	1077.0	695.4	130.6	31.6	11.30	1.68	457.9	41.80	14.30	2.09	1.61	556.5	23
S36	8.12	965.0	622.7	100.1	30.0	7.56	0.62	374.9	19.70	10.09	0.76	2.03	451.6	43
S37	8.51	879.0	566.4	108.6	27.5	9.31	0.85	388.0	29.50	9.06	1.31	5.81	413.4	99
S38	7.92	1011.0	655.7	107.2	21.6	11.61	0.33	362.1	27.50	6.66	1.31	10.44	323.6	28
S39	7.39	978.0	666.4	88.7	32.3	8.47	0.55	351.6	21.80	11.46	0.88	7.47	339.1	32
S40	7.99	943.0	566.7	87.9	28.5	5.46	1.28	333.4	30.00	10.91	0.91	2.65	374.7	27

Appendix IV: CAI 1 and CAI 2 Values of Groundwater Samples for Three Sampling Seasons

	Pre-monsoon (PRM)		Monsoon (MON)		Post-monsoon (POM)	
S. ID	CAI1	CAI2	CAI1	CAI2	CAI1	CAI2
S1	1.436	0.126	1.255	0.107	1.157	0.082
S2	0.802	0.104	0.725	0.089	0.531	0.050
S3	0.838	0.117	0.743	0.099	0.768	0.081
S4	0.989	0.081	0.895	0.072	1.107	0.109
S5	0.631	0.090	0.570	0.077	0.593	0.079
S6	1.105	0.096	1.002	0.085	1.014	0.100
S7	0.825	0.094	0.747	0.082	0.774	0.098
S8	1.114	0.119	1.045	0.107	0.839	0.075
S9	1.087	0.173	1.061	0.161	0.557	0.070
S10	1.710	0.164	1.615	0.150	0.667	0.074
S11	1.064	0.086	1.002	0.080	0.985	0.092
S12	0.403	0.042	0.310	0.031	0.506	0.048
S13	0.444	0.081	0.365	0.063	0.447	0.079
S14	0.660	0.107	0.591	0.092	0.664	0.111
S15	0.739	0.086	0.701	0.079	0.667	0.065
S16	0.832	0.143	0.769	0.125	0.557	0.130
S17	0.870	0.165	0.783	0.142	0.884	0.168
S18	0.856	0.119	0.794	0.106	0.998	0.165
S19	0.827	0.117	0.732	0.099	0.760	0.086
S20	0.538	0.080	0.487	0.070	0.363	0.067
S21	0.635	0.080	0.569	0.070	0.687	0.079
S22	0.597	0.090	0.510	0.074	0.639	0.087
S23	0.659	0.115	0.577	0.095	0.811	0.123
S24	0.712	0.112	0.623	0.092	0.445	0.062
S25	1.103	0.192	1.049	0.173	0.887	0.136
S26	1.034	0.193	0.904	0.161	1.113	0.228
S27	0.771	0.128	0.662	0.105	0.557	0.082
S28	0.715	0.121	0.676	0.109	0.752	0.167
S29	1.155	0.186	1.082	0.166	1.124	0.144
S30	0.804	0.123	0.741	0.107	0.842	0.114
S31	0.806	0.134	0.767	0.121	0.557	0.073
S32	0.828	0.123	0.743	0.105	0.622	0.075
S33	0.812	0.130	0.744	0.113	0.757	0.096
S34	1.358	0.237	1.283	0.211	1.081	0.234
S35	1.233	0.175	1.170	0.157	1.177	0.125
S36	0.642	0.106	0.573	0.092	0.554	0.073
S37	0.971	0.194	0.888	0.166	0.831	0.118
S38	0.881	0.142	0.807	0.125	0.774	0.138
S39	0.626	0.090	0.613	0.084	0.614	0.104
S40	0.998	0.150	0.931	0.133	0.845	0.132

Appendix V: Trace Metal Mean Concentration of Groundwater for Pre-monsoon (PRM) in 2 Years

S. ID	B	Fe	Mn	Cr	Pb	Co	Ni	Cd	As	Cu	Zn
S1	1.105	11.74	4.65	0.083	0.081	0.061	0.390	0.051	0.016	2.13	5.95
S2	1.250	9.95	3.35	0.062	0.052	0.054	0.310	0.003	0.015	2.04	5.17
S3	1.000	8.32	3.02	0.053	0.072	0.057	0.310	0.028	0.000	2.57	4.95
S4	0.805	12.09	4.84	0.088	0.091	0.045	1.095	0.015	0.015	3.11	6.54
S5	0.750	11.66	4.66	0.048	0.071	0.055	0.990	0.013	0.011	2.12	5.53
S6	1.015	11.15	3.84	0.062	0.060	0.045	0.850	0.038	0.019	1.90	5.06
S7	0.650	8.05	2.78	0.041	0.013	0.059	0.060	0.032	0.010	1.77	5.16
S8	0.775	12.62	4.49	0.058	0.074	0.071	0.320	0.017	0.012	1.47	4.94
S9	1.125	10.79	3.42	0.034	0.003	0.061	0.755	0.000	0.011	1.65	4.56
S10	0.095	8.42	2.54	0.037	0.006	0.042	0.090	0.002	0.000	1.44	4.15
S11	0.035	9.65	3.53	0.034	0.023	0.042	0.235	0.002	0.013	1.97	5.01
S12	0.865	7.43	1.87	0.064	0.012	0.000	0.020	0.001	0.012	1.54	5.57
S13	0.710	6.31	1.54	0.013	0.021	0.043	0.010	0.004	0.000	0.98	1.76
S14	1.170	7.10	2.27	0.001	0.004	0.041	0.070	0.000	0.000	0.56	2.23
S15	0.080	5.32	1.41	0.000	0.000	0.041	0.080	0.000	0.000	0.68	2.92
S16	0.010	8.89	1.98	0.071	0.009	0.032	0.015	0.003	0.000	0.98	3.05
S17	0.000	3.67	0.79	0.048	0.009	0.000	0.000	0.000	0.000	0.33	1.65
S18	0.070	10.99	4.66	0.000	0.071	0.056	1.080	0.004	0.005	3.05	4.99
S19	0.650	4.32	1.64	0.043	0.013	0.000	0.080	0.002	0.008	1.76	3.06
S20	0.545	4.21	2.08	0.054	0.005	0.054	0.115	0.002	0.009	0.95	2.87
S21	1.020	7.57	2.81	0.047	0.009	0.066	0.760	0.006	0.010	0.67	2.88
S22	1.090	3.72	1.03	0.016	0.004	0.038	0.090	0.000	0.009	0.53	1.63
S23	0.050	2.15	0.75	0.035	0.000	0.000	0.040	0.000	0.000	0.05	2.91
S24	0.940	10.43	2.67	0.065	0.085	0.031	1.040	0.008	0.000	2.82	4.45
S25	0.675	3.34	1.25	0.053	0.008	0.011	0.090	0.000	0.000	0.07	3.21
S26	0.740	7.95	2.88	0.020	0.000	0.061	0.060	0.014	0.000	1.08	3.22
S27	0.930	2.40	1.16	0.040	0.082	0.015	0.790	0.014	0.007	1.99	3.05
S28	0.870	10.72	3.14	0.081	0.094	0.063	0.000	0.005	0.002	2.11	5.89
S29	0.890	7.65	2.64	0.030	0.059	0.062	0.870	0.000	0.018	1.54	2.77
S30	0.760	6.65	1.56	0.054	0.005	0.048	0.090	0.000	0.019	0.03	0.92
S31	0.490	6.13	1.63	0.064	0.021	0.031	0.050	0.000	0.010	1.57	1.23
S32	0.755	4.43	1.96	0.011	0.006	0.041	0.315	0.000	0.012	1.67	2.08
S33	0.070	1.14	0.66	0.000	0.008	0.000	0.090	0.000	0.000	0.87	0.79
S34	0.000	10.76	2.81	0.000	0.031	0.024	0.330	0.002	0.013	2.17	4.93
S35	0.590	11.84	4.33	0.041	0.061	0.088	0.380	0.001	0.013	1.98	5.55
S36	0.760	10.89	3.45	0.059	0.022	0.071	0.020	0.009	0.008	1.54	5.91
S37	1.115	7.52	2.57	0.073	0.006	0.042	0.080	0.003	0.012	2.08	6.63
S38	0.090	12.26	4.13	0.055	0.096	0.040	0.010	0.009	0.018	2.77	4.96
S39	0.120	10.38	4.65	0.055	0.076	0.072	0.125	0.000	0.002	3.11	5.59
S40	0.605	9.31	3.58	0.051	0.088	0.081	0.110	0.016	0.000	2.06	4.07

Appendix VI: Trace Metal Mean Concentration of Groundwater for Monsoon (MON) in 2 Years

S. No.	B	Fe	Mn	Cr	Pb	Co	Ni	Cd	As	Cu	Zn
S1	1.205	13.73	4.61	0.085	0.081	0.062	0.230	0.041	0.012	2.81	6.21
S2	1.235	10.53	3.40	0.061	0.063	0.056	0.345	0.006	0.016	1.92	5.44
S3	0.950	9.40	3.63	0.048	0.074	0.057	0.360	0.025	0.011	3.13	4.83
S4	1.015	12.23	5.15	0.085	0.076	0.054	1.185	0.014	0.018	3.18	6.46
S5	0.755	12.43	4.92	0.058	0.071	0.055	1.060	0.016	0.012	2.38	5.37
S6	0.995	10.89	3.94	0.074	0.058	0.048	0.805	0.040	0.019	1.89	5.24
S7	0.650	8.20	2.90	0.051	0.015	0.060	0.055	0.042	0.011	1.88	5.31
S8	0.780	13.63	5.66	0.058	0.071	0.084	0.360	0.016	0.013	2.25	4.98
S9	1.225	11.75	4.06	0.040	0.009	0.063	0.790	0.000	0.013	2.28	4.81
S10	0.090	8.44	2.92	0.039	0.008	0.058	0.090	0.002	0.000	1.68	4.25
S11	0.040	8.66	3.52	0.048	0.042	0.044	0.350	0.003	0.014	1.93	5.05
S12	0.800	7.92	0.98	0.082	0.021	0.000	0.050	0.001	0.013	1.65	5.51
S13	0.740	7.32	1.55	0.033	0.022	0.041	0.030	0.002	0.000	0.98	1.98
S14	1.165	7.13	2.12	0.031	0.005	0.039	0.060	0.000	0.000	0.58	2.28
S15	0.185	5.32	1.48	0.000	0.000	0.042	0.085	0.000	0.002	1.48	2.97
S16	0.005	8.85	2.28	0.078	0.021	0.035	0.020	0.004	0.000	1.28	3.17
S17	0.000	3.67	0.70	0.046	0.009	0.000	0.005	0.000	0.000	0.52	1.63
S18	0.090	10.95	4.60	0.000	0.072	0.055	1.140	0.005	0.007	2.99	4.88
S19	0.650	5.32	0.86	0.044	0.023	0.000	0.155	0.004	0.008	1.87	3.54
S20	0.845	5.21	2.54	0.053	0.015	0.065	0.165	0.002	0.009	0.90	2.88
S21	0.990	7.43	2.85	0.057	0.009	0.067	0.775	0.006	0.017	0.62	2.91
S22	1.160	4.82	1.35	0.000	0.008	0.035	0.090	0.000	0.009	0.59	1.69
S23	0.050	1.87	1.08	0.045	0.000	0.000	0.050	0.000	0.000	0.24	2.92
S24	0.970	11.42	2.67	0.065	0.085	0.031	1.060	0.008	0.000	2.80	4.85
S25	0.650	4.33	1.23	0.048	0.008	0.031	0.090	0.000	0.000	0.17	3.33
S26	0.860	7.95	2.88	0.021	0.000	0.061	0.060	0.012	0.000	1.03	3.71
S27	0.940	2.93	2.06	0.042	0.071	0.000	0.795	0.015	0.007	1.89	3.51
S28	0.865	10.78	3.15	0.088	0.092	0.063	0.000	0.006	0.003	2.17	5.69
S29	0.900	8.67	2.72	0.048	0.068	0.061	0.860	0.000	0.018	1.29	2.92
S30	0.750	6.62	1.85	0.054	0.015	0.046	0.170	0.000	0.019	0.14	0.88
S31	0.490	6.46	1.55	0.064	0.027	0.043	0.040	0.000	0.012	1.52	1.64
S32	0.760	4.30	1.97	0.044	0.009	0.045	0.345	0.000	0.013	1.61	2.24
S33	0.175	2.00	0.46	0.000	0.008	0.000	0.090	0.000	0.000	0.89	0.79
S34	0.000	11.74	2.98	0.000	0.038	0.023	0.370	0.002	0.014	2.51	4.91
S35	0.390	10.81	4.82	0.041	0.061	0.090	0.375	0.002	0.016	2.05	5.71
S36	0.760	12.74	4.43	0.059	0.023	0.073	0.060	0.009	0.008	1.53	5.82
S37	1.215	8.58	2.33	0.062	0.016	0.041	0.075	0.002	0.012	2.34	6.72
S38	0.190	13.02	4.08	0.059	0.086	0.032	0.035	0.008	0.018	2.87	5.09
S39	0.315	11.73	4.92	0.065	0.087	0.080	0.185	0.000	0.003	3.32	5.52
S40	0.620	11.08	3.71	0.061	0.077	0.097	0.155	0.015	0.000	2.14	4.28

Appendix VII: Trace Metal Mean Concentration of Groundwater for Post-monsoon (POM) in 2 Years

S. No.	B	Fe	Mn	Cr	Pb	Co	Ni	Cd	As	Cu	Zn
S1	1.295	14.44	4.62	0.088	0.075	0.070	0.215	0.042	0.010	2.86	5.85
S2	1.415	11.20	3.67	0.061	0.062	0.052	0.370	0.006	0.018	2.14	5.56
S3	0.955	10.63	2.85	0.053	0.088	0.058	0.395	0.024	0.012	3.20	4.96
S4	0.925	12.72	4.97	0.093	0.082	0.065	1.045	0.016	0.016	3.26	6.42
S5	0.800	11.42	4.56	0.072	0.071	0.058	1.135	0.019	0.013	2.37	5.83
S6	1.105	11.85	4.84	0.077	0.053	0.055	0.845	0.045	0.020	1.82	5.24
S7	0.645	8.80	3.64	0.052	0.024	0.052	0.060	0.042	0.011	1.92	5.33
S8	0.760	13.27	5.61	0.046	0.072	0.081	0.320	0.017	0.012	2.45	5.19
S9	1.310	11.84	4.05	0.058	0.008	0.066	0.875	0.000	0.012	2.26	4.84
S10	0.115	8.42	2.00	0.045	0.008	0.058	0.090	0.002	0.000	1.77	4.21
S11	0.070	9.00	3.45	0.047	0.053	0.054	0.360	0.002	0.014	1.90	4.87
S12	0.820	8.26	1.20	0.095	0.025	0.000	0.055	0.002	0.014	1.60	5.54
S13	1.010	8.58	1.87	0.042	0.028	0.044	0.040	0.002	0.000	0.99	2.19
S14	1.175	7.77	2.76	0.033	0.003	0.040	0.060	0.000	0.000	0.52	2.27
S15	0.220	4.15	1.48	0.015	0.042	0.048	0.090	0.000	0.002	1.63	2.94
S16	0.000	8.63	2.74	0.078	0.022	0.036	0.025	0.004	0.000	1.42	3.22
S17	0.000	4.14	0.74	0.053	0.014	0.000	0.000	0.000	0.000	0.58	1.66
S18	0.050	11.74	4.65	0.000	0.078	0.057	1.190	0.005	0.006	2.99	4.87
S19	0.605	5.65	0.89	0.051	0.020	0.000	0.180	0.004	0.007	1.87	4.07
S20	0.870	5.09	2.77	0.054	0.011	0.076	0.170	0.002	0.008	0.90	2.88
S21	0.980	7.87	2.82	0.056	0.013	0.061	0.760	0.005	0.016	0.62	2.83
S22	1.450	4.82	1.31	0.000	0.015	0.037	0.085	0.000	0.008	0.59	1.69
S23	0.160	2.10	1.19	0.046	0.000	0.033	0.050	0.000	0.000	0.23	2.91
S24	0.950	10.80	2.83	0.068	0.086	0.039	1.060	0.008	0.000	2.80	4.85
S25	0.660	4.79	1.29	0.065	0.008	0.033	0.090	0.000	0.000	0.17	3.37
S26	0.860	7.92	2.67	0.025	0.000	0.066	0.060	0.009	0.000	1.01	3.70
S27	0.965	2.16	2.34	0.046	0.075	0.000	0.765	0.018	0.006	1.78	3.53
S28	0.870	11.76	3.56	0.092	0.099	0.064	0.000	0.007	0.003	2.17	5.68
S29	0.875	9.63	2.78	0.051	0.056	0.062	0.765	0.000	0.013	1.33	2.92
S30	0.770	8.02	1.81	0.058	0.018	0.053	0.270	0.000	0.018	0.09	0.94
S31	0.455	6.14	1.76	0.068	0.043	0.046	0.010	0.000	0.022	1.54	1.55
S32	0.730	4.86	1.91	0.047	0.014	0.041	0.365	0.000	0.012	1.60	2.08
S33	0.200	1.99	0.54	0.013	0.008	0.000	0.090	0.000	0.000	0.85	0.67
S34	0.000	11.76	2.89	0.000	0.045	0.015	0.370	0.005	0.016	2.51	4.92
S35	0.395	11.32	4.79	0.045	0.066	0.083	0.370	0.002	0.016	1.98	5.78
S36	0.785	12.17	4.65	0.064	0.023	0.078	0.070	0.008	0.007	1.61	5.85
S37	1.260	9.31	2.32	0.062	0.016	0.045	0.080	0.002	0.019	2.38	6.77
S38	0.235	13.74	4.23	0.061	0.086	0.037	0.040	0.008	0.017	2.87	5.57
S39	0.305	11.06	4.98	0.064	0.087	0.085	0.220	0.000	0.005	3.37	5.59
S40	0.680	11.64	3.85	0.062	0.077	0.093	0.155	0.002	0.010	2.18	4.36

Appendix VIII: Mean Fe and Mn Concentrations (mg/L) with Some Selected Water Parameters (Pre-monsoon) of 2 Years

S. ID	pH	EC	TH	Fe	Mn	SO_4	NO_3	H_2CO_3	DO	DOC	log pCO_2	Turbidity
S1	6.63	1043.5	444.7	11.74	4.65	34.70	7.80	643.9	0.96	9.6	−0.877	9.43
S2	6.86	752.5	487.2	9.95	3.35	17.08	1.95	447.7	1.88	4.5	−1.193	4.98
S3	7.06	811.0	121.5	8.32	3.02	18.12	0.93	413.9	1.90	3.6	−1.428	5.87
S4	6.72	1013.0	483.9	12.09	4.84	45.74	13.86	674.5	0.84	10.3	−0.920	19.21
S5	6.75	785.5	411.3	11.66	4.66	12.99	1.42	410.4	1.10	6.6	−1.195	11.7
S6	6.86	925.5	440.1	11.15	3.84	22.15	0.81	675.5	1.2	6.7	−1.195	6.78
S7	6.88	770.0	494.4	8.05	2.78	19.33	2.13	508.4	2.02	5.5	−1.192	5.07
S8	7.19	866.0	440.2	12.62	4.49	12.68	0.94	555.8	0.74	11.7	−1.378	16.4
S9	6.93	717.5	373.5	10.79	3.42	27.71	1.82	345.4	1.22	5.5	−1.451	11.0
S10	6.87	772.0	544.8	8.42	2.54	19.31	8.84	602.1	2.15	4.1	−1.128	5.03
S11	7.12	928.0	474.3	9.65	3.53	31.81	5.33	708.3	2.05	4.8	−1.289	6.65
S12	7.07	975.5	564.0	7.43	1.87	38.47	1.05	542.7	2.25	4.3	−1.321	4.09
S13	6.85	367.5	195.3	6.31	1.54	14.25	1.54	316.0	2.38	4.9	−1.387	2.54
S14	7.03	508.0	301.2	7.10	2.27	17.93	2.11	350.6	2.43	4.6	−1.552	3.16
S15	7.09	571.0	289.7	5.32	1.41	9.75	0.81	510.9	2.92	2.9	−1.419	2.19
S16	7.16	413.0	232.5	8.89	1.98	7.35	1.87	344.0	1.87	5.5	−1.609	6.21
S17	7.16	562.0	276.1	3.67	0.79	8.94	2.12	308.3	3.32	1.98	−1.712	2.00
S18	6.90	871.0	300.3	10.99	4.66	20.98	11.27	402.7	0.88	6.3	−1.324	5.76
S19	6.81	719.0	332.7	4.32	1.64	18.20	5.42	403.1	3.32	1.1	−1.243	3.09
S20	7.24	412.0	241.8	4.21	2.08	5.83	1.20	401.5	3.17	1.2	−1.625	3.12
S21	6.93	542.0	464.5	7.57	2.81	6.92	1.73	473.2	2.10	3.7	−1.276	3.05
S22	7.01	641.0	353.7	3.72	1.03	12.06	1.94	387.4	3.32	1.1	−1.451	2.56
S23	7.09	680.0	361.2	2.15	0.75	17.15	2.42	324.8	3.56	1.0	−1.536	1.93
S24	7.11	675.5	364.5	10.43	2.67	11.59	11.05	362.3	1.25	6.4	−1.590	7.75
S25	7.09	593.5	322.7	3.34	1.25	12.52	8.96	326.2	3.45	1.1	−1.614	2.99
S26	6.97	522.5	291.4	7.95	2.88	9.11	2.41	312.8	1.88	3.8	−1.518	4.21
S27	6.94	425.5	295.6	2.40	1.16	2.77	0.83	363.7	4.12	1.0	−1.399	2.00
S28	6.82	495.0	284.4	10.72	3.14	5.50	0.85	352.2	1.24	4.9	−1.362	8.65
S29	7.25	606.5	403.3	7.65	2.64	11.35	2.50	361.9	1.80	3.5	−1.678	5.85
S30	7.13	591.5	415.4	6.65	1.56	19.52	1.17	373.5	3.12	2.4	−1.550	4.32
S31	6.98	506.0	399.3	6.13	1.63	18.23	6.54	337.2	3.00	2.0	−1.457	3.89
S32	7.28	653.5	381.2	4.43	1.96	9.13	10.82	388.7	3.95	1.7	−1.748	2.12
S33	7.21	565.5	399.4	1.14	0.66	10.64	4.18	364.0	4.35	0.5	−1.678	1.99
S34	6.96	628.5	300.6	10.76	2.81	5.23	1.96	341.0	1.08	6.8	−1.492	7.32
S35	8.04	875.0	401.7	11.84	4.33	19.30	2.69	403.6	0.97	5.6	−2.428	10.43
S36	7.07	658.0	333.2	10.89	3.45	22.69	7.84	332.2	1.45	6.6	−1.612	9.76
S37	6.94	543.5	390.1	7.52	2.57	18.59	10.65	270.8	1.95	5.0	−1.483	4.09
S38	7.30	517.0	333.0	12.26	4.13	11.83	3.91	360.7	0.68	8.7	−1.785	19.56
S39	7.13	544.5	260.5	10.38	4.65	9.63	0.91	411.1	1.12	6.0	−1.422	7.65
S40	6.74	752.5	305.7	9.31	3.58	19.61	7.90	372.7	1.25	4.9	−1.213	6.09

Appendix IX: Mean Fe and Mn Concentrations (mg/L) with Some Selected Water Parameters (Post-monsoon) of 2 Years

S. ID	pH	EC	TH	Fe	Mn	SO_4	NO_3	HCO_3	DO	DOC	Log_{10} pCO_2	Turbidity
S1	7.44	1410.0	603.4	14.44	4.62	31.71	7.09	815.6	0.98	9.5	−1.574	10.76
S2	7.61	1152.5	545.0	11.20	3.67	18.22	1.71	625.7	1.90	8.4	−1.851	6.70
S3	8.14	1114.0	613.3	10.63	2.85	13.09	1.12	559.9	2.01	5.8	−2.394	7.08
S4	7.62	1710.0	453.3	12.72	4.97	41.71	18.34	546.0	0.81	13.7	−1.911	26.10
S5	7.15	885.5	471.0	11.42	4.56	9.99	0.92	446.4	0.98	11.3	−1.556	17.67
S6	7.76	995.5	528.0	11.85	4.84	21.13	0.82	594.0	0.97	10.9	−2.054	11.87
S7	7.47	1066.5	543.3	8.80	3.64	18.09	2.45	457.6	1.99	5.5	−1.837	8.90
S8	7.19	1061.5	465.3	13.27	5.61	10.12	1.88	670.8	1.37	8.9	−1.401	11.76
S9	7.85	995.5	439.5	11.84	4.05	32.73	1.01	445.8	2.00	7.5	−2.240	9.08
S10	7.57	975.5	470.2	8.42	2.00	20.87	4.22	516.2	0.99	8.4	−1.894	7.21
S11	7.15	1231.0	570.3	9.00	3.45	30.38	3.13	610.1	1.09	8.8	−1.251	8.07
S12	7.14	1182.5	571.2	8.26	1.20	41.10	0.92	594.8	2.38	5.3	−1.383	6.32
S13	7.36	663.0	225.8	8.58	1.87	19.80	1.75	317.0	2.48	4.4	−1.886	5.45
S14	8.04	811.0	318.5	7.77	2.76	18.76	2.15	338.3	2.27	4.3	−2.547	4.06
S15	8.19	665.5	396.8	4.15	1.48	8.61	0.80	618.3	2.78	4.2	−2.435	2.89
S16	8.21	915.5	286.6	8.63	2.74	9.32	1.05	249.1	1.79	6.5	−2.851	5.43
S17	8.06	664.0	322.9	4.14	0.74	9.92	1.99	307.2	3.15	3.1	−2.616	3.54
S18	7.39	1166.5	397.0	11.74	4.65	20.64	8.71	334.5	1.21	8.9	−1.897	9.36
S19	7.71	815.0	325.6	5.65	0.89	17.36	5.51	513.5	3.17	1.9	−2.038	5.65
S20	8.29	806.5	292.6	5.09	2.77	6.58	1.85	321.9	3.01	1.8	−2.820	3.54
S21	7.81	838.5	372.7	7.87	2.82	5.90	1.75	518.5	2.72	4.9	−2.132	6.88
S22	8.10	741.5	368.6	4.82	1.31	9.87	1.96	433.1	3.33	0.9	−2.498	2.98
S23	8.19	775.5	321.8	2.10	1.19	19.21	2.53	374.4	3.23	1.2	−2.653	2.78
S24	8.18	1073.0	406.5	10.80	2.83	10.51	10.95	412.4	2.09	5.6	−2.604	6.76
S25	8.09	797.5	346.3	4.79	1.29	9.67	7.95	377.2	3.97	1.8	−2.552	3.32
S26	8.91	827.0	329.8	7.92	2.67	9.91	2.01	282.9	1.97	5.1	−3.490	4.78
S27	7.92	827.0	316.5	2.16	2.34	2.95	0.81	411.5	3.62	1.7	−2.322	2.99
S28	7.67	1192.0	304.2	11.76	3.56	5.18	0.84	266.6	3.18	3.8	−2.281	5.43
S29	8.12	812.0	418.6	9.63	2.78	10.67	1.97	461.5	2.16	3.1	−2.472	5.56
S30	7.39	788.0	466.3	8.02	1.81	17.90	1.10	427.5	2.22	2.9	−1.796	4.00
S31	7.94	891.0	333.0	6.14	1.76	17.14	4.50	437.5	2.22	3.0	−2.294	3.67
S32	8.06	854.0	334.7	4.86	1.91	10.04	8.44	486.8	3.10	1.6	−2.348	4.54
S33	8.11	767.0	390.8	1.99	0.54	10.14	4.17	462.9	4.02	0.6	−2.472	2.29
S34	7.59	934.0	377.1	11.76	2.89	3.87	1.91	274.6	0.97	6.3	−2.188	8.88
S35	8.21	1077.0	457.9	11.32	4.79	14.30	1.61	556.5	0.95	8.6	−2.501	16.43
S36	8.12	965.0	374.9	12.17	4.65	10.09	2.03	451.6	0.99	7.8	−2.482	11.56
S37	8.51	879.0	388.0	9.31	2.32	9.06	5.81	413.4	1.98	4.9	−2.935	7.43
S38	7.92	1011.0	362.1	13.74	4.23	6.66	10.44	323.6	0.84	11.7	−2.442	20.22
S39	7.39	978.0	351.6	11.06	4.98	11.46	7.47	339.1	0.95	5.7	−1.897	11.6
S40	7.99	943.0	333.4	11.64	3.85	10.91	2.65	374.7	0.95	8.1	−2.456	10.6

Appendix X: Statistical Summary of WQI Value of Collected Water Samples for Drinking Purposes

S. ID	CWQI (Drinking)				WWQI (Drinking)			
	PRM	MON	POM	Average	PRM	MON	POM	Average
S1	46.34	43.32	42.56	44.07	322.71	392.20	407.61	374.17
S2	41.09	39.91	39.01	40.00	342.16	422.85	456.88	407.30
S3	44.01	41.06	40.00	41.69	298.34	362.86	372.06	344.42
S4	45.55	43.35	41.88	43.59	311.60	348.42	391.04	350.35
S5	55.65	51.77	50.01	52.48	302.26	392.07	427.95	374.09
S6	44.12	41.88	40.22	42.07	433.51	459.05	478.42	456.99
S7	45.55	42.41	41.08	43.01	377.73	448.88	470.10	432.24
S8	53.73	51.10	50.11	51.65	276.64	375.14	405.92	352.57
S9	45.23	44.03	43.00	44.09	266.62	298.62	371.51	312.25
S10	59.32	55.36	55.05	56.58	304.65	365.91	400.62	357.06
S11	43.01	43.57	41.66	42.75	276.67	329.62	346.63	317.64
S12	40.21	38.35	37.09	38.55	423.10	439.53	479.02	447.22
S13	68.39	66.54	62.55	65.83	176.59	160.18	166.85	167.87
S14	67.23	65.15	63.32	65.23	123.66	166.00	193.74	161.13
S15	71.01	68.88	66.56	68.82	94.84	128.26	141.80	121.63
S16	60.88	59.70	58.07	59.55	233.52	263.85	298.92	265.43
S17	68.06	65.31	64.00	65.79	89.14	93.34	95.51	92.66
S18	43.29	43.63	42.69	43.20	377.65	383.06	379.91	380.21
S19	43.55	42.15	40.78	42.16	411.69	415.07	407.09	411.28
S20	67.42	64.04	63.00	64.82	177.21	255.64	267.85	233.57
S21	65.99	63.85	61.08	63.64	99.13	132.51	142.83	124.82
S22	71.06	68.40	67.00	68.82	88.94	95.01	98.65	94.20
S23	69.44	67.09	68.12	68.22	88.80	113.63	129.06	110.50
S24	45.03	43.01	42.08	43.37	302.44	349.76	376.80	343.00
S25	67.55	65.84	63.12	65.50	123.69	202.48	233.74	186.64
S26	71.76	69.15	70.11	70.34	92.63	111.94	123.13	109.23
S27	73.09	70.57	70.32	71.33	94.20	139.03	159.82	131.02
S28	45.01	42.35	41.55	42.97	354.71	370.18	372.05	365.65
S29	41.88	39.91	38.06	39.95	401.62	489.06	497.88	462.85
S30	42.71	40.42	39.12	40.75	412.08	409.49	397.00	406.19
S31	67.98	66.10	66.00	66.69	149.73	166.87	182.72	166.44
S32	71.03	69.39	67.43	69.28	95.64	159.03	179.13	144.60
S33	67.90	67.85	66.00	67.25	111.54	144.79	156.54	137.62
S34	66.74	64.06	64.02	64.94	188.66	222.25	232.87	214.59
S35	47.32	45.51	44.23	45.69	408.70	509.38	516.56	478.21
S36	66.88	65.41	63.55	65.28	319.72	330.33	343.76	331.27
S37	67.09	66.66	65.78	66.51	175.68	200.61	208.19	194.83
S38	51.55	49.94	49.00	50.16	400.51	434.76	456.05	430.44
S39	45.01	41.53	40.09	42.21	488.62	531.17	543.3`	461.05
S40	51.05	49.19	48.11	49.45	311.62	194.54	422.04	309.40

Appendix XI: Irrigation Water Quality Parameter Values for Groundwater Samples in Pre-monsoon (PRM) Season

S. ID	EC	TDS	TH	Na%	SAR	SSP	RSBC	PI	MAR	KR	Mg: Ca	Na: Ca	Hazard class
S1	1043.5	663.4	444.7	17.50	0.316	17.93	4.27	57.75	29.33	0.212	0.249	0.345	C3S1
S2	752.5	421.5	487.2	8.92	0.154	9.32	2.00	37.22	45.98	0.098	0.511	0.209	C3S1
S3	811.0	530.1	121.5	10.30	0.192	10.81	0.67	34.73	45.43	0.115	0.500	0.242	C3S1
S4	1013.0	676.5	483.9	7.69	0.131	8.11	3.30	41.97	21.50	0.083	0.164	0.122	C3S1
S5	785.5	482.3	411.3	12.30	0.200	12.86	1.51	45.97	35.71	0.140	0.333	0.251	C3S1
S6	925.5	589.5	440.1	4.53	0.074	4.85	4.68	39.14	33.94	0.047	0.308	0.083	C3S1
S7	770.0	487.5	494.4	5.51	0.092	5.69	2.76	34.82	44.02	0.058	0.472	0.120	C3S1
S8	866.0	533.0	440.2	5.98	0.095	6.16	3.29	40.33	34.51	0.064	0.316	0.112	C3S1
S9	717.5	446.9	373.5	6.04	0.089	6.19	0.58	37.56	33.47	0.064	0.302	0.111	C2S1
S10	772.0	474.4	544.8	4.83	0.084	5.24	4.27	33.86	48.70	0.051	0.570	0.114	C2S1
S11	928.0	596.6	474.3	4.69	0.076	5.32	5.35	40.33	34.90	0.049	0.322	0.087	C3S1
S12	975.5	618.5	564.0	4.20	0.074	4.53	2.18	30.31	41.58	0.044	0.427	0.086	C3S1
S13	367.5	220.3	195.3	11.64	0.132	12.41	2.31	70.20	27.97	0.132	0.233	0.210	C2S1
S14	508.0	316.2	301.2	5.85	0.076	6.34	1.66	46.49	31.36	0.062	0.274	0.104	C2S1
S15	571.0	366.1	289.7	5.10	0.065	5.42	3.97	54.73	24.87	0.054	0.199	0.082	C2S1
S16	413.0	240.9	232.5	9.73	0.116	10.16	2.82	62.18	38.96	0.108	0.383	0.203	C2S1
S17	562.0	360.0	276.1	12.57	0.170	12.99	0.63	54.68	20.76	0.144	0.157	0.209	C2S1
S18	871.0	555.1	300.3	30.48	0.582	31.00	1.71	80.35	30.54	0.438	0.264	0.726	C3S1
S19	719.0	455.0	332.7	11.30	0.164	11.86	2.18	51.66	32.93	0.127	0.295	0.218	C2S1
S20	412.0	239.6	241.8	11.39	0.151	11.51	2.71	59.41	29.79	0.129	0.255	0.211	C2S1
S21	542.0	321.2	464.5	3.87	0.057	4.05	1.59	38.15	24.40	0.040	0.194	0.061	C2S1
S22	641.0	410.6	353.7	4.96	0.069	5.13	1.70	41.02	33.86	0.052	0.307	0.091	C2S1
S23	680.0	403.8	361.2	4.05	0.057	4.19	0.63	35.51	36.39	0.042	0.343	0.076	C2S1
S24	675.5	412.3	364.5	3.36	0.047	3.51	0.46	36.76	25.15	0.035	0.202	0.053	C2S1
S25	593.5	373.4	322.7	3.40	0.045	3.55	0.55	39.22	25.98	0.035	0.211	0.055	C2S1
S26	522.5	321.0	291.4	5.37	0.069	5.54	1.27	44.08	34.62	0.057	0.318	0.100	C2S1
S27	425.5	273.0	295.6	5.11	0.065	5.45	1.87	46.81	30.53	0.054	0.264	0.089	C2S1
S28	495.0	320.3	284.4	5.02	0.059	5.49	2.24	53.13	29.70	0.053	0.254	0.086	C2S1
S29	606.5	371.2	403.3	3.99	0.061	4.59	0.65	32.04	39.50	0.042	0.392	0.079	C2S1
S30	591.5	357.2	415.4	5.63	0.092	6.02	0.38	32.21	39.14	0.060	0.386	0.113	C2S1
S31	506.0	324.2	399.3	9.62	0.143	9.88	1.06	43.27	38.04	0.106	0.368	0.198	C2S1
S32	653.5	409.1	381.2	7.61	0.106	8.03	2.61	46.07	43.58	0.082	0.463	0.168	C2S1
S33	565.5	364.1	399.4	4.90	0.066	5.28	2.52	42.79	46.85	0.052	0.529	0.112	C2S1
S34	628.5	307.8	300.6	4.85	0.067	5.18	1.52	39.15	41.40	0.051	0.424	0.100	C2S1
S35	875.0	421.0	401.7	6.04	0.095	6.55	1.04	36.14	35.61	0.064	0.332	0.115	C3S1
S36	658.0	456.3	333.2	5.67	0.078	5.98	1.51	40.72	41.40	0.060	0.424	0.118	C2S1
S37	543.5	321.1	390.1	6.12	0.091	6.45	−0.52	33.30	36.96	0.065	0.352	0.119	C2S1
S38	517.0	356.9	333.0	4.55	0.062	4.84	2.57	40.34	51.07	0.048	0.626	0.112	C2S1
S39	544.5	307.9	260.5	4.64	0.055	4.95	4.03	55.80	46.93	0.049	0.530	0.105	C2S1
S40	752.5	430.0	305.7	8.232	0.111	8.62	2.42	49.45	39.57	0.090	0.393	0.171	C3S1

Appendix XII: Irrigation Water Quality Parameter Values for Groundwater Samples in Monsoon (MON) Season

S. ID	EC	TDS	TH	Na%	SAR	SSP	RSBC	PI	MAR	KR	Mg: Ca	Na: Ca	Hazard class
S1	1237.0	802.3	487.89	15.34	0.284	15.73	3.56	43.92	24.67	0.181	0.196	0.277	C3S1
S2	927.0	554.3	518.42	8.10	0.142	8.45	1.72	32.67	41.95	0.088	0.434	0.175	C3S1
S3	1034.0	675	539.48	10.03	0.184	10.55	0.49	32.14	38.94	0.111	0.383	0.210	C3S1
S4	1193.0	815.9	550.66	6.44	0.114	6.80	2.23	34.94	17.73	0.069	0.129	0.096	C3S1
S5	976.5	624.5	451.15	9.36	0.155	9.86	0.80	35.97	30.61	0.103	0.265	0.171	C3S1
S6	1119.5	734.7	499.21	3.62	0.060	3.90	4.50	36.01	31.57	0.038	0.277	0.063	C3S1
S7	966.5	633.2	492.22	4.53	0.075	4.69	2.65	32.9	39.39	0.047	0.390	0.090	C3S1
S8	1045.5	668.6	468.12	4.88	0.079	4.99	3.06	36.09	31.27	0.051	0.273	0.086	C3S1
S9	906.5	589.6	398.28	5.18	0.077	5.28	0.32	34.09	29.58	0.055	0.252	0.089	C2S1
S10	963.0	616.3	615.10	3.85	0.070	4.18	3.31	28.66	44.30	0.040	0.477	0.083	C2S1
S11	1209.5	744.5	505.03	3.84	0.064	4.44	4.88	36.45	31.37	0.040	0.274	0.067	C3S1
S12	1172.0	758.7	592.28	3.64	0.065	3.95	1.97	28.27	38.76	0.038	0.380	0.071	C3S1
S13	560.5	369	200.48	10.06	0.112	10.63	2.47	62.67	23.60	0.112	0.185	0.168	C2S1
S14	706.0	461.7	322.53	4.52	0.060	4.92	1.40	40.9	27.53	0.047	0.228	0.075	C2S1
S15	753.5	503.7	332.65	3.50	0.047	3.73	3.73	46.15	25.45	0.036	0.205	0.056	C2S1
S16	600.5	382.5	240.04	8.71	0.105	9.05	2.84	55.11	34.20	0.095	0.312	0.167	C2S1
S17	745.0	498.4	362.67	9.15	0.136	9.50	−0.34	37.98	21.98	0.101	0.169	0.148	C2S1
S18	1154.5	747.8	381.60	28.47	0.551	28.92	1.21	52.97	25.88	0.398	0.210	0.617	C3S1
S19	909.5	591.4	355.09	9.67	0.143	10.15	1.84	42.98	28.86	0.107	0.243	0.173	C2S1
S20	596.5	384.5	303.47	9.87	0.135	9.95	2.28	48.69	24.34	0.109	0.193	0.166	C2S1
S21	725.0	461.2	437.61	2.89	0.044	3.00	1.08	34.21	20.87	0.030	0.158	0.043	C2S1
S22	828.0	550.7	383.44	4.03	0.058	4.12	1.20	36.15	29.32	0.042	0.249	0.068	C2S1
S23	862.5	542	385.20	3.09	0.044	3.16	0.47	32.89	32.83	0.032	0.293	0.055	C2S1
S24	865.0	551.5	391.91	2.39	0.034	2.46	0.14	33.57	21.50	0.024	0.164	0.036	C2S1
S25	787.5	511.3	350.56	3.45	0.047	3.57	0.16	36.14	21.55	0.036	0.165	0.052	C2S1
S26	722.0	473.3	323.06	4.13	0.055	4.21	0.78	38.44	28.89	0.043	0.244	0.070	C2S1
S27	635.0	421.1	324.44	4.20	0.056	4.46	1.39	40.97	25.33	0.044	0.203	0.068	C2S1
S28	681.0	463.8	282.12	3.53	0.043	3.92	1.68	45.56	22.22	0.037	0.171	0.054	C2S1
S29	802.5	511.5	462.51	2.14	0.033	2.68	0.27	28.47	35.58	0.022	0.331	0.039	C2S1
S30	777.5	496.8	484.60	4.80	0.079	5.14	0.32	29.77	36.26	0.050	0.341	0.091	C2S1
S31	685.0	462.4	384.96	8.52	0.130	8.72	0.67	37.22	32.53	0.093	0.289	0.159	C2S1
S32	844.0	550.7	348.22	6.22	0.088	6.55	2.45	40.93	39.03	0.066	0.384	0.125	C2S1
S33	856.5	587.1	352.40	2.80	0.038	3.08	2.05	37.23	39.97	0.029	0.400	0.055	C2S1
S34	811.0	442.4	400.46	3.48	0.051	3.70	1.14	32.75	40.04	0.036	0.401	0.069	C2S1
S35	1072.0	568.5	456.96	5.22	0.084	5.66	0.80	32.58	32.21	0.055	0.285	0.093	C3S1
S36	842.0	598.1	386.22	4.14	0.060	4.36	0.99	33.53	39.69	0.043	0.395	0.082	C2S1
S37	744.0	467.7	404.46	5.33	0.080	5.59	−0.56	30.83	34.13	0.056	0.311	0.098	C2S1
S38	702.5	492.8	345.28	3.27	0.045	3.48	2.10	37.75	41.99	0.034	0.434	0.067	C2S1
S39	734.0	443.9	286.41	5.24	0.066	5.49	3.59	49.15	38.63	0.055	0.378	0.104	C2S1
S40	945.0	576.2	335.50	6.44	0.089	6.76	2.02	41.82	33.49	0.069	0.302	0.119	C3S1

Appendix XIII: Irrigation Water Quality Parameter Values for Groundwater Samples in Post-monsoon (POM) Season

S. ID	EC	TDS	TH	Na%	SAR	SSP	RSBC	PI	MAR	KR	Mg: Ca	Na: Ca	Hazard class
PIS1	1410.0	891.2	593.1	10.66	0.206	11.00	4.16	38.09	22.61	0.119	0.175	0.177	C3S1
S2	1152.5	744.3	538.8	4.96	0.086	5.33	3.08	33.07	33.7	0.052	0.305	0.091	C3S1
S3	1114.0	690.8	607.7	8.65	0.166	8.98	1.20	31.31	34.72	0.095	0.319	0.167	C3S1
S4	1710.0	1108.1	454.8	5.56	0.089	5.79	1.82	36.51	21.91	0.059	0.168	0.087	C3S1
S5	885.5	550.8	467.5	9.91	0.169	10.42	0.97	35.85	32.46	0.110	0.288	0.187	C3S1
S6	995.5	624.9	521.6	2.90	0.048	2.99	2.51	31.81	31.01	0.030	0.270	0.050	C3S1
S7	1066.5	683.7	543.8	5.19	0.090	5.37	0.95	28.93	40.07	0.055	0.401	0.105	C3S1
S8	1061.5	594.2	474.9	4.96	0.081	5.10	4.13	38.00	28.04	0.052	0.234	0.083	C3S1
S9	995.5	658.3	441.3	4.35	0.068	4.46	0.88	33.52	27.45	0.045	0.227	0.072	C2S1
S10	975.5	625.0	470.5	4.29	0.069	4.59	1.80	33.75	29.54	0.045	0.252	0.073	C2S1
S11	1231.0	783.7	571.6	4.13	0.073	4.56	1.80	30.53	28.59	0.043	0.240	0.069	C3S1
S12	1182.5	757.2	570.5	3.98	0.070	4.30	2.19	30.12	34.08	0.041	0.310	0.072	C3S1
S13	663.0	466.0	225.7	9.40	0.110	9.73	1.83	54.97	25.74	0.104	0.208	0.161	C2S1
S14	811.0	511.7	316.7	4.35	0.057	4.44	0.93	39.76	27.38	0.046	0.226	0.072	C2S1
S15	665.5	449.0	391.1	4.75	0.070	4.92	3.89	43.38	20.38	0.050	0.154	0.072	C2S1
S16	915.5	567.8	288.2	6.26	0.080	6.38	0.16	38.97	32.23	0.067	0.285	0.113	C2S1
S17	664.0	453.3	322.9	9.80	0.138	10.17	0.09	41.02	23.65	0.109	0.186	0.164	C2S1
S18	1166.5	763.4	396.0	21.56	0.387	22.07	−0.86	44.67	20.23	0.275	0.152	0.396	C3S1
S19	815.0	559.8	327.2	10.95	0.158	11.47	3.42	50.28	23.97	0.123	0.189	0.186	C2S1
S20	806.5	500.2	292.7	4.67	0.059	4.77	0.88	41.93	25.23	0.049	0.203	0.075	C2S1
S21	838.5	534.2	375.8	2.97	0.042	3.14	2.78	40.46	24.19	0.031	0.191	0.046	C2S1
S22	741.5	500.4	367.0	2.76	0.039	2.91	1.47	37.92	23.52	0.028	0.185	0.043	C2S1
S23	775.5	501.8	325.0	3.90	0.052	4.07	1.98	40.33	36.34	0.041	0.343	0.073	C2S1
S24	1073.0	685.8	406.1	3.11	0.046	3.18	0.29	34.02	20.55	0.032	0.155	0.046	C2S1
S25	797.5	503.9	352.5	3.06	0.042	3.24	0.89	37.12	25.19	0.032	0.202	0.049	C2S1
S26	827.0	540.3	329.4	2.51	0.033	2.65	−0.23	34.25	26.46	0.026	0.216	0.040	C2S1
S27	827.0	528.9	315.9	5.05	0.067	5.38	2.13	43.92	27.19	0.053	0.224	0.084	C2S1
S28	1192.0	755.8	306.1	4.40	0.057	4.82	−0.16	36.92	26.3	0.046	0.214	0.072	C2S1
S29	812.0	520.5	419.0	3.54	0.053	4.11	1.77	35.05	31.18	0.037	0.272	0.061	C2S1
S30	788.0	499.7	467.6	4.78	0.077	5.19	0.96	31.60	35.64	0.050	0.332	0.090	C2S1
S31	891.0	546.8	333.6	5.01	0.068	5.34	3.23	42.91	41.22	0.053	0.421	0.103	C2S1
S32	854.0	555.5	334.4	7.55	0.106	7.92	3.91	46.39	39.41	0.082	0.390	0.155	C2S1
S33	767.0	489.6	393.1	3.53	0.051	3.78	2.63	37.15	37.32	0.037	0.357	0.067	C2S1
S34	934.0	600.6	384.7	3.89	0.056	4.16	−0.57	30.26	34.38	0.040	0.314	0.071	C2S1
S35	1077.0	695.4	456.1	5.09	0.081	5.51	2.59	36.37	28.74	0.054	0.242	0.087	C3S1
S36	965.0	622.7	373.4	4.20	0.060	4.39	2.40	38.93	33.31	0.044	0.300	0.076	C2S1
S37	879.0	566.4	384.3	4.98	0.073	5.24	1.35	37.02	29.68	0.052	0.253	0.086	C2S1
S38	1011.0	655.7	356.6	6.59	0.094	6.69	−0.06	36.64	25.14	0.071	0.201	0.108	C2S1
S39	978.0	666.4	354.4	4.91	0.069	5.09	1.12	36.37	37.77	0.052	0.364	0.095	C2S1
S40	943.0	566.7	336.8	3.39	0.046	3.84	1.75	38.76	35.08	0.035	0.324	0.062	C3S1

Appendix XIV: Statistical Summary of IWQI (CWQI and MWQI) Value of Collected Water Samples

	CWQI (Irrigation)				MWQI (Irrigation)			
S. ID	PRM	MON	POM	Average	PRM	MON	POM	Average
S1	66.3	62.7	61.4	63.5	81.8	78.4	79.3	79.8
S2	67.0	64.0	62.9	64.6	82.9	81.6	83.2	82.6
S3	73.7	70.6	70.0	71.4	85.3	82.8	81.0	83.0
S4	72.9	68.9	67.0	69.6	86.0	84.4	83.1	84.5
S5	73.7	71.3	70.6	71.9	84.1	78.1	80.9	81.0
S6	70.0	67.3	65.9	67.7	78.9	79.5	74.2	77.5
S7	74.2	70.6	69.1	71.3	85.4	79.2	77.0	80.5
S8	69.5	66.4	65.1	67.0	86.3	78.0	80.6	81.6
S9	65.4	61.7	60.0	62.4	78.0	74.5	75.4	76.0
S10	80.0	75.4	73.8	76.4	85.1	88.0	88.8	87.3
S11	69.9	68.5	67.2	68.5	82.0	79.6	80.4	80.7
S12	74.1	70.0	68.7	70.9	81.0	78.9	77.9	79.3
S13	80.8	78.4	78.0	79.1	84.0	81.9	80.8	82.2
S14	77.8	74.1	72.6	74.8	85.9	83.0	82.1	83.7
S15	76.9	74.0	73.7	74.9	88.8	84.6	82.2	85.2
S16	80.1	76.6	74.2	77.0	85.1	82.5	81.0	82.9
S17	77.7	75.7	73.7	75.7	89.0	83.0	81.3	84.4
S18	74.3	71.9	70.2	72.1	83.5	79.5	77.9	80.3
S19	81.1	78.9	77.0	79.0	85.7	83.1	82.0	83.6
S20	76.2	73.2	71.5	73.6	86.0	84.2	82.5	84.2
S21	79.0	76.4	75.0	76.8	90.3	86.4	84.3	87.0
S22	78.4	75.0	73.4	75.6	91.8	80.0	82.0	84.6
S23	78.9	76.2	74.9	76.7	86.3	84.1	82.5	84.3
S24	77.7	75.3	74.4	75.8	88.9	83.8	81.3	84.7
S25	81.2	78.2	77.0	78.8	85.1	82.0	81.5	82.9
S26	74.6	71.8	70.1	72.2	87.5	83.6	81.0	84.0
S27	75.0	72.0	70.7	72.6	90.4	84.0	85.1	86.5
S28	70.2	67.8	66.9	68.3	81.4	77.4	75.9	78.2
S29	72.8	70.5	68.1	70.5	83.6	80.9	80.1	81.5
S30	73.5	69.0	67.7	70.1	81.1	77.5	75.0	77.9
S31	74.6	71.6	69.9	72.0	89.5	87.8	86.5	87.9
S32	82.1	79.0	77.1	79.4	91.8	87.1	85.1	88.0
S33	74.6	72.6	70.5	72.6	89.4	86.0	84.3	86.6
S34	75.3	70.9	69.9	72.0	87.3	84.7	82.8	84.9
S35	73.1	69.0	65.4	69.2	82.6	79.0	77.5	79.7
S36	77.5	75.0	74.8	75.8	87.3	82.8	84.0	84.7
S37	75.9	73.8	74.9	74.9	91.0	87.0	85.9	88.0
S38	68.8	66.9	68.0	67.9	81.8	80.3	79.1	80.4
S39	66.5	63.9	63.0	64.5	76.1	75.1	72.4	74.5
S40	67.6	65.0	64.1	65.6	72.5	69.4	68.2	70.0

Index

A
Aquifer, 2, 18, 40, 66, 136, 189, 224, 288, 302
As contamination, 176, 179–182
Atomic adsorption spectroscopy (AAS), 49, 50, 53–58

B
Bedrock, 6, 7, 9, 18, 19, 25, 41, 94, 109, 305
Bicarbonate index (BCI), 98–102
Boilers, 284, 288, 289

C
Carcinogenic and noncarcinogenic, 27, 144, 191, 198–201, 211–219
Climate change, 13, 14, 19, 145, 188, 301–307
Cooling systems, 282, 291–293
Corrosivity, 285, 287, 289–291, 294–298

F
Fe-Mn chemistry, 155–174

G
Groundwater, 2, 18, 40, 66, 136, 188, 224, 283, 302, 309–310
Groundwater availability, 22–23, 302–303
Groundwater recharge, 7, 8, 19, 21, 22, 41, 107, 127, 128, 161, 302, 303

H
Hazard class, 225, 226, 235–237, 240, 251–254, 256–266, 268, 270–272, 320–322
Health risk assessment (HRA), 27, 63, 137, 190–192, 198–202, 211–218
Heavy metal contamination index, 192, 210
Hydrogeochemical process, 2–15, 25, 26, 48, 66–131, 181
Hydrogeochemistry, 4–6, 84, 88, 110, 126

I
Industrial water, 62, 63, 72, 76, 282–298, 306
Industrial water quality indices, 284–289
Integrated Irrigation Water Quality Index (IIWQIndex), 63, 225, 252, 259–275, 277
Irrigation water quality index (IWQI), 225, 242–245, 252–259, 277, 323
Irrigation water quality parameters, 27, 63, 224, 225, 227–229, 231, 233–242, 249, 254–257, 270, 276, 320–322
Isotope dilution, 3

M
Mineral dissolution, 15, 78, 79, 93, 102, 105, 109, 110, 112, 116, 122, 123, 130, 131, 151, 165, 170, 259
Multimeter, 47, 49–51

M. S. Islam, *Hydrogeochemical Evaluation and Groundwater Quality*, https://doi.org/10.1007/978-3-031-44304-6

Multivariate analysis, 85, 105, 146–153, 157–160
Multivariate techniques, 66, 85, 137, 146, 157, 182

P
Permeability index diagram, 277
Physicochemical parameters, 38, 44, 48, 68, 72, 146, 150, 165, 194, 209, 227, 288, 289

R
Rock weathering, 14, 26, 89, 91, 99, 103, 110–116, 125, 130, 145, 170, 206

S
Saturation index (SI), 63, 78, 84, 88, 122–126, 130, 194, 284, 285, 288, 294–298
Scale formation, 72, 74, 284, 286, 288–290, 292, 294, 297, 298

T
Trace metals, 25–27, 31, 48–50, 53, 62, 63, 66, 112, 131, 136–183, 190–192, 197, 198, 211–213, 215, 218, 225, 229–231, 233, 236, 254, 314–316
Trace metal sources, 136–183

U
US salinity hazard diagram, 248–249, 276
UV Spectrophotometer, 49, 60, 61

W
Water chemistry, 4, 5, 14, 25, 26, 44, 66–82, 89, 106, 110, 129, 130, 154, 161, 162, 224, 235, 237, 245, 298, 302
Water contamination, 42, 174
Water facies, 5, 105–109, 129
Water management, 22, 42, 176, 188, 191, 305–307
Water quality index (WQI), 27, 28, 63, 189, 190, 192, 194–196, 202–210, 218, 225, 227, 229, 231, 232, 242–245, 251–253, 257, 260, 263, 265, 275, 276, 295, 319
Water resource, 131, 188, 229, 282, 305, 306
Water sampling, 48
Water uses, 205, 224, 233, 276
WHO guideline, 143, 204
Wilcox diagram, 248–250, 277

SPRINGER NATURE

GPSR Compliance

The European Union's (EU) General Product Safety Regulation (GPSR) is a set of rules that requires consumer products to be safe and our obligations to ensure this.

If you have any concerns about our products, you can contact us on ProductSafety@springernature.com

In case Publisher is established outside the EU, the EU authorized representative is:

Springer Nature Customer Service Center GmbH
Europaplatz 3
69115 Heidelberg, Germany

The manufacturer's authorised representative in the EU is Springer Nature Customer Service Centre GmbH, Europaplatz 3, 69115 Heidelberg, Germany. If you have any concerns regarding our products, please contact ProductSafety@springernature.com

Printed and bound by CPI Group (UK) Ltd, Croydon, CR0 4YY
23/09/2026
02221541-0005